W9-CDT-700

Michael Foster and the Cambridge School of Physiology

Sir Michael Foster (1836-1907)

Probably taken in 1900 in the Adirondack mountains in upstate New York, during Foster's visit to the United States. See Henry Dale, "Sir Michael Foster. . . ," *Notes and Records of the Royal Society of London, 19* (1964), 10-32, on 29-30.

Photograph courtesy of the Library, Trinity College, Cambridge.

Michael Foster and the Cambridge School of Physiology

The Scientific Enterprise in Late Victorian Society

Gerald L. Geison

Princeton University Press
Princeton, New Jersey

Published by Princeton University Press, Princeton, New Jersey
In the United Kingdom: Princeton University Press, Guildford, Surrey

Library of Congress Cataloging in Publication Data will be
found on the last printed page of this book
Publication of this book has been aided by the Louis A. Robb
Fund of Princeton University Press.
This book has been composed in Linotype Baskerville.
Printed in the United States of America by Princeton
University Press, Princeton, New Jersey

For My Family

Contents

Preface

[The concept of a research school] is most easily seen in terms of the work done on a family of problems, usually descended from some seminal ancestor problem. The problems are related not only by descent, but also by their objects of investigation, a battery of tools, and a body of methods. In these, criteria of value demarcate schools most sharply; and a well-developed school will also have its social side, in the network of friendships and loyalties built up in the course of co-operative work; and an institutional aspect, perhaps in the place of employment of its nucleus . . . as well as some elements of the formal apparatus of a field, such as journals and societies.

J. R. Ravetz (1971)[1]

In the history of science, despite great ferment and activity in recent years, many large areas remain virtually unexplored. Conspicuous among them is the history of physiology after 1850. This book is designed to fill one or two of the gaps in that story. It does not, however, attempt to trace the general development of physiological thought during this period or any part of it. My aim is in some respects more limited and in others more expansive than that. It is more limited in that attention is focused on one small group of English physiologists, the Cambridge School between 1870 and 1900; and chiefly on their efforts to resolve one important physiological problem, the origin of the rhythmic heartbeat. It is more expansive in that considerable attention is given to the social and institutional context within which this school of physiologists emerged and developed its special approach and ideas. In other words, this book examines the role both of internal scientific issues and of external, socioinstitutional factors in the rise and early development of the Cambridge School.

The respective roles of internal and external factors in the development of science have long been a subject of spirited and sometimes acrimonious debate.[2] To state the issue at its crude extremes, the "internalist" portrays science as a socially insular and essentially intellec-

[1] Jerome R. Ravetz, *Scientific Knowledge and its Social Problems* (Oxford, 1971), p. 224.

[2] For interesting discussions of this debate and access to the pertinent literature, see Arnold Thackray, "Science: Has Its Present Past a Future," in *Historical and Philosophical Perspectives of Science*, ed. R. H. Stuewer (Minneapolis, 1970), pp. 112-127; J. E. McGuire, "Newton and the Demonic Furies: Some Current Problems and Approaches in the History of Science," *History of Science, 11* (1973), 21-48; and Barry Barnes, *Scientific Knowledge and Sociological Theory* (London, 1974), pp. 99-124.

tual enterprise, while the "externalist" insists that sociopolitical forces and material conditions play the determining role in scientific activity, not excluding scientific thought. Few if any historians of science now actively defend either of these extreme positions, and most probably agree that the internalist and externalist approaches are complementary rather than opposed. As early as 1968, in fact, Thomas S. Kuhn suggested that the greatest challenge facing historians of science is precisely that of integrating the two approaches into a single enriched perspective.[3] More recently, in a subtle variation on this theme, we have been urged to discard or transcend the very categories in which the debate has been conceived. Merely to posit an internal-external dichotomy, it is said, predisposes one to draw anachronistic or artificial distinctions. Especially when or where clearly differentiated scientific communities do not exist, it is often ahistorical to pretend that purely "scientific" issues can be teased out of their "nonscientific" context. And even where sharply defined scientific communities do exist, it is vital to recognize the extent to which ostensibly value-free scientific ideas are inextricably bound up with the larger sociocultural setting. In particular, scientific concepts are embodied in metaphors that may tacitly serve special ideological interests and that certainly are limited by the cultural resources of a given time and place.[4]

Only in the concluding chapter does this book make any direct effort to meet these deep historiographic concerns. Up to that point, my very organizational strategy—with the institutional framework of the Cambridge physiologists (Part Two) being discussed separately from their research (Part Three)—suggests the extent to which I have felt able to draw (or unable to deny) a "natural" distinction between internal and external factors in the case of this highly specialized group. In its narrative core, this study seeks only to describe the rise of the Cambridge School and to account for its early success. Those aims, I believe, absolutely demand a consideration of both internal

[3] Thomas S. Kuhn, "The History of Science," in *International Encyclopedia of the Social Sciences*, ed. David L. Sills (New York, 1968), *14*, 74-83, on 76.

[4] See, e.g., Robert M. Young, "Malthus and the Evolutionists: The Common Context of Biological and Social Theory," *Past and Present*, *43* (1969), 109-145, esp. 109-110, 140-141; Young, "The Historiographic and Ideological Contexts of the Nineteenth-Century Debate on Man's Place in Nature," in *Changing Perspectives in the History of Science: Essays in Honour of Joseph Needham*, ed. Mikulas Teich and Robert M. Young (London, 1973), pp. 344-438; Steven Shapin and Arnold Thackray, "Prosopography as a Research Tool in History of Science: The British Scientific Community, 1700-1900," *History of Science*, *12* (1974), 1-28, esp. 22, n. 9; Barnes, *Scientific Knowledge*, esp. Chapters 3, 5-6; and Karl Figlio, "The Metaphor of Organization: An Historiographical Perspective on the Bio-Medical Sciences of the Early Nineteenth Century," *History of Science*, *14* (1976), 17-53.

and external factors, but no systematic attempt is made to articulate the ways in which the two sorts of factors relate to each other. For the most part, they are treated here as literally complementary, as two separate elements that are mutually required to complete the story.

But I would insist that yet a third element must be included in any proper account of a research school: individual human personality. In the wake of the internal-external debate and the recent "revolution" in the historiography of science,[5] the biographical approach may have lost much of its appeal for younger and more sophisticated historians of science.[6] It seems important, therefore, to emphasize a truism whose truth may now be too little honored—namely, that neither scientific ideas nor scientific institutions float in ethereal isolation from the men and women who give them life. The degree of attention hereafter given to the life, work, and influence of one now-almost-forgotten man, Michael Foster, is in part an expression of that belief. Indeed, while I do not generally subscribe to the Emersonian dictum that "an institution is the lengthened shadow of one man," I am frankly astonished at the extent to which it applies to the embryonic Cambridge School.

That is not to say that what follows is a full-scale biography of Foster. While I am eager to establish his roots and background before he arrived at Cambridge, I largely ignore the later (and more public) phase of his career. That period of his life, though vitally important in other ways, can shed little light on the hitherto mysterious process by which a "second-rate" scientist came to found an undeniably great research school. As a matter of fact, this study originated in the belief that much too little attention had been paid to scientists like Foster, who played a crucial role in the diffusion and organization of scientific knowledge without himself contributing to it in any monumental or permanent fashion. In other words, I sensed a need for a more systematic examination of those "great teachers" who seem to have left little in their wake except celebrated students. Among physiologists, Foster stood out as one whose reputation was that of "a discoverer of men rather than of facts."[7] I began this study in the hope that something could be done to specify the nature and sources of his influence

[5] See Gerd Buchdahl, "A Revolution in Historiography of Science," *History of Science, 4* (1965), 55-69.

[6] I refer here to biographies of individuals, not to the collective biographical approach of the "prosopographers," which now seems to be gaining a solid foothold in the history of science and which can be seen as a potentially fruitful attempt to supplement the older tradition of scientific biography. Cf. McGuire (n. 2); and Shapin and Thackray (n. 3).

[7] W. H. Gaskell, "Sir Michael Foster, 1836-1907," *Proc. Roy. Soc., B80* (1908), lxxi-lxxxi, on lxxviii.

on English physiology—something more informative than the vague tributes of his former students and something more satisfying than Abraham Flexner's judgment that Foster produced his achievement at Cambridge "in subtle ways that defy expression."[8]

From the Cambridge School of Physiology Foster's influence radiated out to the general biomedical enterprise there and elsewhere in the English-speaking world. Some attention is given below to these broader results of Foster's efforts, and some attempt is made to establish their dependence on his evolutionary and broadly biological approach to physiology. Inevitably, however, the nature of his influence becomes harder to specify the further it extends beyond its institutional base in the Cambridge School. By focusing so minutely on that school, and on Foster's role in it, I have sought to meet the demand that an assertion of influence be based on something beyond mere coincidence in time and space,[9] and to urge that, so long at least as our attention is focused on small groups, we need not be excessively skeptical about the possibility of doing meaningful influence studies.[10]

Historians of science have thus far produced exceedingly few full-length studies of laboratory-based research schools, wherein small groups of scientists work side by side in a common institutional setting. In fact, Maurice Crosland's book on the Society of Arcueil is the only example that comes readily to mind.[11] Even that example might have escaped attention had not Owen Hannaway stressed, in his review of Crosland's book, that the Society of Arcueil should be seen not as a scientific society in the usual sense, but as a private research school which adopted the Newtonian style and research interests of its leaders, Berthellot and Laplace.[12] More directly pertinent here, though it is an article and not a full-length study, is J. B. Morrell's attempt to construct a model to explain the success of Justus Liebig's school of chemistry at Giessen.[13] Comparing Liebig's success with Thomas Thomson's failure

[8] Abraham Flexner, *Universities: American, English, German* (New York, 1930), p. 7.

[9] Dennis M. McCullough, "W. K. Brooks' Role in the History of American Biology," *J. Hist. Biol.*, *2* (1969), 411-438, on 413.

[10] Cf. Quentin R. D. Skinner, "The Limits of Historical Explanation," *Philosophy*, *41* (1966), 199-215. I am indebted to Robert M. Young for introducing me to this paper.

[11] M. P. Crosland, *The Society of Arcueil: A View of French Science at the Time of Napoleon I* (Cambridge, Mass., 1967). I deliberately exclude the several existing books on the Cavendish Laboratory of Physics at Cambridge. In my view, none of these works is entirely adequate, and none of them even tries to enlarge our general understanding of "the research school."

[12] Owen Hannaway, *Isis*, *60* (1969), 578-581.

[13] J. B. Morrell, "The Chemist Breeders: The Research Schools of Liebig and Thomas Thomson," *Ambix*, *19* (1972), 1-46.

to establish a flourishing research school at Glasgow, Morrell draws attention to a variety of interwoven internal and external factors, including the prescience of the director's research program, the quality of his students, the "ripeness" of his field of study, the nature and simplicity of the techniques employed, access to channels of publication, the extent of a director's charismatic powers, and the degree of institutional support he enjoyed. Though this study was not (and is not) designed as an explicit test of Morrell's model, I do invoke many if not all of the same factors to describe the rise of the Cambridge School and to account for its success.

But there is one element in the story that I single out for special and detailed attention: the problem of the heartbeat. In fact, if this book has any one central theme, it is that the institution known as the Cambridge School of Physiology descended from and crystallized around the problem of the heartbeat. In pursuit of this problem, the members of the Cambridge School displayed a common style, adopting the evolutionary approach of their leader, Michael Foster, as well as his commitment to the generally unpopular myogenic theory of the heartbeat. And it was their successful resolution of this problem, particularly through the work of W. H. Gaskell, that launched the Cambridge School on its path to greatness. This conception of the story requires that the discussion in Part Three be quite resolutely technical. For that, there can ultimately be no apology, though I have tried to make the discussion as accessible as possible to the nonspecialist.

Gerald L. Geison
Princeton, New Jersey
17 March 1977

Acknowledgments

It is a decade since I began this book, and I have accumulated a large store of personal and intellectual debts. None of the later debts could have accrued without the initial support and encouragement of two superb and selfless advisers: Leonard G. Wilson, who led me into the topic; and Frederic L. Holmes, who directed my Ph.D. dissertation for the Department of the History of Science and Medicine at Yale University. Now that I am privileged to call them colleagues, I have an even larger appreciation and an even deeper gratitude for all they did for me during our years at Yale, and all they have done for me since.

Also while at Yale, from 1965 through 1969, I benefited greatly from stimulating exchanges with my fellow graduate students, particularly Jerry Bylebyl, my brother Roger Geison, Alan Kay, Peter Niebyl, Charley Sawyer, and Polly Winsor. The Yale Medical Historical Library would have been a much less enjoyable place to work without the ready wit and wisdom of Professor Lloyd G. Stevenson or the splendid library staff, led by Madelaine Stanton and Ferenc Gyorgyey. In somewhat different ways, Professors Asger Aaboe, Bernard Goldstein, and Derek Price also helped to make the Department of the History of Science and Medicine a generally agreeable and always challenging place to be.

In the summer of 1968, while still at Yale and in the midst of writing my dissertation, I began to correspond with Richard D. French, then a Rhodes Scholar at Oxford who was cataloguing the papers of E. A. Sharpey-Schafer at the Wellcome Institute of the History of Medicine. Rick's knowledge of that and other archives, together with his impressive grasp of the published literature on Victorian physiology, made him a valuable resource for me from the outset. He sent me photocopies of several important letters and offered countless bibliographical suggestions. The remarkable similarity of our research interests soon became apparent, and more than once Rick and I must have been reading the same papers on the same day. To some extent we were in a race, but it is surely the least trying race I have ever run. At that time we exchanged information and ideas as fully as sensible transatlantic correspondence would allow. The rate of exchange increased enormously between September 1970 and June 1972, when we worked side by side in the Program in the History and Philosophy of Science at Princeton University. I have made some effort to indicate my indebtedness to Rick in the text and footnotes below, but a brief attempt at a more general assessment seems in order here.

Rick's recent book, *Antivivisection and Medical Science in Victorian Society* (Princeton University Press, 1975), contains important material of use to me in this study, but the two books only rarely overlap. Even when they do, as in my brief account of antivivisection sentiment (Chapter 2), our views are not always identical and were developed for the most part independently. I owe a considerably larger debt to Rick's 1970 article, "Darwin and the Physiologists."[1] He sent me an earlier version in July 1969, as I was writing that chapter of my dissertation which ultimately became Chapters 8 and 9 of this book. To the best of my recollection, I had already independently established Foster's active role in the early cardiological research of the Cambridge School, and more especially the intimate link between his work and Gaskell's. I had also, I think, fully recognized the broadly biological and evolutionary tone of Foster's approach, though perhaps without yet seeing it as part of the "evolutionary climate of opinion" that Rick perceived in English physiology as a whole. Of one large and specifiable debt I am sure: Rick alerted me to the significance of G. J. Romanes' work for my own themes. He certainly also articulated several points that had been only vaguely formulated in my own mind—how many, precisely, I am at a loss to say. Rick was, in any case, the most immediate intellectual influence on my Ph.D. dissertation. It is a pleasure to thank him here for that and other things, including his general support and continuing friendship.

This book differs from my dissertation in several ways, most of them owing to the insights or suggestions of others. The works of J. R. Ravetz and J. B. Morrell helped to sharpen my conception of a research school and to expand my discussion of the institutional advantages Foster enjoyed at Cambridge (Chapters 6 and 10). A new section on the general rise of physiology in Victorian England (Chapter 6) grows out of a suggestion by Paul Cranefield. Most importantly, the chapter below on university reform (Chapter 4) differs radically in tone and thesis from the counterpart chapter in my dissertation. For helping me to see my way through to this change in perspective, I am much indebted to conversations with Charles Gillispie, Arthur Engel, and Lawrence Stone, and to a bruising session at the Davis Center for Historical Studies at Princeton, where Professor Stone and his associates convinced me of the need for drastic recasting.

[1] R. D. French, "Darwin and the Physiologists, or the Medusa and Modern Cardiology," *J. Hist. Biol.*, *3* (1970), 253-274.

Several colleagues have read the manuscript versions of this book in whole or in part and have offered useful suggestions. In addition to Leonard Wilson, Larry Holmes, Rick French, Paul Cranefield, and Charles Gillispie, they include Thomas S. Kuhn, Robert Stauffer, John Parascandola, and an anonymous referee for Princeton University Press. John Hannon and Gail Filion, successively my editors at the Press, have aided and encouraged me at several crucial stages in the process of getting the book into print. As copy editor, Judith May has done much to improve the clarity of my prose and argument. In giving everyone above my sincere thanks, let me also release them from any responsibility for my errors of omission or commission.

Dozens of other colleagues, friends, and correspondents have contributed to this book in one way or another. In the footnotes I have tried to remember those who helped me to locate and secure pertinent materials, often at expense to themselves. At long last I have an opportunity to acknowledge my deep indebtedness to Joseph Barrell, who first excited my interest in the history of science when I was an undergraduate at Beloit College. In 1970 Nobel laureates Lord Adrian and A. V. Hill graciously allowed me to interview them about the Cambridge School of Physiology as it was when they were trained there in the early years of this century. Richard Dupuis, who also supplied several references and documents, willingly shared with me his wide knowledge of the nineteenth-century evangelical revival. Drs. William Bynum and Arthur Rook clarified certain aspects of Victorian medical education for me. Dr. Robert Robson of Trinity College, Cambridge, sent me several pertinent documents and made it possible for me to gain access to the Trinity College Minute Books. J. E. T. Hales-Tooke, solicitor in Cambridge, gave freely of his time and energy to do several personal favors for me. Edwin Clarke, now director of the Wellcome Institute, has several times interceded on my behalf. At one particularly hectic stage in the rewriting process, Mark Adams of the University of Pennsylvania found me a quiet place to stay and work. Barbara Suomi did a simply brilliant job of typing the final manuscript, offered general good humor and support, and handled with skill and dispatch all matters relating to the book while I was in Paris last year. Dorothy Ebersole—in addition to many other valuable contributions—arranged and typed my Appendices.

I owe a special debt of gratitude to the late Sir Robert Foster, grandson of Sir Michael, who granted me access to the remaining family papers and gave me permission to copy and to quote from some twenty letters that Michael sent home to his family during his voyage to the Red Sea (Chapter 3). Sir Robert and Lady Foster twice extended to me the full hospitality of their charming home, and they have consistently encouraged me in my work.

A large and inevitably anonymous corps of librarians has earned my heartfelt thanks for their general efficiency and cordial helpfulness. In addition to the libraries of the universities with which I have been formally associated (Yale, Minnesota, and Princeton), I have used the archives of the American Philosophical Society, the Royal Society of London, the Royal College of Physicians, the University of London, University College London, the University of Cambridge, the Johns Hopkins University, the Bayerische Staatsbibliothek in Munich, the Royal Botanic Gardens in Kew, the Wellcome Institute of the History of Medicine in London, and Trinity College, Cambridge. To the last four of these institutions, and to the Imperial College of Science and Technology, I am grateful for permission to quote from archives in their possession or under their control.

Several agencies and organizations have contributed financially to the realization of this enterprise. A National Science Foundation Graduate Fellowship supported my work at Yale from 1966 through 1969. As a research fellow at the University of Minnesota in 1969-1970, I was given generous time and facilities for research. A travel grant from the American Philosophical Society enabled me to go to England in the summer of 1970, at which time I examined most of the manuscript material used in the preparation of this book.[2] At Princeton, the History Department awarded me two summer research grants, and the University Committee on Research in the Humanities and Social Sciences defrayed the cost of typing the final manuscript.

Chapter 2 below is a somewhat revised and enlarged version of my "Social and Institutional Factors in the Stagnancy of English Physiology," *Bulletin of the History of Medicine, 46* (1972), 30-58. For permission to repeat so much of it here, I am grateful to the *Bulletin* and its editor, Lloyd G. Stevenson.

In dedicating this book to my family, I mean to express my profound gratitude for their varied and vital contributions over the years. No member of that family will be surprised or disappointed that I should mention Lynne especially. It is easy, but it is the merest beginning, to thank her for her superb job of typing my dissertation. The rest is harder to express. From the day I began this study, she has been my chief source of sustenance and support, and too often—when things have not gone well—the victim of my somber moods. For her sake, if for no other reason, I am glad at last to let this book go.

[2] See the *American Philosophical Society Year Book 1971* (Philadelphia, 1972), pp. 588-589.

Abbreviations

Brit. Assoc. Rep.	*Report of the . . . Meeting of the British Association for the Advancement of Science*
Brit. J. Hist. Sci.	*British Journal for the History of Science*
Brit. Med. J.	*British Medical Journal*
Bull. Hist. Med.	*Bulletin of the History of Medicine*
Centralblatt	*Centralblatt für die medicinischen Wissenschaften*
CUR	*Cambridge University Reporter*
DNB	*Dictionary of National Biography*
DSB	*Dictionary of Scientific Biography*
J. Anat. Physiol.	*Journal of Anatomy and Physiology*
J. Hist. Biol.	*Journal of the History of Biology*
J. Hist. Med.	*Journal of the History of Medicine and Allied Sciences*
J. Physiol.	*Journal of Physiology*
Med. Hist.	*Medical History*
Müller's Archiv	*Archiv für Anatomie, Physiologie, und wissenschaftliche Medicin*
Phil. Trans.	*Philosophical Transactions of the Royal Society of London*
Proc. Camb. Phil. Soc.	*Proceedings of the Cambridge Philosophical Society*
Proc. Roy. Soc.	*Proceedings of the Royal Society of London*

PART ONE

The Background: Foster and English Physiology, 1840-1870

1. Introduction

In truth, the more closely men are studied, the more elusive they become. The visual field narrows in proportion to the sharpness of our gaze. The forest fades when we regard a tree. The particular inevitably destroys the general. Sir Michael Foster was a power in the development of physiology, but the secret of his influence is not to be found in the data of the biographical dictionary. His strength lay rather in temperament, in felicity of choice, and in the admirable co-ordination of qualities in themselves of sound but not surpassing excellence.

W. T. P[orter] (1907)[1]

Original thinkers and investigators do not therefore represent the only type of university professor. They will always be the distinguished figures; theirs will usually be the most profound and far-reaching influence. But even universities, modern universities, need and use men of different stamp —teachers whose own contributions to learning are of less importance than their influence in stimulating students or their resourcefulness in bringing together the researches of others. Michael Foster was not the less a great university professor, though he was not himself a great original thinker: in subtle ways that defy expression, he created the great Cambridge school of physiology.

Abraham Flexner (1930)[2]

Just over a a century ago, in May 1870, the Master and Senior Fellows of Trinity College, Cambridge, appointed Michael Foster praelector in physiology. Although Foster may have been wanted chiefly for collegiate rather than university purposes, he brought with him a broader vision and more ambitious aims. During the nineteenth century, no single appointment was to have greater importance for the development of the biological sciences in the English universities. It was the leading event in the revival of the great English physiological tradition inaugurated by William Harvey in the seventeenth century and represented by Stephen Hales in the next, but which in the course of the nineteenth century had faded into dormancy.

By the middle of the last century, English physiology had become a stagnant backwater rather than a major tributary flowing into the mainstream. Physiology remained in the hands of amateurs, outside

1 W. T. P[orter], "Sir Michael Foster," *Boston Medical and Surgical Journal, 156* (1907), 309-310.

2 Abraham Flexner, *Universities: American, English, German* (New York, 1930), p. 7.

the universities, while in France and especially in Germany it was becoming established as a professional pursuit in university laboratories. England simply had no physiologists to compare with Claude Bernard in France or Carl Ludwig in Germany.

At the time of Foster's appointment, there was in England no independent professorship in physiology, no journal devoted exclusively to it, and no independent recognition of it in the various sections of the British Association for the Advancement of Science. Still the leading English textbook on the subject was William Benjamin Carpenter's *Principles of Human Physiology,* first published in 1842. Although generally sensible, this work never attained the international reputation or influence of Johannes Müller's *Handbuch der Physiologie des Menschen* (1834-1840) or of Carl Ludwig's *Lehrbuch der Physiologie des Menschen* (first edition, 1852-1854).

While on the Continent physiology had become a rigorous experimental science, it remained in England largely submerged in anatomy and in essentially religio-philosophical concerns. Antivivisection sentiment also ran peculiarly high there, further contributing to the stagnancy of physiology. In all of England, there was but one small and inadequate laboratory where experimental physiology was regularly taught. And even this laboratory—at University College London—had been organized only in the late 1860's, largely under the influence of Foster himself, who had been brought there as instructor in practical physiology and histology just four years before accepting the praelectorship at Trinity College. With the partial exception of William Sharpey, Foster's mentor at University College, no one in England was teaching physiology as it was then being taught on the Continent.[3]

Yet by the time Foster died in 1907, English physiology had not only rejoined the mainstream, but in several important areas actually led the way. On a visit to England in 1895, the German-trained physiologist T. W. Engelmann perceived a new "young species of physiologist," a species that would soon repair the neglect into which English physiology had fallen during the first seventy-five years of the century. "We will," he wrote, "learn much from them."[4] In 1902, the future Nobel

[3] See Edward Sharpey-Schafer, *History of the Physiological Society During Its First Fifty Years, 1876-1926* (London, 1927), pp. 1-5; Sharpey-Schafer, "Developments of Physiology," *Nature, 104* (1919), 207-208. More generally, see Karl E. Rothschuh, *History of Physiology,* trans. and ed. G. R. Risse (Huntington, N.Y., 1973).

[4] See Helmut Kingreen, *Theodor Wilhelm Engelmann (1843-1909): Ein bedeutender deutscher Physiologe an der Schwelle zum 20. Jahrhundert,* no. 6 in Münstersche Beiträge zur Geschichte und Theorie der Medizin, ed. K. E. Rothschuh et al. (Münster, 1972), pp. 49-50. I am most grateful to Prof. Rothschuh for sending me a copy of this and other volumes in the series free of charge.

laureate Otto Loewi came from Marburg to England in order to study with the "world-renowned physiologists" now abundant there, believing that Germany had "little by little lost its leadership to England."[5] This renaissance can scarcely be ascribed entirely to Foster's influence, but he was its most important leader. Among contemporaries, only John Scott Burdon Sanderson even approached him in the breadth and strength of his influence on English physiology.

Foster exerted his most profound influence at Cambridge. Under his inspiration and direction, the Cambridge School of Physiology became the leading center for physiological research in the English-speaking world and a school of international renown. H. Newell Martin, Walter Holbrook Gaskell, John Newport Langley, Sir Walter Morley Fletcher, Sir Joseph Barcroft, T. R. Elliott, and Nobel laureates Sir Charles Sherrington and Sir Henry Dale—these are only the most famous of the physiologists trained at Cambridge during Foster's years there. Foster retired in 1903, but his influence persisted long after through his students, notably Gaskell, who served until 1914 as university lecturer in physiology, and Langley, who succeeded Foster as professor of physiology and held the chair until 1925, to be succeeded in turn by Barcroft. Throughout this period, the continued success and growing fame of the Cambridge School of Physiology depended very largely on Foster's intellectual sons and grandsons. As founder and director of this school, Foster surely deserves to be ranked with Johannes Müller and Carl Ludwig as one of the three greatest teachers of physiology in the nineteenth century.

But Foster's influence was by no means confined to physiology. In other branches of biology, too, he inspired or encouraged men who later occupied important Cambridge posts. They include embryologist Francis Maitland Balfour; pathologist Charles Smart Roy; morphologists Adam Sedgwick and Arthur Shipley; psychologist James Ward; anthropologist A. C. Haddon; botanists Sydney Vines, Marshall Ward, and Francis Darwin; and physiological chemists Arthur Sheridan Lea and Sir William Bate Hardy. In the 1890's Foster led the fight for the establishment of a scientific school of agriculture at Cambridge. In 1898, he persuaded Frederick Gowland Hopkins to bring the new biochemistry. Hopkins soon produced important work on muscular metabolism and vitamins, winning a Nobel prize for the latter, and went on to become founder and director of the Cambridge School of Biochemistry, serving as professor from 1914 to 1943.

As the fame of its scientific faculty spread, the Cambridge Medical

[5] Otto Loewi, "An Autobiographical Sketch," *Perspectives in Biology and Medicine, 4* (1960), 1-25, on 10.

School grew apace. By 1900, from a moribund state in 1870, it had become probably the largest medical school in England. For his part in this spectacular rise, Foster has been included among "the great triumvirate" of the late nineteenth-century revival.[6] Across the entire range of biomedical studies, much of the luster of this golden era at Cambridge was due to Foster and to those who were in one way or another his students and protégés.

Those whom Foster sent out into the larger world were scarcely less distinguished. Archibald Liversidge left Cambridge to become professor of chemistry at the University of Sydney from 1872 to 1909. Henry Head became a renowned practitioner and teacher of neurology. From 1892 to 1919, John George Adami enjoyed a distinguished career as professor of pathology and bacteriology at McGill University in Montreal. In 1876, H. Newell Martin became first professor of biology at newly established Johns Hopkins University in Baltimore and thus carried to the United States the methods and style of teaching he had learned from Foster and T. H. Huxley. The Hopkins became in turn the training ground for many of the leaders of the next generation of American biologists, physiologists, and geneticists. Even today many British and American biologists and physiologists can trace their intellectual ancestry back to Foster.

In addition to his work as teacher, Foster made immense contributions to the organization and professionalization of physiology and other sciences. He wrote the useful section on nerve and muscle for Burdon Sanderson's *Handbook for the Physiological Laboratory* (1873), the first such work in English and a favorite target of the antivivisectionists.[7] A leading force in the founding of the British Physiological Society (1876), the first such society in the world, Foster also founded in 1878 and edited until 1894 the still active *Journal of Physiology*, the first English-language periodical devoted exclusively to physiology. Of his several manuals, the best known and most influential was *A Textbook of Physiology* (first edition, 1876). Distinguished by its graceful style, evolutionary perspective, and balanced discussion of unsettled physiological problems, Foster's textbook quickly became the leading work of its kind in English. Before 1900 it had gone through six English editions and part of a seventh and had been translated into Italian, German, and Russian.

[6] Walter Langdon-Brown, *Some Chapters in Cambridge Medical History* (Cambridge, 1946), pp. 88-99. See also Arthur Rook, ed., *Cambridge and Its Contribution to Medicine* (London, 1971), passim.

[7] Richard D. French, *Antivivisection and Medical Science in Victorian Society* (Princeton, 1975), pp. 47-50.

Long an outspoken advocate of greater public support for science, Foster became biological secretary to the Royal Society in 1881. Like Huxley, whom he succeeded, Foster used this powerful post to promote a closer partnership between the Society and the government and to secure research grants for those he thought worthy. Not everyone, not even all scientists, shared Foster's priorities and some disliked his manner of achieving his aims. Perhaps too fond of the trappings of power, he held onto his post until 1903, and the belief that he had served beyond the point of effectiveness was general enough for the Royal Society to pass a bylaw limiting the tenure of its secretaries.[8]

Foster sat on several important governmental commissions, chairing the Royal Commission on Tuberculosis in 1901. He was president of the British Association for the Advancement of Science in 1899, the same year he was made baronet. He played an important role in the founding of the International Association of Academies, the *International Catalogue of Scientific Papers*, and the International Physiological Congresses, the first of which was held in 1889 and at the fifth of which (in 1901) he was elected permanent honorary president. The English public perhaps knew him best as Member of Parliament for the University of London from 1900 to 1906. In this capacity, as in his post at the Royal Society, Foster did not win universal admiration. Whether out of principle or political naiveté, he switched parties in the middle of his term. Having run as a Liberal Unionist because of that party's opposition to Irish home rule, he crossed the aisle to join the Liberals because of their position on education. For this act of desertion, he was criticized, indeed caricatured, as a muddle-headed man of divided loyalties,[9] and he lost his bid for reelection in 1906.

Few men can have led so interesting a career and yet received so little attention. Since Foster was certainly one of the most important and powerful forces in Victorian science, it is astonishing that he escaped enshrinement in the *Life and Letters* genre so popular in his time. He is perhaps best remembered today for his biography of Claude Bernard (1899) and for his book of lectures on the history of physiology (1901). Never a prolific contributor to the physiological literature, Foster published no research papers after 1876, and his name is attached to no significant scientific discovery. He is somewhat better known for his achievements as teacher, organizer, and administrator; but these are elusive talents, and it has in the end remained something

[8] Henry Lyons, *The Royal Society, 1660-1940: A History of Its Administration under Its Charter* (Cambridge, 1944), p. 300.

[9] See, e.g., *The Westminster Gazette*, 12 January 1903.

of a mystery how he came to found such an impressive school of research biologists and physiologists.

To solve this mystery, it is necessary first to understand the intellectual climate and the social and institutional setting in which Foster lived and worked. It would be impossible to recognize what was distinctive in his approach without some awareness of the state of English physiology before 1870. And since Foster was above all else an educator, it is particularly important to understand the nature of English medical and scientific education before he came upon the scene. If the students who took Foster's courses found them a revelation, it can only have been by comparison with what had gone before.

But medical and scientific education in Victorian England were part of a more general educational setting, so that the history of teaching in physiology is but a chapter in the history of English education. The celebrated "reform" of the English universities had really only begun when Foster arrived at Cambridge in 1870. Scientific education, in particular, had been surprisingly little affected by the Royal Commission on the Universities in 1850. It was at least partly for this reason that Foster's approach seemed so new, and that his achievement belongs properly to the wider history of university reform.

In the words of English botanist W. T. Thiselton-Dyer, "Foster stood at the parting of the ways in biological methods of teaching."[10] But this is far from saying that he owed nothing to his predecessors or his age. To William Sharpey, his mentor at University College, and especially to T. H. Huxley, Foster owed a great deal indeed. Besides offering general inspiration and encouragement, both interceded on his behalf to arrange posts where he could exercise his talents and both taught him to view physiology as but one branch of the broad science of biology.

What Foster may have owed to his era is obviously more difficult to specify. His life (1836-1907) coincided almost precisely with the reign of Queen Victoria (1837-1901). Perhaps only in the seventeenth century has England, and especially English science, known a more vital period in its history. Whether one calls this period "the age of reform" or, in Gertrude Himmelfarb's more accurate phrase, "the age of conservative revolution,"[11] it was an age when success and influence came to depend almost as much on talent as on the social status into which one was born. Although the Anglican aristocracy remained strongly entrenched, political power was extended to other groups by the fran-

[10] W. T. Thiselton-Dyer, "Michael Foster—A Recollection," *Cambridge Review*, *28* (1907), 439-440, on 439.

[11] Gertrude Himmelfarb, *Victorian Minds* (New York, 1968), p. xi.

chise bills of 1832 and 1867. Religious dissenters achieved full civil rights, and a position in government no longer required subscription to the Thirty-nine Articles of the Church of England. In 1870, when Foster went to Cambridge, the benefits of elementary education were made available to all as a result of the Elementary Education Act. Democracy and meritocracy were on the ascendant in England, even if social and economic distinctions generally remained very sharp. A generation earlier, middle-class religious dissenters such as Huxley and Foster would have found it vastly more difficult to rise to positions of national prominence and influence.

Like the Anglican leadership they served and supplied, the venerable universities at Oxford and Cambridge long resisted this democratic and reforming impulse. By the 1870's, however, Cambridge in particular was prepared to undertake substantial reform. This change of attitude was crucial not only to the success of the Cambridge School of Physiology but also to that of the even more famous Cambridge School of Physics. That these two schools paralleled each other almost exactly in their rise and development is a curiously neglected fact. It serves to emphasize that Foster's achievement at Cambridge was not entirely isolated and in fact depended on a new spirit already gaining ground there before he arrived.

Whatever light these social and institutional factors may shed on it, Foster's achievement remains mysterious until one recognizes how much it depended on his conviction that physiology belonged to the rest of biology and until one emphasizes the little known fact that he was, in the early part of his career, a competent if unprolific research physiologist. The former conviction not only informed his teaching in physiology but also allowed him to exert a tremendous influence on the more general development of the biomedical sciences at Cambridge. In 1873 he introduced a complete course in elementary biology, modeled on one that Huxley had established in 1871 at South Kensington for selected schoolmasters. Foster's version seems to have been the first such course taught in any university. Its impact on the future development of English biology and physiology was enormous. Together with Foster's strong commitment to evolutionary principles, it expressed and reinforced a broadly biological perspective then becoming rare among Continental physiologists.

Even more crucial to the success of the Cambridge School was Foster's brief career as a research physiologist. From the beginning to the end of that career, the problem of the origin of the heartbeat held a special fascination for him, and experimental physiology got its start at Cambridge when Foster directed his own energies and those of his students toward a resolution of the problem. That this fundamental fact should have escaped earlier attention is not especially surprising. For one thing, Foster and his students concentrated on the problem of the heartbeat only very briefly (mostly between 1874 and 1877) and no one, including Foster himself, could have realized that a great research school was then and thus in the making. The entire process was very probably unconscious, informal, and unplanned.

Moreover, once Foster abandoned active research, his direct influence on the work carried out in his laboratory faded rapidly. By 1883, when he was promoted from praelector at Trinity College to professor of physiology in the university, Gaskell and Langley had assumed supervision of the day-to-day activities of the physiological laboratory. Insofar as the work done there during the next two decades had any focus, it no longer concerned the problem of the heartbeat, but the structure and functions of the nervous system. Indeed, the fundamental contributions of Gaskell and Langley to the field made Cambridge famous chiefly as a center for the study of the involuntary or autonomic nervous system.

Meanwhile, Foster had stepped into his more familiar role as lecturer, administrator, editor, and organizer of activities that often took him from Cambridge. Already in the 1880's he had begun to fit the image that Henry Dale recalled so vividly from his student days in the late 1890's:

> The outside claims on his time and activities . . . had already reached the stage at which it seemed to us normal, as we surged out of the theatre after a lecture, to see the Professor climbing into a waiting hansom cab, to catch the train which would take him to duties in London for the Royal Society and, in increasing measure, for the Government of the day.[12]

As his own research faded farther into the past, and as his impact on the new work disappeared, everyone naturally tended to forget that Foster had once been a competent researcher and an influential director of research. By the time of his death in 1907, even Gaskell and Langley had apparently forgotten how important their mentor had once been in determining their own initial interests and approach. In separate obituaries, Langley discussed Foster's work on the heart without recalling that his own first excursion into physiological research was related to it, while Gaskell scarcely referred at all to Foster's research and

[12] Henry Dale, "Sir Michael Foster, K.C.B., F.R.S., A Secretary of the Royal Society," *Notes and Records of the Royal Society of London, 19* (1964), 10-32, on 25.

introduced the phrase, often repeated in later years, that Foster was "a discoverer of men rather than of facts."[13]

However understandable, the tendency to forget Foster's short career as research physiologist and research director has helped to obscure some of the key elements in his achievement at Cambridge. Like Liebig before him, Foster spent several years working cheek by jowl with a small band of students in an undersized and underequipped laboratory. Quite apart from the opportunity it gave him to exert a direct influence on his students, this informal and physically inauspicious atmosphere promoted a sense of esprit de corps and offered an optimal setting for the exercise of the charismatic powers that Foster's students ascribed to him.[14] Indeed, Foster virtually required such a setting to be effective, for not even his most avid admirers went so far as to call him a magnetic lecturer; others found his lectures quite simply boring.[15] Only in small groups, it seems, and preferably in informal settings, could Foster communicate his subtle brand of inspiration. As the increasingly preoccupied and remote professor, his influence over young students waned. Accordingly, the years of his professorship (1883-1903) receive much less attention below than do the years of his praelectorship (1870-1883), which were in every sense the crucial and formative years in the rise of the Cambridge School.

The resulting emphasis on the problem of the heartbeat has several advantages. Besides permitting, indeed requiring, a thorough historical analysis of an important but hitherto neglected problem in nine-

[13] See J. N. Langley, "Sir Michael Foster—In Memoriam," *J. Physiol., 35* (1907), 233-246; and W. H. Gaskell, "Sir Michael Foster, 1836-1907," *Proc. Roy. Soc., B80* (1908), lxxi-lxxxi.

[14] On Foster's charismatic power to inspire students, see Chapter 6 below, "'Work, finish, publish.'" For an instructive comparison, see the discussion of Liebig's charisma in J. B. Morrell, "The Chemist Breeders: The Research Schools of Liebig and Thomas Thomson," *Ambix, 19* (1972), 1-46, esp. 6, 36-37.

[15] See e.g., Dale, "Sir Michael Foster," p. 28; and F. J. Allen, "Evolution of the Cambridge Medical School," *Brit. Med. J.*, 1920 (1), 505ff., 651-653, on 652-653. C. S. Sherrington went so far as to call Foster an "appalling lecturer," whose large classes were poorly attended. See Judith P. Swazey, *Reflexes and Motor Integration: Sherrington's Concept of Integrative Action* (Cambridge, Mass., 1969), p. 7. For the following piece of additional evidence, I am indebted to P. A. Merton of the Physiological Laboratory at Cambridge, who sent it to me in a private letter of 28 October 1972: "I once talked to (H. W.?) Mills (now dead), the Emeritus Reader in Stereochemistry, who had attended Michael Foster's lectures as an undergraduate. He looked, Mills said, exactly like the picture of him that shows him holding a piece of chalk. He spoke deliberately, slowly waving the chalk backwards and forwards, with a very soporific effect. It used to be said by the undergraduates of the day that you could take perfect notes of his lectures by buying his text book and crossing out the bits he didn't read."

teenth-century physiology, it gives the book an internal focus it might otherwise lack. Moreover, by emphasizing the contribution of the evolutionary approach to the successful resolution of the problem, this analysis may provide some clues to the difference in approach between English and Continental physiology during the late nineteenth century, and some insight into the reasons why at least some physiologists felt that Germany "little by little lost its leadership to England" during that period.

2. The Stagnancy of English Physiology, 1840-1870

Speaking of emancipation, don't you think science wants a little heroic striking off of fetters. I mean of course you do, but don't you think something useful might be done by comparing [in print] the scientific work done by Englishmen with that done by furriners during the last 30 years, shewing the influence of fetters? In *physiology* at least we should look *very* small.

Michael Foster to T. H. Huxley (1865) [1]

By the early 1870's Foster's lament had been taken up by others, as there emerged among English students of physiology a general awareness that their country, the richest and most powerful in the world, had contributed little to the recent spectacular developments in physiology. In September 1870 Peter Braidwood, editor of the *Liverpool Medical and Surgical Reports,* sent a letter to *Nature* deploring the depressed state of British physiology. "Though foremost in many things," he wrote, "Britain is far behind Continental countries in the field of physiological science."[2] In 1872, when the British Association met at Brighton, John Scott Burdon Sanderson told the section on anatomy and physiology that "we English physiologists . . . must admit with regret that we have had very little to do with the unprecedented development of our science during the last two decades." He predicted that it would take years of effort "to regain the position which we in England once had and ought never to have lost."[3] One year later, the same audience heard William Rutherford say how strange it seemed to him that "the country of Harvey, John Hunter, Charles Bell, Marshall Hall and John Reid should not always have been in the front rank as regards physiology."[4]

That these statements had real justification is clear from a variety

[1] Foster to Huxley, 1 July 1865, Huxley Papers, IV. 150. The Huxley Papers are deposited in the archives of the Imperial College of Science and Technology, South Kensington, London. I am grateful to the Governors of Imperial College for allowing me to quote from these papers, and to the staff of the American Philosophical Library, Philadelphia, whose microfilm copy of the papers I used.

[2] P. M. Braidwood, "English Physiology," *Nature, 2* (1870), 413-414, on 413.

[3] J. S. Burdon Sanderson, "Address to the Department of Anatomy and Physiology," *Brit. Assoc. Rep., 42* (1872), Transactions, pp. 145-150, on 145, 147.

[4] William Rutherford, "Address to the Department of Anatomy and Physiology." ibid., *43* (1873), Transactions, pp. 119-123, on 122.

of sources recently brought together by Joseph Ben-David.[5] Especially striking evidence comes from Karl Rothschuh's catalog of physiological discoveries, which lists almost four hundred German contributions for the period 1840-1870. For the same period, the number of English contributions is a paltry twenty-two. English contributors fared even less well during the last two of these three decades. Between 1850 and 1870, while German physiologists made almost three hundred contributions, English workers made but nine, including none at all between 1860 and 1865.[6] If Rothschuh has overlooked any really significant English contributions, they are apparently also unknown to other historians of physiology. His compilation provides impressive support for Edward Sharpey-Schafer's judgment that in England "physiology had ceased to exist as an active science" during the 1860's.[7]

Strictly speaking, of course, English physiology had not really ceased to exist. It only seemed that way by comparison with the remarkable developments on the Continent. During the middle years of the nineteenth century, physiology was pursued in England much as it had always been pursued, while in France and even more dramatically in Germany it underwent immense qualitative change and quantitative expansion. To understand the factors contributing to the relative stagnancy of English physiology during this period, we must first gain some insight into physiology as it was then being pursued in Europe.

[5] See Joseph Ben-David, *The Scientist's Role in Society, A Comparative Study* (Englewood Cliffs, N.J., 1971), pp. 188-190, tables 1-3.

[6] See A. Zloczower, "Career Opportunities and the Growth of Scientific Discovery in 19th Century Germany with Special Reference to the Development of Physiology," unpublished M.A. thesis (Hebrew University, Jerusalem, 1960), p. 7. I am grateful to the authorities of Hebrew University for making a copy of this thesis available to me. Zloczower's data, which are also reproduced in Ben-David, *Scientist's Role*, p. 188, table 1, are based on an analysis of K. E. Rothschuh, *Entwicklungsgeschichte physiologischer Probleme in Tabellenforme* (Berlin, 1952). Under "English" contributions, Zloczower (and thus Ben-David) includes a very few by Scottish or Irish workers. I have subtracted these non-English contributions in arriving at the figures reported in my text.

It should be emphasized that Rothschuh's catalogue is heavily biased in favor of German contributions. For the period 1800-1840 he cites 92 German contributions, compared to 53 French and 29 English. For the period 1870-1900, he gives Germans credit for 362 discoveries, compared to 36 by French and 67 by English physiologists. On other grounds, one would expect to find the French more prominent between 1800 and 1840, and the English much nearer in number to the German discoveries between 1870 and 1900. Nonetheless, Rothschuh's compilation offers a valuable rough guide to the shifting fortunes of physiology in the three countries.

[7] E. A. Sharpey-Schafer, "Developments of Physiology," *Nature, 104* (1919), 207-208.

National styles in Continental physiology, 1840-1870

Most of what has been written about this period in physiology has focused on Claude Bernard in France and the "1847 Group" in Germany. Among other things, Bernard contributed fundamentally to our understanding of the functions of the pancreas in digestion, the glycogenic function of the liver, the vasomotor system, and the action of poisons, especially curare and carbon monoxide.[8] The 1847 Group included Hermann Helmholtz, Emil DuBois Reymond, Ernst Brücke, and Carl Ludwig. Its name derives from the year in which these four men met in Berlin and issued a famous pledge to reconstitute physiology by explaining vital phenomena in terms of physical and chemical principles.[9] The important contributions of Helmholtz and DuBois Reymond to neurophysiology and of Ludwig to the mechanics of the circulatory system are well known in their general features, and are often regarded as evidence of the value of their program.

What has especially attracted the interest of historians is the apparent difference in approach between Bernard and the German group, and the possible derivation of this difference from a dissimilarity in philosophical outlook. There has been a tendency to place Bernard in the tradition of his teacher, Magendie, and to assume that the strong commitment of both to vivisection was somehow related to their more or less vitalistic or at least organismic point of view. The Germans, by comparison, have been portrayed as the leaders of the new physicochemical revolution, as men whose commitment to the techniques of physics and chemistry depended on their thoroughgoing mechanism. Thus Owsei Temkin has proposed a distinction between the French "vitalistic materialists" and the German "mechanistic materialists," and Everett Mendelsohn has suggested that the

[8] The literature by and about Bernard is extensive. For a comprehensive guide, see M. D. Grmek, *Catalogue des manuscrits de Claude Bernard avec la bibliographie de ses travaux imprimés et des études sur son oeuvre* (Paris, 1967). The standard biography is J.M.D. Olmsted, *Claude Bernard, Physiologist* (New York, 1938). See also Joseph Schiller, *Claude Bernard et les problèmes scientifiques de son temps* (Paris, 1967); and M. D. Grmek, "Claude Bernard," *DSB*, II (1970), 24-34. For two important recent studies, which exhaustively examine Bernard's contributions to toxicology and digestive physiology respectively, see M. D. Grmek, *Raisonnement expérimental et recherches toxologiques chez Claude Bernard* (Genève, 1973); and Frederic Lawrence Holmes, *Claude Bernard and Animal Chemistry: The Emergence of a Scientist* (Cambridge, Mass., 1974).

[9] See Paul F. Cranefield, "The Organic Physics of 1847 and the Biophysics of Today," *J. Hist. Med., 12* (1957), 407-423.

French and German physiologists of the nineteenth century adopted different "national styles" of investigation and explanation.[10]

But if scientific methodology and thought often do reflect different national styles, we need to know much more than we now do about the possible sources of such differences. In the case of nineteenth-century physiology, the possible contribution of institutional differences to national styles has scarcely been explored. In placing so much emphasis on such quasi-philosophical currents as vitalism or mechanism, historians have sometimes ignored the complexity of the relationship between philosophical outlook and physiological methodology. As Paul Cranefield points out, physiologists who are indifferent or even opposed to philosophical mechanism may nonetheless employ purely mechanistic methods of investigation.[11] Moreover, the differential appeal of mechanism or vitalism may itself depend in part on institutional factors. It may depend, for example, on the nature of scientific education in each country and on the availability of laboratories, instruments, and opportunities for research.[12]

In terms of these institutional factors, the differences between French and German physiology during the mid-nineteenth century seem to be differences of degree rather than of kind.[13] One may wonder, therefore, whether the distinction between the French and German approaches to physiology has not been somewhat exaggerated. We certainly need to know more about the work and thought of lesser known physiologists in both countries who seem not to conform to these supposed national distinctions. If, for example, we knew more about the work and influence of such French "mechanists" as Henri Dutrochet (1776-1847), J. L. M. Poiseuille (1799-1869), J. B. August Chauveau (1827-1917), and E. Jules Marey (1830-1904), or such

[10] See Owsei Temkin, "The Philosophical Background of Magendie's Physiology," *Bull. Hist. Med.*, *20* (1946), 10-35, esp. 22-23; Temkin, "Materialism in French and German Physiology of the Early Nineteenth Century," ibid., *20* (1946), 322-327; Everett Mendelsohn, "The Biological Sciences in the Nineteenth Century; Some Problems and Sources," *History of Science*, *3* (1964), 39-59, on 45, 48-49; and esp. Mendelsohn, "Physical Models and Physiological Concepts: Explanation in Nineteenth Century Biology," *Brit. J. Hist. Sci.*, *2* (1965), 201-219.

[11] Cranefield, "Organic Physics," p. 422.

[12] Some preliminary evidence for the views expressed in this paragraph can be gleaned from ibid.; Schiller, *Bernard*; and F. L. Holmes, untitled book review, *Bull. Hist. Med.*, *39* (1965), 489-491. For an enriched perspective on the complexity of the relationship between philosophical outlook and physiological methodology, see also Russell C. Maulitz, "Schwann's Way: Cells and Crystals," *J. Hist. Med.*, *26* (1971), 422-437; and the remarkable posthumous essay by Charles Culotta, "German Biophysics, Objective Knowledge, and Romanticism," *Historical Studies in the Physical Sciences*, *4* (1975), 3-38.

[13] See Joseph Ben-David, "Scientific Productivity and Academic Organization in Nineteenth-century Medicine," *American Sociological Review*, *25* (1960), 828-843.

German "neo-vitalists" as Rudolf Heidenhain (1834-1897) and E. F. W. Pflüger (1830-1904), then the supposed distinction between "vitalistic" France and "mechanistic" Germany might begin to break down. It is particularly interesting in this connection that Emil DuBois Reymond of the 1847 Group regarded his move toward mechanism as a move toward Dutrochet's point of view.[14] Even where Bernard and the leading German physiologists are concerned, the tendency to draw sharp distinctions may obscure how important chemical analysis was to Bernard's work, and how indispensable vivisection was to the German group, or at least to Ludwig.[15]

Bernard himself, despite his well-known antipathy to the approach of certain German physiologists and animal chemists, did not separate French and German physiology into sharply distinct traditions. For him the essential achievement of nineteenth-century physiology was simply its systematic introduction of "the experimental method in the science of vital phenomena." Though this method had originated in French laboratories, it was "perfected" and "brought forth more fruit" in other countries, especially Germany, because of the more extensive development there of laboratories or physiological institutes, "admirably endowed for the experimental study of vital phenomena."[16] For Bernard, apparently, the ascendancy of German over French physiology had nothing to do with a commitment to philosophical mechanism. In his view, "men of science achieve their discoveries, their theories and their science apart from philosophers." Rather, the crucial factor was the nature and extent of laboratory development, "for scientific methods are learned, in fact, only in laboratories."[17] If we pursued the implications of Bernard's argument and concentrated more on what French and German physiologists actually did in their laboratories and less on their programmatic and retrospective statements, then we might be surprised at the degree of similarity between their approaches.[18]

[14] See Vladislav Kruta, "Rene-Joachim-Henri Dutrochet," *DSB*, IV (1971), 263-265, on 265.

[15] Cf. the sources in n. 12 above.

[16] Claude Bernard, *An Introduction to the Study of Experimental Medicine*, trans. by Henry Copley Green, with an introduction by Lawrence J. Henderson and a new foreword by I. Bernard Cohen (New York, 1957), pp. 147-148.

[17] Ibid., p. 225. For further evidence that Bernard attributed the triumph of German physiology to its laboratories, see Claude Bernard, *Rapport sur les progrès et la marche de la physiologie générale en France* (Paris, 1867).

[18] Promising attempts to separate what was actually done in the laboratory from programmatic statements of vitalism or mechanism have been made by Schiller, *Bernard*; Cranefield, "Organic Physics;" and Culotta, "German Biophysics." Although Schiller obviously believes that Bernard's approach places him apart from and above other contemporary physiologists, he tends otherwise to narrow the

The anatomical bias of English physiology

Certainly by comparison with the situation in England, the general similarity between French and German physiology is more striking than any differences. On the Continent physiology had become an increasingly rigorous and more broadly based experimental science, while in England it remained for the most part subsumed in anatomy and in religious and philosophical issues. By 1860 the leading English textbooks must have sounded very strange indeed to a Continental physiologist. William Carpenter, whose textbook was the standard work on the subject in Britain, continued to combine moral with physiological instruction, warning of the dangers of excessive sexual intercourse, and speaking in raptures about the Creator and man's immortal soul. A very large portion of his book was devoted to what can perhaps best be described as moral psychology, his chief aim being to preserve the notion of free will.[19] And at the very time when Continental physiologists sought to free their discipline from its subservience to anatomy, Englishmen Robert Todd and William Bowman spoke of their desire to give anatomy "a greater degree of prominence than had been usual in Physiological works."[20] By this point, whatever their differences, French and German physiologists would have agreed that moral pronouncements were out of place in a textbook of physiology, and that what was needed was not more anatomy, but more experimentation.

The most obvious obstacle to physiological experimentation in nineteenth-century Britain was the strength of antivivisection sentiment there.[21] What is not so obvious is the very real possibility that

---

more general distinction between French and German physiology. So, too, does Cranefield, who concludes (p. 421) that "the great success of the 1847 group was based on methods which may justly be called experimental, biological, and vivisectional, rather than on a theoretical bias toward the physical and chemical." Although Culotta confines his attention to German physiology, and though his results point toward a rather more nuanced conclusion, he clearly reveals the breadth of research concerns among the practitioners (as distinct from the philosophical spokesmen) of German biophysics.

[19] W. B. Carpenter, *Principles of Human Physiology*, a new American from the last [fifth] London edition, ed. Francis Gurney Smith (Philadelphia, 1856), esp. pp. 45, 435-437, 517-649, 752-753.

[20] Robert Bentley Todd and William Bowman, *The Physiological Anatomy and Physiology of Man*, complete in one volume (Philadelphia, 1857), p. vii.

[21] For the most complete and probing account of the English antivivisection movement, see Richard D. French, *Antivivisection and Medical Science in Victorian Society* (Princeton, 1975). See also Hubert Bretschneider, *Der Streit um die Vivisektion im 19. Jahrhundert: Verlauf, Argumente, Ergebnisse* (Stuttgart, 1962); Mark N. Ozer, "The British Vivisection Controversy," *Bull. Hist. Med., 40* (1966), 158-

this sentiment exerted its most deleterious influence on British physiology between 1840 and 1870, even though no organized antivivisection movement existed until the 1870's. In that decade, agitation spearheaded by the newly formed antivivisection societies led to the appointment of a Royal Commission and to the enactment of the world's first legislation to control vivisection. T. H. Huxley had hoped that Parliament's fox-hunting proclivities might prevent any such "fatal legislative mistake" and predicted that its enactment would mean that "we shall import our physiology as we do our hock and our claret from Germany and France. . . ."[22]

Curiously enough, however, the agitation for and passage of the Vivisection Act of 1876 may have done British experimental physiology more good than harm. The advocates of animal experimentation made a better case in Parliament than their opponents had expected, and the act as finally passed by no means satisfied the more fervent antivivisectionists. It stipulated fairly severe restrictions on animal experimentation, but these restrictions could be removed by statutory certificates, awarded to qualified experimentalists at the discretion of governmental officials in the Home Office. For six years, the administration of the act did impede a few would-be vivisectors, especially in Scotland, but it most assuredly did not eliminate vivisection experiments. And by 1882, the act was being administered in such a way as virtually to guarantee certificates to qualified physiologists.[23] Furthermore, the emergence of an organized antivivisection movement awakened a sense of esprit de corps among the small band of British physiologists, and led to the formation in 1876 of the still active Physiological Society of Great Britain. At the time, no such society existed even in France or Germany, and the antivivisection movement must assume inadvertent responsibility for this incrongruous act of leadership on the part of British physiology. E. A. Sharpey-Schafer (né Schäfer), a charter member of the Society, later expressed this ironic state of affairs with the phrase "ex malo bonum."[24] After 1876, British

167; Lloyd G. Stevenson, "Religious Elements in the Background of the British Anti-Vivisection Movement," *Yale Journal of Biology and Medicine, 29* (1956), 125-127; and Stevenson, "Science Down the Drain: On the Hostility of Certain Sanitarians to Animal Experimentation, Bacteriology and Immunology," *Bull. Hist. Med., 29* (1955), 1-26.

22 Leonard Huxley, *Life and Letters of Thomas Henry Huxley*, 2 vols. (London, 1900), I, pp. 434-435. Huxley sat on the Royal Commission on Vivisection, and few subjects produced so many examples of his celebrated wit in prose style. Cf. ibid., I, pp. 427-441.

23 See French, *Antivivisection*, esp. Chapter 7.

24 See Edward Sharpey-Schafer, *History of the Physiological Society during its First Fifty Years, 1876-1926* (Cambridge, 1927), p. 5.

physiologists may even have felt somewhat less vulnerable to innuendoes against their personal character, since their experiments in a sense now carried the sanction of Parliament.

None of the paradoxical advantages stemming from the organized antivivisection movement of the 1870's was available to earlier British physiologists. Not even the introduction of anaesthetics in the 1840's could stem the flow of antivivisection sentiment. It is hard to explain why this sentiment developed such strength in the nineteenth century and why it centered so preeminently in England (or Britain), but Lloyd Stevenson and Richard D. French have implicated a wide variety of factors, including xenophobia, the classical emphasis of British education, the dominance of aristocratic values in a class-conscious society, the special appeal of the pastoral ideal during a period of rapid urbanization, the meliorist spirit of the "age of reform," a widespread irritation at the growing hubris of science and "scientific medicine," the social and psychological implications of women's circumscribed role in Victorian society, and certain themes in British religion and philosophy.[25] The German physiologist Rudolf Heidenhain captured part of the explanation when he linked antivivisection sentiment with the English campaign against formal gardens. The English, he said, looked upon both vivisection and formal gardens as artificial and unjustified intrusions on God's natural creations.[26]

Whatever its sources, antivivisection sentiment increased in strength as animal experiments increased in frequency. Marshall Hall, the most eminent of England's tiny band of mid-century experimental physiologists, was one who felt the force of the rising tide. In 1847 the *Medico-Chirurgical Review* accused him of unjustified cruelty in some of his experiments on living animals. Hall vehemently denied the charge and resuscitated a proposal of his, first made in 1831, that there be formed in England a Physiological Society, "the objects of which should be . . . to prevent cruelty, to promote physiological science, and, as has now become requisite, to protect individual character."[27] Thirty years later, similar objectives animated the founders of the Physiological Society of Great Britain, but in Hall's day the supporters and practitioners of experimental physiology were so few that he spoke to no avail. Other evidence amply confirms the presence and strength of English antivivisection sentiment during the 1840's and

[25] See Stevenson, "Religious Elements"; and French, *Antivivisection*, esp. Chapters 2 and 9-11.

[26] See Bretschneider, *Streit um Vivisektion*, pp. 16-17.

[27] Marshall Hall, "On Experiments in Physiology as a Question of Medical Ethics," *Lancet*, 1847 (1), 161. Cf. ibid., pp. 58-60, 135.

1850's,[28] and when Claude Bernard succeeded Magendie at the Collège de France in 1856, he suggested that the persistence of this "prejudice" was "one of the great obstacles retarding the development of experimental physiology in [England]."[29]

Another obstacle, by no means unrelated to antivivisection sentiments,[30] was the powerful grip that natural theology exerted on the English mind in the nineteenth century. Few works indeed were so widely read or so influential in early Victorian England as William Paley's *Natural Theology*, which students like Charles Darwin learned almost by heart,[31] and the *Bridgewater Treatises*, in which the "power, wisdom, and goodness" of God were to be demonstrated by examples drawn from nature.[32] Viewed from the perspective of natural theology, the only legitimate goal of physiology was to demonstrate that each and every structure had been perfectly designed by the Creator to subserve a function. This is the attitude expressed in the 1850's by Todd and Bowman in their guiding principle that "each organ has its proper use in the animal or plant," a principle in which they saw the "strongest evidence of Design" and "incontestable proof that the whole was devised by One Mind, infinite in Wisdom, unlimited in resource."[33] Virtually all British students of physiology shared this attitude. To them, function apart from structure was inconceivable, and experiment served at best to confirm functions more properly deduced from anatomy. In France, meanwhile, Magendie and Bernard criticized the tendency to correlate function so exclusively with the structure of particular organs; function, they emphasized, often transcends the activity of individual organs and can even exist in the absence of anatomical structure.[34] In their view, physiology could advance only by securing its emancipation from anatomy, and experiment was for them the key to this emancipation.

During the middle years of the nineteenth century, then, the gen-

[28] See French, *Antivivisection*, esp. Chapter 2.

[29] Claude Bernard, *Fr. Magendie: Leçon d'ouverture du cours de médecine du Collège de France* (Paris, 1856), p. 14.

[30] Cf. French, *Antivivisection*, pp. 22, 36, 307-308, 350, 352-355, 358, 362-364, 368.

[31] See Francis Darwin, ed., *The Life and Letters of Charles Darwin*, 2 vols. (New York, 1904-1905), II, p. 15; I, pp. 40-41. The full title of Paley's work was *Natural Theology; or, Evidences of the Existence and Attributes of the Deity, Collected from the Appearances of Nature* (London, 1802). By 1850 this work had passed through thirty reprints or editions!

[32] On the *Bridgewater Treatises* in general, see Charles C. Gillispie, *Genesis and Geology* (Cambridge, Mass., 1951), esp. pp. 209-216, 246-248, 298.

[33] Todd and Bowman, *Physiological Anatomy*, p. 30.

[34] See Schiller, *Bernard*, pp. 42-48; J. M. D. Olmsted, *François Magendie: Pioneer in Experimental Physiology and Scientific Medicine in XIXth Century France* (New York, 1944), p. 31; and Bernard, *Introduction*, pp. 105-112.

eral British attitude toward physiology was little more than a sterile imitation of the views expressed a generation earlier by Sir Charles Bell (1774-1832). Largely by ignoring or manipulating strong evidence to the contrary, Bell tried to claim priority over Magendie for the most important physiological discovery of the nineteenth century —namely, the distinction between motor and sensory nerves.[35] Insofar as Bell did approach this discovery, vivisection experiments played a significant role. But Bell evidently abhorred such experiments, for he wrote to his brother on 1 July 1822:

> I should be writing a third paper on the Nerves, but I cannot proceed without making some experiments, which are so unpleasant to make that I defer them. You may think me silly, but I cannot perfectly convince myself that I am authorized in nature, or religion, to do these cruelties—for what?—for anything else than a little egotism or self aggrandisement.[36]

Within two months, Magendie, who had no such qualms about vivisection, announced his discovery of the sensory-motor distinction. Thereafter Bell undertook a prolonged and systematic public campaign to denigrate the role of experiments, both Magendie's and his own, in this and all other physiological discoveries. In 1823, Bell went so far as to claim that "experiments have never been the means of discovery; and a survey of what has been attempted of late years in physiology will prove that the opening of living animals has done more to perpetuate error than to confirm the just views taken from the study of anatomy and natural motions."[37] For sentiments such as these, Bell became a darling of the antivivisectionists, and he did not recoil from the embrace.

Not surprisingly, Bell also shared with his compatriots a strong commitment to natural theology. In fact, his work on the human hand was one of the most famous of the *Bridgewater Treatises*.[38] He also reflected and exploited a widespread English xenophobia, espe-

[35] For an exhaustive discussion of the priority dispute, with reproductions of passages from the pertinent primary and secondary sources, see Paul F. Cranefield, *The Way In and the Way Out. François Magendie, Charles Bell, and the Roots of the Spinal Nerves* (New York, 1974).

[36] Charles Bell, *Letters of Sir Charles Bell Selected from his Correspondence with his Brother George Joseph Bell* (London, 1870), pp. 275-276.

[37] Charles Bell, "Second Part of the Paper on the Nerves of the Orbit," *Phil. Trans.*, *113* (1823), 298-307, on 302.

[38] Charles Bell, *The Hand, Its Mechanism and Vital Endowments, as Evincing Design*, *Bridgewater Treatises* IV (London, 1833). This work reached its seventh edition in 1870. For examples of Bell's attitude toward natural theology, see his *Letters*, pp. 260, 424-425. For his connection with the *Bridgewater Treatises*, see Benjamin Spector, "Sir Charles Bell and the Bridgewater Treatises," *Bull. Hist. Med.*, *12* (1942), 314-322.

cially toward postrevolutionary France, by insisting on "the superiority of English physiology to French, which is so improperly popular,"[39] and by exhorting his disciples to continue in the path of "their own great Countrymen":

> Surely it is time that the schools of this kingdom should be distinguished from those of France. Let physiologists of that country borrow from us, and follow up our opinions by experiments but let us continue to build up that structure which has been commenced in the labours of [our great anatomists].[40]

In the 1820's, when Bell took his stand, his denigration of experiment and exaltation of anatomy could have had only minimal affect on the international standing of English physiology. By the 1850's, however, his continuing influence was yet another factor contributing to its isolation and stagnancy.

Owsei Temkin has already suggested that antivivisection sentiment, natural theology, and chauvinism "gave English physiology and biology a setting different from that of continental Europe."[41] That the most specific common effect of these factors was to foster a deep and persistent anatomical bias in English physiology seems also to have been recognized, especially by Lloyd Stevenson, but the point deserves greater emphasis.[42] What needs further to be appreciated is that this same bias was also a natural consequence of the organization and goals of medical and scientific education in nineteenth-century England.

Anatomical bias and the structure of English medical education[43]

Still the most striking feature of the English medical profession in the nineteenth century was its division into three distinct orders—the physician, the surgeon, and the apothecary (as well as a fourth, hybrid

39 Bell, *Letters*, p. 263.

40 As quoted in Olmsted, *Magendie*, p. 43.

41 Owsei Temkin, "Basic Science, Medicine, and the Romantic Era," *Bull. Hist. Med.*, *37* (1963), 97-129, on 114. Many of the same issues are raised in June Goodfield-Toulmin, "Some Aspects of English Physiology: 1780-1840," *J. Hist. Biol.*, 2 (1969), 283-320.

42 See Lloyd G. Stevenson, "Anatomical Reasoning in Physiological Thought," in *The Historical Development of Physiological Thought*, ed. Chandler McC. Brooks and Paul F. Cranefield (New York, 1959), pp. 27-38, esp. 33. As early as 1873, William Rutherford (n. 4) deplored the anatomical bias of English physiologists, associating it with their lack of training in the physical sciences and their consequent inability to think in submicroscopic or "molecular" terms.

43 On the general state of the medical profession and medical education in nineteenth-century England, my views owe much to Charles Newman, *The Evolution of Medical Education in the Nineteenth Century* (London, 1957).

order, the surgeon-apothecary). These distinctions resulted from the historical circumstance that there had grown up a multiplicity of licensing bodies, of which the three most important were the Royal College of Physicians, the Royal College of Surgeons, and the Worshipful Society of Apothecaries. A license from the Royal College of Physicians permitted one to prescribe medicines and, legally, to perform surgery as well, though a physician rarely so demeaned himself. Licentiates of the College of Surgeons were supposed to practice surgery only, and those from the Society of Apothecaries were expected only to prepare and dispense internal remedies.[44] Since, however, physicians were so few and so expensive, these guidelines were frequently ignored in practice, and the apothecaries and surgeon-apothecaries performed, in reality, the functions of the general practitioner. In consequence, the lines separating the three orders became increasingly blurred as the century progressed, with the rapid growth of the urban middle classes increasing greatly the demand for medical care of all kinds.[45] Nonetheless, the basic pattern of medical education changed only slowly in response to the new realities of medical practice.

Of the three orders of medical practitioners, only the rare and elite physicians were expected to have a university education or an M.D. degree. The surgeon and apothecary classes were trained instead by apprenticeship and (with increasing regularity) in the proprietary medical schools and teaching hospitals concentrated in London. The faculties in these schools were drawn almost exclusively from the hospital staffs, and they were concerned above all with matters of practical value. Chemistry was given an important curricular place, but it more nearly resembled eighteenth-century mineralogy than nineteenth-century chemistry—perhaps out of concern for such utilitarian matters as urinogenital calculi. Botany, or more properly materia medica, was also prominent in the curricula of these schools because it too was presumed to possess therapeutic value.[46]

Most obviously practical of all the sciences was anatomy, more particularly human anatomy, which was "less pursued and cultivated for itself than in its applications to surgery." Nothing so distinguished

[44] With the passage of the Apothecaries Act of 1815, apothecaries won the right to prescribe as well as to dispense medicines, but it remained illegal for them to charge a fee for prescribing. See, e.g., William H. McMenemy, "Education and the Medical Reform Movement," in the *Evolution of Medical Education in Britain*, ed. F. N. L. Poynter (London, 1966), pp. 136-137. To this act, however, can largely be traced the rising influence of the apothecaries on medical practice and education, and the concomitant decline of the influence of the university-educated physicians.

[45] S. W. F. Holloway, "Medical Education in England, 1830-1858: A Sociological Analysis," *History, 49* (1964), 299-324.

[46] See Newman, *Medical Education*, pp. 97-100.

the London medical schools as their "anatomical-chirurgical character."[47] The physicians and surgeons of this period expected that the art of healing could best be served by following the lead of the French clinical school—that is, by concentrating on morbid anatomy and by utilizing the new methods of physical diagnosis (notably the stethoscope). As, therefore, the emphasis shifted from patient-reported symptoms to physical signs, "the practice of medicine," writes Charles Newman, "came to be more and more concerned with structural changes and their detection."[48] Together with the traditional dependence of surgery on anatomy, it was this shift toward the structural in medical practice that largely accounted for the "anatomical-chirurgical" emphasis of the London hospital schools. Institutions whose sole educational aim was to train competent medical practitioners quite naturally insisted on the utility of the subjects included in their curricula.

For experimental physiology, the yoke of utility was a burdensome one. It is difficult indeed to think of any properly physiological discovery that made a significant direct impact on the art of healing before 1870. One may wonder whether Claude Bernard's sole candidate, the experimental eradication of "the itch," seemed a very impressive argument for the utility of experimental physiology.[49] In 1874 Foster attempted to justify vivisection by claiming that it had already played an important role in the advance of the medical art, but he could offer only two specific examples of its direct usefulness. He claimed that animal experiments had been crucial to the advance of methods of ligature, and to the treatment in particular of aneurysms, and he emphasized that Claude Bernard's work on the glycogenic function of the liver had provided "the only gleam of light into [diabetes] which we possess."[50] But Foster simplified and probably exaggerated the role of animal experiments in the discovery of "Hunter's method" of treating aneurysms,[51] and he openly admitted that even Bernard's

[47] Adolph Muehry, *Observations on the Comparative State of Medicine in France, England, and Germany, during a Journey into these Countries in the Year 1835*, trans. by E. G. Davis (Philadelphia, 1838), pp. 94-99.

[48] Newman, *Medical Education*. pp. 83-96, quote on 85. On the French clinical school, see Erwin H. Ackerknecht, *Medicine at the Paris Hospitals, 1798-1848* (Baltimore, 1967).

[49] Bernard, *Introduction*, pp. 214-215.

[50] Michael Foster, "Vivisection," *Macmillan's Magazine, 29* (1874), 337-376, on 374-375.

[51] See Lloyd G. Stevenson, "The Stag of Richmond Park: A Note on John Hunter's Most Famous Animal Experiment," *Bull. Hist. Med., 22* (1948), 467-475; and Stevenson, "A Further Note on John Hunter and Aneurism," ibid., *26* (1952), 162-167.

work had accomplished little in a therapeutic sense; diabetes remained incurable. If metabolic studies had contributed some dietary rules of marginal value, and if the work of Marshall Hall permitted a more accurate diagnosis of certain nervous disorders,[52] nonetheless medical practice before 1870 followed a path quite independent of that followed by experimental physiology, or even experimental science in general.[53] Precisely because experimental science seemed so irrelevant to medical practice, it was scarcely represented at all in the curricula of the London hospital schools until very near the end of the century. And when it was finally introduced, it was not for its own sake or for its methodology, but for its immediate practical value. In physics, for example, the principles of heat and mechanics were imparted so that students might later be able to give useful advice on such domestic matters as ventilation.[54]

When taught at all in the London hospital schools, physiology was taught in conjunction with anatomy or from an anatomical point of view,[55] and almost without exception by a medical practitioner recruited from the hospital staff. At St. Mary's and the London Hospital medical schools, physiology was taught by the clinical staff into the 1880's.[56] At Guy's, it was not until the 1870's that Henry Pye-Smith urged the appointment of "a competent Physiologist, unconnected with the practice of Medicine or Surgery," and even then the proposal was rejected out of fear that "such a teacher would be very apt

[52] See Charlotte Hall, *Memoirs of Marshall Hall, M.D., F.R.S.* (London, 1861), pp. 479-489. From the point of view of *therapy*, however, Hall's work offered very little help; the therapeutics of the spinal system was still "in its infancy." Ibid., p. 489.

[53] Probably the first major impact of experimental science on medical practice is to be found in Lister's use of Pasteur's work for the introduction of antiseptic surgery in the late 1860's and early 1870's. On the whole, the separation between medical science and medical practice before about 1870 seems to conform to the more general pattern of the relationship between science and technology. I hope to develop this theme more completely in a separate study. A preliminary airing of some of the issues can be found in Derek J. De Solla Price, "Is Technology Historically Independent of Science? A Study in Statistical Historiography," *Technology and Culture, 6* (1965), 553-568, on 566-567.

[54] Newman, *Medical Education*, pp. 208-209.

[55] As Newman (ibid., p. 108) should have recognized when he claimed, on the authority of James Paget, that physiology was separated from anatomy in most of the London medical schools in the 1840's. Paget's own assessment of his course and of physiological teaching in the other hospital schools shows how exaggerated Newman's claim is. See Stephen Paget, ed., *Memoirs and Letters of Sir James Paget* (London, 1902), pp. 130-131, 218-219.

[56] See Zachary Cope, *The History of St. Mary's Hospital Medical School, or a Century of Medical Education* (London, 1954), pp. 22, 80-82; and A. E. Clark-Kennedy, *The London: A Study in the Voluntary Hospital System*, 2 vols. (London, 1963), II, pp. 75, 78, 165.

to overlook the special bearings of Physiology on the practice of Medicine and Surgery and to treat the subject as one of abstract science." The teaching of physiology was entrusted instead to a surgeon until 1898.[57]

In their aims and organization, then, the London hospital schools fostered a utilitarian and correspondingly anatomical conception of physiology. In no way did they provide an institutional setting conducive to the liberation of physiology from anatomy. But not all of English medical education was confined to the hospital schools. Medical faculties had existed for centuries at the ancient universities of Oxford and Cambridge, and they were joined in the late 1820's by the faculties of newly founded University and King's Colleges in London. What was the fate of physiology in these faculties? Did they, by virtue of their association with institutions of wider aims than the hospitals, promote a broader vision of physiology?

Graduates of Oxford and Cambridge dominated the elite physician class of medical practitioners. An Oxford or Cambridge M.D. automatically carried with it the right to practice as a physician outside the immediate vicinity of London. In London (and for seven surrounding miles), the licensing and regulation of physicians was the exclusive right of the Royal College of Physicians; but for much of the century, only graduates of the ancient universities could belong to the ruling body of the College, so that in London too the physician class was dominated by Oxford and Cambridge men.[58]

If these special privileges seem to favor the development of major medical centers at Oxford and Cambridge, there were far stronger forces against it. For one thing, the medical faculties of these universities (until 1871) graduated none but members of the Church of England, so that Catholics and dissenters were automatically excluded. Even among Anglicans, few could afford the tuition fees, and those who could faced a long and expensive course of study before getting the medical degree. Clinical material was also far less abundant in the small university towns than in London. But the chief obstacle to the effective study of medicine at the ancient universities lay in the general goals and character of the education available there. Medicine suffered the same fate as the sciences and all other studies that could be taught properly only on a university-wide basis.[59] Especially as it

[57] See Hujohn A. Ripman, ed., *Guy's Hospital, 1725-1948* (London, 1951), p. 70; and H. C. Cameron, *Mr. Guy's Hospital, 1726-1948* (London, 1954), p. 284.

[58] George Clark, *A History of the Royal College of Physicians of London*, 2 vols. (Oxford, 1964-1966), II, pp. 671-696 et passim.

[59] For a remarkably pugnacious expression of this view, see E. Ray Lankester, "The Relation of Universities to Medicine," *Brit. Med. J.*, 1878 (2), 501-507.

came to require expensive laboratory facilities, medicine was not the sort of subject that could be taught in small residential colleges,[60] and these colleges utterly dominated the educational system at both universities. During the first seventy years of the nineteenth century, the number of medical degrees awarded annually by Oxford or Cambridge rarely exceeded three or four, and medical education was moribund at both places throughout the period.[61]

Insofar as medical education did take place at Oxford or Cambridge, it differed markedly from the blatantly utilitarian training offered in the London hospital schools. Although the ultimate motive was scarcely less pragmatic, medical education at the ancient universities was distinguished by its classical emphasis. Among the Anglican elite, it was a self-serving article of faith that only a classical education guaranteed learning, culture, and character; and it was largely because they had satisfied this condition that Oxford and Cambridge men enjoyed their special privileges in the Royal College of Physicians.[62] As a group, the physicians displayed far greater concern for protecting their privileges than for the state of medical and scientific education. Not even the quality of clinical instruction in the London hospitals could arouse their interest.[63] Nothing in their training qualified or disposed them to direct the attention of their professional brethren away from anatomy and toward the experimental sciences, especially since they looked with disdain on manual procedures and shared with their patrons an antipathy to animal experimentation.

The religious, social, and economic exclusiveness of Oxford and Cambridge led in the late 1820's to the foundation of two new middle-class institutions in London—University and King's Colleges. From the beginning, both new colleges emphasized medical education, and both played an important role in its reformation, especially University College which (unlike King's) welcomed religious dissenters. These were the first institutions in England to provide a university education for intending general practitioners, and the first to chal-

[60] In an earlier and simpler era, the colleges apparently did serve as an adequate center for medical education. Cf. Arthur Rook, "Medicine at Cambridge, 1660-1760." *Med. Hist., 13* (1969), 107-122; Rook, "Medical Education at Cambridge, 1600-1800," in *Cambridge and Its Contribution to Medicine*, ed. Rook (London, 1971), pp. 49-63; and A. H. T. Robb-Smith, "Cambridge Medicine," in *Medicine in Seventeenth Century England* (Berkeley, 1974), pp. 327-369.

[61] See Humphry Davy Rolleston, *The Cambridge Medical School: A Biographical History* (Cambridge, 1932), esp. pp. 21-31. As late as 1878, Lankester (n. 59), p. 504, claimed that there was "not a single medical student" at Oxford.

[62] McMenemy, "Medical Reform," p. 140. Cf. Clark, *College of Physicians*, passim.

[63] R. S. Roberts, "Medical Education and the Medical Corporations," in *Evolution of Medical Education*, ed. Poynter, p. 80.

lenge the principle that university education was the exclusive preserve of the elite physician class.[64] A development of great importance, it did not, however, immediately produce any fundamental change in the teaching of the medical sciences or of physiology in particular.

For most of the century at least, neither King's nor University College was a university in the sense that we understand the word today. Though radically different in some ways from Oxford and Cambridge, they were like them in one crucial respect: original research was deemphasized if not actively discouraged. Both of the new London colleges saw themselves as teaching institutions, as places where the existing knowledge was to be transmitted for its practical value. For some of the chairs at University College, authorship was considered "a positive disqualification." To its founders, "research was no part of the duty of a teacher."[65] King's College, in essence an Anglican version of University College, made it clear in its founding regulations that "it was concerned solely with the dissemination of existing knowledge and not at all with the extension of the bounds of knowledge."[66] Even so brilliant an ornament as physicist Clerk Maxwell may have been asked to resign from the faculty at King's (in 1865) because of his inability to perform his primary task of teaching "raw youth."[67] Obviously, neither University nor King's College sought to mimic the research orientation of the German universities.

So far as medical education is concerned, the result was not strikingly different from that in the best of the hospital schools in London. As in those schools, the medical faculties at the two new London colleges were composed almost exclusively of active medical practitioners. Although the new colleges seemed able to attract the most original minds in the English medical profession, there were no endowments that allowed them to devote a significant amount of their time to research.[68] Their job was to teach and to teach in such a way that students would be able to pass the qualifying examinations of the Royal College of Surgeons or the Society of Apothecaries.

What particularly encouraged this narrow mode of teaching was the fact that until 1854 medical graduates from King's or University College could not use their university degrees (including the M.D.) as licenses to practice medicine. Unlike their counterparts from Oxford

64 Cf. Holloway, "Sociological Anaylsis," p. 323.

65 H. Hale Bellot, *University College London, 1826-1926* (London, 1926), pp. 51-52.

66 F. J. C. Hearnshaw, *The Centenary History of King's College London, 1828-1928* (London, 1929), p. 80.

67 Ibid., pp. 247-248.

68 Cf. Lankester (n. 59), pp. 502-503.

or Cambridge, they were required to pass a licensing examination given by one of the old medical corporations.[69] Since students who had had no university education were equally eligible for these examinations, and since the hospital schools were less expensive, the overwhelming majority of intending practitioners continued to be trained in the hospital schools. As late as 1850, eight of every nine practitioners in England remained unscathed by university education.[70]

In effect, then, the two new London colleges were placed in competition with the hospital schools, and this tended to encourage a leveling of standards among the rivals. In both types of institutions, the curricula and mode of teaching were naturally framed in keeping with the traditional requirements of the medical licensing bodies. Since these bodies did not include laboratory physiology in their examinations until after 1870,[71] it is not surprising that physiology was almost as anatomical in its orientation at the two new colleges as it was in the hospital schools. So long as University of London degrees carried no real legal advantages, so long would the education of medical students at University and King's Colleges tend to approximate that given in the older institutions.

Until about 1870, the major achievement of the physiologists at University and King's was to show that microscopic anatomy deserved a more important place in the English medical curriculum than it had hitherto been granted. Between 1838 and 1870 physiology was taught at King's College by Robert Bentley Todd, William Bowman, and Lionel Smith Beale. All three were skilled microscopists. In their two major collaborative works, Todd and Bowman incorporated the results of important new research, especially by Bowman, but neither of them escaped the tendency to correlate function in every case with anatomical structure.[72] Beale studied under and later assisted Todd and Bowman, and he remained faithful to their tradition in his own published works. A master of microscopic technique, he wrote a series of

[69] In 1854 an Act of Parliament conferred on medical graduates of the University of London "the same privileges in the practice of medicine as belonged to those of Oxford and Cambridge." Clark, *College of Physicians*, II, p. 696.

[70] See A. H. T. Robb-Smith, "Medical Education at Oxford and Cambridge," in *Evolution of Medical Education*, ed. Poynter, pp. 49-51. It should further be emphasized that of those who did have university degrees, the overwhelming majority had Scottish degrees, which required at most three years of study. From Robb-Smith's data, it follows that of the practitioners with university degrees in 1850, about 9% came from Oxford or Cambridge, 11% from the University of London, and nearly all the rest from Scottish universities.

[71] The Royal College of Surgeons and the University of London led the way by requiring "practical" physiology of their medical students after 1870. See Chapter 6 below, "Seizing the transient moment."

[72] See above, pp. 18, 21.

highly successful manuals on the use of the microscope. But he was also an outspoken vitalist, who ridiculed the "physicalists" for their belief that experimental analysis could lead to an understanding of physiological processes.[73]

During this same period, University College entrusted its physiological teaching to William Sharpey, professor of physiology and general anatomy from 1836 to 1874. In the England of his day, he was the only teacher of physiology who did not simultaneously practice medicine or serve on a hospital staff. Perhaps this helps to explain why he also became his country's most influential physiological mentor. When after 1870 English physiology enjoyed a great renaissance, it was led by men who had once studied or worked under Sharpey: Michael Foster, John Scott Burdon Sanderson, and Edward A. Schäfer. Nonetheless, even Sharpey was mainly an anatomist by training and inclination. He may have been the first teacher in England to separate microscopic from gross anatomy and to connect microscopic anatomy with physiology,[74] but he did not himself teach experimental physiology. If he appreciated its importance and the principles on which it was based, he remained without experience in its methods. It is said that he did not even have a simple kymograph. Not until the late 1860's, when Foster came to University College to assist Sharpey, was experimental physiology taught there or anywhere else in England.[75]

The relative unimportance of the anatomical bias for the position of English physiology, 1800-1840

During the first half of the nineteenth century, roughly speaking, the anatomical bias of English physiology was relatively unimportant for its international position. To be sure, the most exciting physiological discoveries during this period resulted from the experimental investigation of the nervous system, and the work of Magendie and Pierre Flourens, among others, placed France at the forefront of this field. But during most of this period at least, British physiologists continued to make important contributions to the science. Indeed, W. B. Carpenter could insist in 1855 that the British achievements, if less frequent,

73 See Gerald L. Geison, "The Protoplasmic Theory of Life and the Vitalist-Mechanist Debate," *Isis*, *60* (1969), 272-292; and Geison, "Lionel Smith Beale," *DSB*, I (1970), 539-541.

74 Cf. *Proc. Roy. Soc.*, *52* (1893), i-vii, on ii. More generally on Sharpey, see D. W. Taylor, "The Life and Teaching of William Sharpey (1802-1880): 'Father of Modern Physiology' in Britain," *Med. Hist.*, *15* (1971), 126-153, 241-259; and Chapter 3 below, "William Sharpey."

75 See Chapter 3 below, "University College."

were actually of more fundamental importance than their Continental counterparts. His argument depended in part on the misguided assumption that Charles Bell had at least co-discovered the distinction between motor and sensory nerves, but it also rested on the more defensible claim that Marshall Hall and other British workers were chiefly responsible for the establishment and development of the basic concept of the reflex. In his textbook of physiology, Carpenter sounded the theme:

> It is a circumstance not devoid of interest, that, during the present century, notwithstanding the large amount of anatomical and experimental inquiry which has been directed to the Nervous System both in France and in Germany, and the vast additions to our knowledge of *details* which has hence arisen, the great advances in the *general doctrines* of this department of the science should have been made by British physiologists.[76]

At the time, most of his compatriots probably shared Carpenter's view. The increasing stagnancy of British physiology had not yet become obvious to them.

Part of the reason was that anatomical reasoning remained an extremely valuable guide to physiological investigation. If this were true in the case of such organs as the kidney and skeletal muscle, as William Bowman's work in the 1840's clearly suggests,[77] it was even more true in the case of the nervous system. Nowhere are structure and function more intimately related. The nervous system, as Joseph Schiller points out, is the most "anatomical" of physiological systems.[78] Indeed, E. G. T. Liddell has suggested that throughout much of the nineteenth century, confusion reigned in neurophysiology precisely because "knowledge from experiment was growing faster than knowledge from the microscope."[79] And for this reason, anatomically oriented British physiologists could continue to make important contributions to neurophysiology.[80]

[76] Carpenter, *Human Physiology* (5th ed.), p. 649.

[77] William Bowman, "On the Minute Structure and Movements of Voluntary Muscles," *Phil. Trans.*, 1840, pp. 457-501; and Bowman, "On the Structure and Use of the Malpighian Bodies of the Kidney, with Observations on the Circulation through that Gland," ibid., 1842, pp. 57-80. See also L. G. Wilson, "The Development of Knowledge of Kidney Function in Relation to Structure—Malpighi to Bowman," *Bull. Hist. Med.*, *34* (1960), 175-181.

[78] Joseph Schiller, "Claude Bernard and Vivisection," *J. Hist. Med.*, 22 (1967), 246-260, on 250-251.

[79] E. G. T. Liddell, *The Discovery of Reflexes* (Oxford, 1960), passim, quote on 77.

[80] See, e.g., R. D. Grainger, *Observations on the Structure and Functions of the Spinal Cord* (London, 1837). Cf. Liddell, ibid., pp. 73-77.

Moreover, even the experimental investigation of the nervous system did not yet require extensive laboratories or unusual techniques. To the extent that the new experimentalists relied exclusively on vivisection, their approach was more adapted to *localizing* function than to *explaining* it, and in this sense remained an essentially anatomical approach.[81] Operative skill was essential, as was the willingness to perform the operations, but even in Britain there were always a few investigators willing and able to perform them, most notably Marshall Hall. No more than a few such men were needed, as yet, to keep at least partially abreast of French and German developments.

At a more general level, and independently of the methods used to investigate it, the nervous system seems to have held a special attraction for British workers, perhaps because of its pertinence to some of the religio-philosophical issues that loomed so large at the time. Because of the strength of natural theology, physiology in general was perceived as an essentially moral exercise. Early in the century, William Lawrence was one Englishman who discovered the hazards of attempting to draw a distinction between religion and the science of physiology.[82] And of all physiological systems, the nervous system had the greatest religious significance. Its large role in human life, especially through the brain, was looked upon as the most persuasive natural evidence of man's moral nature and of his exalted place in the Great Chain of Being.[83] While on the one hand such concerns tended to isolate British physiology from Continental developments, they also tended to make the nervous system a subject of special investigation in Britain. The peculiarly moral tone of Carpenter's textbook certainly coincides with and perhaps underlies a distinct emphasis on the nervous system.

The stagnancy becomes apparent: experimental science and the English universities, 1840-1870

As late as 1874, Carpenter continued to cite the contributions of British neurophysiologists as evidence against the increasingly vocal complaint that British physiology had fallen behind Continental, and

[81] Cf. Schiller, *Bernard*, pp. 56-57.

[82] See Temkin, "Basic Science"; Goodfield-Toulmin, "Some Aspects"; and Karl Figlio, "The Metaphor of Organization: An Historiographical Perspective on the Bio-Medical Sciences of the Early Nineteenth Century," *History of Science, 14* (1976), 17-53, esp. 32ff.

[83] This attitude is at least latent in much of Carpenter's writing. See, e.g., William Carpenter, *Principles of Human Physiology*, first American ed., with additions by the author, and notes and additions by Meredith Clymer (Philadelphia, 1843), pp. 69-71, 88.

especially German, physiology. He did so in a review of David Ferrier's important experimental work on the cerebral localization of motor phenomena.[84] Yet Carpenter's review reveals his own inability or unwillingness to perceive the full implications of Ferrier's work,[85] and he was in any case fighting a losing battle. By the 1870's the general stagnancy of British physiology had become so obvious that many began to consider it a matter for national shame. Insofar as it was valid at all, Carpenter's defense of British physiology was limited to neurophysiology and no such narrow defense could carry conviction in the 1870's. For three decades, the domain of physiology had been undergoing tremendous expansion.

More or less crude investigations of the nervous system lost their privileged place in the world of physiological research; they became just one part of a more sophisticated and more broadly based study of physiological processes in general. In this extension and diversification of the subject, the techniques of physics and chemistry played a central role. They had in fact long been used in physiology—by Stephen Hales, Lavoisier, and Magendie himself, for example[86]—but during the middle years of the nineteenth century they became an indispensable and everyday part of the methodological armament of Continental physiologists. Especially between 1850 and 1870 the application of physical and chemical methods, and of physiological instruments based upon them (most notably Ludwig's kymograph), opened up whole new vistas.[87] Whether or not this development really deserves to be called a "physicochemical revolution," there can be no doubt that it profoundly altered the character and direction of physiological research. In concert with the technique of vivisection, the methods of physics

[84] W. B. Carpenter, "On the Physiological Import of Dr. Ferrier's Experimental Investigations into the Functions of the Brain," *West Riding Lunatic Asylum Medical Reports, 4* (1874), 1-23, on 2.

[85] See Robert M. Young, *Mind, Brain and Adaptation in the Nineteenth Century: Cerebral Localization and Its Biological Context from Gall to Ferrier,* (Oxford, 1970), esp. p. 214.

[86] I refer to Hale's work on blood pressure and fluid dynamics in plants and animals, Lavoisier and Laplace's work on animal heat at the end of the eighteenth century, and to Magendie's work on non-nitrogeneous diets and on the mechanics of the circulation.

[87] On the evolution of the kymograph, see Hebbel E. Hoff and L. A. Geddes, "Graphical Registration before Ludwig: The Antecedents of the Kymograph," *Isis, 50* (1959), 5-21; and Hoff and Geddes, "The Technological Background of Physiological Discovery: Ballistics and the Graphic Method," *J. Hist. Med., 15* (1960), 345-363. For a useful recent contribution to the more general and pitifully neglected history of physiological techniques, see Dietmar Rapp, *Die Entwicklung der physiologischen Methodik von 1784 bis 1911: Eine quantitative Untersuchung,* no. 2 in Münstersche Beiträge zur Geschichte und Theorie der Medizin, ed. K. E. Rothschuh et al. (Münster, 1970). Once again, my warm thanks to Prof. Rothschuh for sending me several issues of this series.

and chemistry rendered physiology a broadly experimental science and brought into view a wide range of hitherto inaccessible physiological phenomena.

The more physiology came to depend on this experimental approach, the more obvious became the stagnancy of English physiology. Part of the problem, of course, was the persistence and increasing influence of antivivisection sentiment. But antivivisection had always been just one contributing factor in the nonexperimental tradition of English physiology. More fundamental social and institutional circumstances had also contributed importantly to that tradition and indeed to antivivisection sentiment itself. Chief among these had been the peculiarities of English medical and scientific education. What happened between 1840 and 1870 was that these peculiarities became even more apparent and even more exaggerated by comparison with the dramatic institutional developments then taking place in Europe, and particularly in Germany. Because English universities did not offer professional opportunities for research scientists of any kind, English physiologists were faced not only with antivivisection sentiment, but also with newer and even more serious obstacles isolating them from the physiological mainstream.

For one thing, the increasing reliance on techniques borrowed from physics and chemistry came at a time when English physicists and chemists were themselves failing to keep pace with their Continental counterparts. Although theoretical or mathematical physics was well represented, especially by Clerk Maxwell, experimental physics fared less well. No longer was there an English physicist doing experimental work comparable in importance with that of Michael Faraday in the 1830's. Not until the late 1860's was an organized course in laboratory physics taught anywhere in England, and not until after 1870 did there exist facilities reasonably adequate to the task.[88] The relative decline of experimental physics in England during the mid-nineteenth century is clear from the table of physical discoveries compiled in 1929 by T. J. Rainoff: between 1840 and 1870, the number of English contributions was 261, while French physicists made slightly over 400 discoveries and German physicists slightly over 600.[89]

[88] See D. M. Turner, *History of Science Teaching in England* (London, 1927), pp. 124-129; and Romualdas Sviedrys, "The Rise of Physics Laboratories in Britain," *Historical Studies in the Physical Sciences,* 7 (1976), 405-436, which discusses a number of earlier private laboratories in Britain, while confirming the point that the organized teaching of physics in university-based laboratories effectively dates from 1870.

[89] T. J. Rainoff, "Wave-like Fluctuations of Creative Productivity in the Development of West-European Physics in the Eighteenth and Nineteenth Centuries," *Isis, 12* (1929), 287-319, on 311-313. Data reproduced in Ben-David, *Scientist's Role,* p. 192.

British chemistry, meanwhile, after what Edward Thorpe called the "momentous" and "brilliant" decades from 1800 to 1820, had entered its "low ebb" in the 1830's. Thorpe attributed this low ebb in part to the resistance of British chemists to the mathematical notation by then in use on the Continent, but more especially to the "pervasive dilettantism" of which Liebig had complained during a visit to England in 1837.[90] Thorpe emphasized that there was at the time no chemical laboratory at Oxford or Cambridge, and that professional opportunities were extremely limited. There was, in fact, "nothing to tempt men to take up chemistry as a means of livelihood."[91] By 1870, considerable progress had been made toward providing teaching laboratories in chemistry, but even then English chemistry continued to be "conspicuous for [its] position in the rear."[92] In 1871, in a long letter to *Nature,* Edward Frankland emphasized the magnitude of England's inferiority. Such chemical laboratories as did exist in England were "adapted for beginners, there is not in any of them a separate department constructed and fitted for original research." Small wonder that of 1273 important papers in chemistry published in 1866, only 127 (10%) were contributed by English authors. Germans, on the other hand, had contributed 777 (61%) and the French 245 (19%).[93] The situation was particularly serious in those areas of chemistry most relevant to physiology. The great impulse given to organic and physiological chemistry by Liebig's *Organic Chemistry* (1840) and *Animal Chemistry* (1842) seems scarcely to have crossed the English Channel during the nineteenth century.[94]

As a matter of fact, the stagnancy of English physiology between about 1840 and 1870 was only a particularly striking example of the

[90] Edward Thorpe, "On the Progress of Chemistry in Great Britain and Ireland during the Nineteenth Century," in *Essays in Historical Chemistry* (London, 1923), pp. 551-589, on 557, 587-588.

[91] Ibid., p. 598.

[92] E. Frankland, "Chemical Research in England," *Nature, 3* (1871), 445.

[93] Ibid.

[94] See F. Gowland Hopkins, "Huxley Memorial Lecture: On Biochemistry, Its Present Position and Outlook," *Lancet,* 1924 (1), 1247-1252, esp. 1247; and F. G. Young, "The Rise of Biochemistry in the Nineteenth Century, with particular reference to the University of Cambridge," in *Cambridge and Its Contribution,* ed. Rook, pp. 155-172. To be sure, the founding of the short-lived Royal College of Chemistry (1845-1853) has traditionally been ascribed to the impact of Liebig's work, but that interpretation is clearly simplistic. See Gerrylynn K. Roberts, "The Establishment of the Royal College of Chemistry: An Investigation of the Social Context of Early-Victorian Chemistry," *Historical Studies in the Physical Sciences,* 7 (1976), 437-485, which also reveals the pragmatic (most frequently, medical) orientation of Victorian chemical laboratories, including the Royal College of Chemistry. The *theoretical* dimension of Liebig's work in organic and physiological chemistry fell on particularly fallow ground in England.

general stagnancy of experimental science in England during that period. Thus Foster proposed to Huxley a whole series of articles on all the sciences, "each singing a mournful song." "This," he thought, "would be more likely to shock the public than a single article on physiology alone."[95] If physiology suffered somewhat more than other fields, it did not suffer alone. Its state and prospects were closely bound up with those of the other sciences, and in particular with their position in the English universities. For what permitted the remarkable developments in Continental science, and what prevented the English from keeping pace, was the whole interrelated pattern of the institutionalization of science and the state of the universities in each country. Attention must be confined here to the institutionalization of physiology, but the pattern that emerges is roughly representative of the situation in other disciplines as well.

Although physiology achieved some recognition as an independent science in France as early as the 1820's,[96] its emergence as a full-fledged profession took place between 1850 and 1870. This was true to some extent in France, where Bernard was now able to make a decent living while engaged exclusively in the teaching and pursuit of physiology, but more especially in the German *Sprachgebiet,* where the discipline was propelled into full professional status by the unique and extensive German university system.

Consisting by the 1860's of some twenty universities in active competition with each other, the German system encouraged specialization and research. The goal of each university was to attract and to hold the most productive scholar in each field. Since appointments depended on a demonstrated capacity to make and to publish original contributions, anyone aspiring to a professorship devoted much of his time to the establishment of his credentials in a hitherto unexplored area of knowledge. During the 1850's and 1860's, physiology was among the most promising of the relatively unexplored fields, and many students chose to concentrate on it. By 1864, separate professorships in physiology had been created at nearly all the German universities; and by 1876, only Giessen retained the old union between anatomy and physiology. A particularly important feature of the German system was the inter-

95 Foster to Huxley [1865], Huxley Papers, IV. 148.

96 The chief step was the founding in 1821 of Magendie's *Journal de physiologie expérimentale et pathologique* (1821-1831). In 1823, the Faculty of Medicine, University of Paris, appointed André Dumeril to what Olmsted calls "the first independent chair of Physiology in any university." J. M. D. Olmsted, "Physiology as an Independent Science," in *Science in the University* (Berkeley, 1944), p. 296. More generally, see Joseph Schiller, "Physiology's Struggle for Independence in the First Half of the Nineteenth Century," *History of Science,* 7 (1968), 64-89.

university mobility, which increased the bargaining power of the professoriate. Typically, a leading physiologist could and would insist upon a new and independent laboratory (or institute) as a condition of appointment. Naturally, then, the creation of independent professorships in physiology tended to go hand in hand with the establishment of new laboratories in the field.[97]

In England, meanwhile, physiology remained in the hands of amateurs, usually physicians or surgeons who pursued physiological research in leisure moments snatched from the demands of a busy medical practice. Before 1870 there was no English professorship in physiology. Nor was there an independent laboratory in the subject, unless one wishes so to describe a little three-room affair at University College, which was organized only in the late 1860's and paled by comparison with the magnificent institute built almost simultaneously for Carl Ludwig in Leipzig.[98]

Those few English physiologists who did make important contributions to the subject during this period inevitably found their hopes frustrated if they sought a career in physiological research. Perhaps the best known English contribution between 1850 and 1870 was A. V. Waller's work on nerve degeneration. But Waller had been trained in Paris and in Bonn, and he returned to these more sympathetic centers in the early 1850's in order to continue his investigations. He taught physiology only briefly in an English university, and that was at little known Queen's College, Birmingham. Even there, he retained clinical duties as physician to the Birmingham hospital.[99]

Another important English contributor was Edward Smith (1818?-1874), whose significant work on human metabolism and nutrition has recently been rescued from oblivion by Carleton B. Chapman.[100] In the 1850's, Smith's imaginative study of metabolism during exercise attracted considerable attention and praise in Germany, where it was recognized as an impressive challenge to Liebig's dictum that the

[97] The most thorough analysis of the professionalization of German physiology is Zloczower, "Career Opportunities." Cf. Ben-David, "Scientific Productivity." For an extensive discussion of physiological research schools, especially in Germany, with helpful genealogical charts, see Karl E. Rothschuh, *History of Physiology*, trans. and ed. G. R. Risse (Huntington, N.Y., 1973), Chapters 5-7. See also Rothschuh, "Ursprünge und Wandlungen der physiologischen Denkweisen im 19. Jahrhundert," *Technikgeschichte, 33* (1966), 329-355.

[98] See H. P. Bowditch, "The Physiological Laboratory at Leipzig" *Nature, 3* (1870), 142-143; and J. S. Burdon Sanderson, "Physiological Laboratories in Great Britain," *Nature, 3* (1871), 189.

[99] *DNB, 20*, 579-580.

[100] Carleton B. Chapman, "Edward Smith (?1818-1874), Physiologist, Human Ecologist, Reformer," *J. Hist. Med., 22* (1967), 1-26.

energy for muscular exercise comes solely from protein. But in England, Smith's work made little impression. None of the English universities offered him the opportunity to fulfill his promise as a physiologist, and he made his living from a medical practice he seems not to have enjoyed.

Rothschuh's catalog of physiological discoveries includes Waller and Smith among the nine English workers who made significant contributions during the period 1850 to 1870.[101] Rothschuh also cites Henry Gray (of *Gray's Anatomy*) for his suggestion that the spleen functions as an organ of storage; Joseph Toynbee for his investigation of the mechanism by which the Eustachian tube is opened and closed;[102] Clerk Maxwell for work that belongs more properly to theoretical physics than to physiology; and Thomas Addison for his description of the disease of the suprarenal capsules that has since borne his name.[103] Thomas Graham is recognized for his pioneering efforts in colloid chemistry, and Edward Frankland for his work of 1866 on the heats of combustion of the major classes of nutrient compounds.[104] Finally, Joseph Lister, who studied under William Sharpey at University College, is cited for his investigation of the role played by the vasomotor system in the frog's peripheral circulation.[105]

The work and careers of these nine men provide a suggestive index to the state of English physiology between 1850 and 1870. Two (Gray and Toynbee) were essentially anatomists; two (Graham and Frankland) foreign-trained chemists; one (Maxwell) a theoretical physicist; and one (Addison) a clinician. Of the three who demonstrated greatest

101 Rothschuh, *Entwicklungsgeschichte*, item 1367, p. 70 and item 596, p. 33. According to Zloczower, Rothschuh's catalogue contains ten English contributions between 1850 and 1870, but I can find only nine. I suspect that Zloczower has inadvertently included the contribution of William Rutherford in 1868 (Rothschuh, item 1441, p. 74), when Rutherford was still in Scotland, though he published his paper in an English periodical, *The Journal of Anatomy and Physiology*. Rothschuh also cites one other Scottish contribution between 1850 and 1870 (Playfair's work on the nutritional requirements of the active human body: item 699, p. 38) and one Irish contribution (Stokes's description of the normal respiratory curve: item 187, p. 13).

102 Joseph Toynbee has apparently fallen into complete obscurity, but his extensive anatomical and pathological studies of the organs of hearing played an important role in the emergence of medical ear specialists. See Prof. von Tröltsch, "Joseph Toynbee, ein Nekrolog," *Archiv für Ohrenheilkunde, 3* (1867), 230-239.

103 Rothschuh, *Entwicklungsgeschichte*, item 483, p. 28 (Gray); item 1964, p. 99 (Toynbee); item 1828a, p. 92 (Maxwell); and item 1121, p. 58 (Addison).

104 Ibid., item 855, p. 46 (Graham); and item 577, p. 32 (Frankland). On Graham, see R. Angus Smith, *The Life and Works of Thomas Graham* (Glasgow, 1884); On Frankland, *DNB, 22* (Supplement), 662-665.

105 Rothschuh, *Entwicklungsgeschichte*, item 1530, p. 78. On Lister, see Claude Dolman, "Joseph Lister," *DSB*, VIII (1973), 399-413.

promise as physiologists, one (Lister) went on to a famous career in surgery, while the other two (Waller and Smith) remained largely unappreciated at home and depended primarily on medical practice for their income.

Equally suggestive is the career of George Harley (1829-1896), who returned to England in the 1850's after having acquired an impressive Continental training in histology and physiology.[106] Harley testified that "the mere reporting of medical cases in the journals" was a practice he despised, "yet the journals far preferred medical cases to scientific truths."

> For example I read . . . three scientific medical papers at the Royal Medico-Chirurgical Society, none of which appeared in their Transactions; they were too scientific, I suppose . . . [but] after sending in the report of a case on intermittent haematuria, which required but a modicum of brains to work out, the . . . Society deemed it sufficiently meritorious to find a place in their Transactions![107]

Harley's experience is particularly interesting because it illustrates at once four of the problems faced by Englishmen attracted to physiological research before the 1870's: (1) the utilitarian attitude of the medical profession, (2) the absence (until 1876) of a physiological society to which Harley might have communicated the results of his more scientific research, (3) the absence (until 1878) of an English journal in which those results might have been published,[108] and (4) the absence of professional opportunities in general. In the end, Harley too earned his livelihood chiefly from medical practice.

The failure of English society to provide professional opportunities to its most promising physiologists was part of a more general indifference to basic research of any kind. In the 1850's, during his early struggles to make a living in science, T. H. Huxley complained:

> Nothing but what is absolutely practical will go down in England. A man of science may earn great distinction, but not bread. He will get invitations to all sorts of dinners and conversaziones, but not enough income to pay his cab fare. A man of science in these times is like an Esau who sells his birthright for a mess of pottage.[109]

During a visit to England in 1850, the Swiss histologist Albert von Koelliker was also struck by the utilitarianism of English society, which

[106] Mrs. Alec Tweedie, ed. *George Harley, F.R.S., The Life of a London Physician* (London, 1899).

[107] Ibid., pp. 159-160.

[108] In 1878 Michael Foster founded the still active *Journal of Physiology.*

[109] Huxley, *Life and Letters,* I, p. 66.

presented such a marked contrast to the situation in the German-language universities.[110] Despite the remarkable developments on the Continent, the English attitude toward science had changed very little during the nineteenth century. For the most part, scientific research was still considered a harmless pastime, a gentlemanly avocation. At least until 1870, England continued to rely for its science on the peculiar genius of its traditional band of independent amateurs, and English scientific societies remained under the sway of a pervasive dilettantism.

This is not to deny that some English thinkers had made strenuous efforts to alter the situation. To professionalize science was the central aim of Charles Babbage's celebrated *Reflections on the Decline of Science in England* (1830). The founding in 1831 of the British Association for the Advancement of Science was part of this same movement, as was the growing agitation for a reform of the venerable Royal Society. But these efforts were only partly effective, and they took time. Not until the 1860's did serious scientific investigators predominate in the Royal Society over dilettantes from the clergy, peerage, and legal profession.[111] In a social environment that extolled the virtues of self-help, voluntarism, and individualism, the government quite naturally eschewed large-scale support of science.[112] By German or even French standards, the British government gave meager support to the Royal Society, and failed to give the Royal Institution the funds it sought to fulfill its stated aim of promoting knowledge by experiments and original investigations.[113] By no means did all English scientists deplore this governmental restraint, and some positively applauded the separation of English science from the state.[114] Because of this sentiment, the British Association long relied almost solely on private resources, though these fell far short of the wants of the emerging breed of active researchers, particularly in physiology, which was allotted a diminutive portion of the total.[115]

110 A. Koelliker, *Erinnerungen aus meinem Leben* (Leipzig, 1899), passim, esp. p. 123. I am grateful to Owsei Temkin for bringing this source to my attention.

111 Henry Lyons, *The Royal Society, 1660-1940: A History of Its Administration under Its Charter* (Cambridge, 1944), Chapter 7.

112 J. B. Morrell, "Individualism and the Structure of British Science in 1830," *Historical Studies in the Physical Sciences, 3* (1971), 183-204.

113 See *Report on the Past, Present and Future of the Royal Institution, Chiefly in Regard to its Encouragement of Scientific Research*, by the Honorary Secretary (London, 1862).

114 See Roy M. MacLeod, "Resources of Science in Victorian England: the Endowment of Science Movement," in *Science and Society, 1600-1900*, ed. Peter Mathias (Cambridge, 1972), pp. 111-116.

115 S. F. Mason, *A History of the Sciences*, new revised ed. (New York, 1962), p. 446.

During the first thirty or forty years of the nineteenth century, British science did not suffer because of its amateur tradition as much as Babbage tried to pretend it had. His self-serving and narrowly focused critique ignored many of the important contributions made by British workers during that period—by Humphry Davy, Michael Faraday, Marshall Hall, and Charles Lyell, for example.[116] The valuable and original work done after mid-century by such independent workers as Charles Darwin and Alfred Russel Wallace suggests that even then the amateur tradition of English science was not inimical to works of genius in some areas. But the gradual transformation of science, especially in the state-supported German universities, into a full-blown professional activity—pursued in a highly competitive environment by many investigators, instead of by the few who could afford to do it for nothing—eventually left much of British science at a relative standstill. This was especially true in those sciences in which large and expensive laboratories had become increasingly important, notably physics, chemistry, and physiology.

The situation was especially serious in physiology because it required laboratories as elaborate and costly as those in physics or chemistry, while its immediate practical benefits were even less apparent. The utilitarianism that kept experimental physiology out of the London hospital schools was by no means confined to them. John Scott Burdon Sanderson faced the problem directly when he said in 1872 that the revival of British physiology depended above all on overcoming "that practical tendency of the national mind which leads us Englishmen to underrate or depreciate any kind of knowledge which does not minister directly to personal comfort or advantage."[117]

What English physiology would ultimately require to compete with its Continental rivals was a setting where it could be pursued for its own sake, relatively free from utilitarian demands. In short, it required a home in universities which honored and encouraged the pursuit of fundamental scientific research. If this lesson was not clear enough from the spectacular success of physiologists in the German universities, the French experience only reinforced it. In France, as in England, medical education had evolved essentially out of medical practice and was conducted in the hospitals, which were associated only nominally with the University of Paris. But it was not in the medical faculties that the leaders of French physiology were able to make their way. As Erwin Ackerknecht has emphasized, Magendie, Flourens, and Claude Bernard, among others, were never professors in a French med-

[116] Cf. Nathan Reingold, "Babbage and Moll on the State of Science in Great Britain: A Note on a Document," *Brit. J. Hist. Sci.*, *4* (1968), 58-64.

[117] J. S. Burdon Sanderson, "Address," p. 147.

ical school or faculty.[118] Rather, they found their natural home in institutions where the pursuit of knowledge was honored for its own sake—at the Collège de France, the Sorbonne, and the Muséum d'Histoire Naturelle.[119]

As a matter of fact, this same lesson emerges when one distinguishes between English and Scottish physiology. It is a remarkable fact, hardly to be dismissed as mere coincidence, that every really noteworthy British physiologist of the early to mid-nineteenth century had been trained initially in Scotland, almost always at the University of Edinburgh. Charles Bell, Marshall Hall, John Reid, William Sharpey, and William Carpenter were all graduates of Edinburgh. So was A. P. Wilson Philip (1770-1847), a controversial experimental physiologist whose work on nervous influences in digestion and heart action attracted Johannes Müller's attention, if not always his admiration.[120] Yet another medical graduate of Edinburgh was William Prout, the chemist who identified hydrochloric acid in the gastric juice and who was the leading practitioner of physiological chemistry in early nineteenth-century Britain.[121] Even at the second level, it is hard to find a physiologist who had received all of his training in England.[122]

118 See E. H. Ackerknecht, "Medical Education in 19th Century France," *Journal of Medical Education, 32* (1957), 148-153, on 150-151. Cf. Ackerknecht, *Paris Hospitals*, p. 12.

119 Magendie was appointed to the chair in medicine at the Collège de France in 1831. Despite the chair's official title, he was allowed to teach experimental physiology. He was also given considerable freedom with the limited financial resources at his disposal, and a new laboratory was built for him in 1840. See Olmsted, *Magendie*, esp. pp. 220, 228, 231. When Bernard succeeded Magendie in 1856, he was given similar freedom. It was to give greater scope to Bernard's talents that chairs in general physiology were created both at the Sorbonne (in 1854) and at the Muséum d'Histoire Naturelle (in 1868). See Joseph Schiller, "Claude Bernard and Brown-Séquard: The Chair of General Physiology and the Experimental Method," *J. Hist. Med., 21* (1966), 260-270.

120 Biographical sketches of all the men named in this paragraph can be found in the appropriate volumes of the *Dictionary of National Biography*. For a detailed examination of Philip's life and career, see William H. McMenemy, "Alexander Philips Wilson Philip (1770-1847), Physiologist and Physician," *J. Hist. Med., 13* (1958), 289-328.

121 See W. H. Brock, "The Life and Work of William Prout," *Med. Hist., 9* (1965), 101-126; and Anthony M. Kasich, "William Prout and the Discovery of Hydrochloric Acid in the Gastric Juice," *Bull. Hist. Med., 20* (1946), 340-358.

122 John Bostock (1773-1846), whose three-volume *Elementary System of Physiology* (1824-1826) was for a brief period the standard textbook in England, was also a graduate of Edinburgh. So was James Blundell, who attracted at least some attention as a research physiologist. See J. H. Young, "James Blundell (1790-1868), Experimental Physiologist and Obstetrician," *Med. Hist., 8* (1964), 159-169. The only exceptions that come to mind are the surgeons John Abernethy (1783-1867), Benjamin Brodie (1783-1862), and William Lawrence (1783-1867). All three were trained entirely in England, and all three exerted some influence on English physiology during the first third of the century.

Although a number of factors contributed to the superiority of Scottish over English physiology, surely the most important was the very different character of medical and scientific education in Scotland. There, as in Germany, the medical schools had evolved from the universities, from learning, rather than from the hospitals and practice.[123] Furthermore, because they cost relatively little to attend and imposed no religious tests, the Scottish universities attracted the scientifically most progressive students in Britain. At least until 1870, as George Haines has impressively documented, experimental science found virtually no advocates or practitioners among the Anglicans who had control over and access to Oxford, Cambridge, and King's College in England.[124] By mid-century, to be sure, the Scottish universities had lost much of their distinctively Continental flavor,[125] but they had by then trained what few competent physiologists Britain could claim. Even as late as 1870, when a controversy arose over the state of laboratory facilities for physiology in Britain, the University of Edinburgh was better equipped than any place in England.[126]

Conclusion

Although the stagnancy of English physiology between 1840 and 1870 depended importantly on antivivisection sentiment and on the utilitarian and correspondingly anatomical bias of English medical education, it was even more strongly correlated with the general state of higher education in England. First France, then Germany, led the advance of nineteenth-century experimental physiology precisely because, and to the degree that, their universities proved conducive to its development. England fell behind, meanwhile, chiefly because its universities did not yet view fundamental research in general, and scientific research in particular, as one of their central functions.

Until 1870, at least, physiology as taught at University College London was the best England had to offer, primarily because of the influence of William Sharpey and because physiology there was separated from surgical anatomy and brought into a more productive union with microscopic anatomy. Indeed, Sharpey's efforts have earned him recognition as the "father of modern physiology" in Britain,[127] and Univer-

123 See Vern and Bonnie Bullough, "The Causes of the Scottish Medical Renaissance of the Eighteenth Century," *Bull. Hist. Med., 45* (1971), 13-28.

124 George Haines, *German Influence upon English Education and Science, 1800-1866* (New London, Conn., 1957).

125 See George Elder Davie, *The Democratic Intellect: Scotland and Her Universities in the Nineteenth Century* (Edinburgh, 1961), esp. pp. 44-47, 68-73.

126 J. S. Burdon Sanderson, "Physiological Laboratories."

127 Cf. Taylor, "Sharpey."

sity College therefore deserves to be called the cradle of the modern revival. For all his achievements, however, Sharpey had only meager resources for research, did virtually no original work himself, and "was certainly not . . . a great research worker gradually building up a school."[128] And University College, despite its curricular emphasis on medicine and science, was not especially impressed by abstract or "disinterested" research on the German model. Like its Anglican rival King's College, it viewed as its chief function the production of "practically" educated men destined for careers in medical practice or business.

About twenty years after the establishment of these two new London colleges, another institution was launched with a similar conception of its purpose—Owens College, Manchester, founded on the bequest of an industrialist in 1851 and later incorporated into the University of Manchester. It transcended its utilitarian origins rather more quickly than its London analogues, chiefly through the influence of the German-trained chemist Henry Roscoe, but the "Germanization" of Owens College was not yet complete in 1870. In any case, physiology was not taught there until 1873.[129]

At Oxford and Cambridge, meanwhile, physiology shared the fate that there befell science and modern studies in general. Despite certain religious and political differences, early Victorian Oxford and Cambridge shared a quite similar educational philosophy and educational goal. This goal differed markedly from that of the middle-class, industrial institutions in London and Manchester, but its implications for scientific research were much the same. For though Oxford and Cambridge did not aspire to produce "practical" men in the mold of these middle-class institutions, neither did they aim to produce productive scholars. They sought instead to produce disciplined Christian gentlemen capable of intelligent and responsible leadership in the clergy, law, and public service. In the revealing if hyperbolic language of an Oxford flyleaf of 1830, their goal was

> to keep up a succession of sound orthodox Protestants in Church and State; men of true English growth . . . ; men who despise the

[128] Ibid., p. 138.

[129] The chief sources for the history of Owens College are *Essays and Addresses by Professors and Lecturers of the Owens College, Manchester*, published in commemoration of the opening of the new college buildings October 7th, 1873 (London, 1873); Joseph Thompson, *The Owens College, Its Foundation and Growth* (Manchester, 1886); P. J. Hartog, ed., *The Owens College, Manchester (Founded 1851), A Brief History of the College and Description of Its Various Departments* (Manchester, 1901); Edward Fiddes, *Chapters in the History of Owens College and of Manchester University, 1851-1914* (Manchester, 1937); and H. B. Charlton, *Portrait of a University, 1851-1951* (Manchester, 1951).

> frivolous refinements of the age, and are determined to put a stop to all the dangerous innovations of the boasted march of intellect, as destructive of the stout-hearted complacency of the true British character, and utterly useless to the man whose mind has been cast in the mould of our ancient institutions.[130]

Because they viewed a disciplined mind as the intellectual portion of the ideal total product—a disciplined man—and because they believed that classics and mathematics could better do the work of "mental discipline," most early nineteenth-century Oxford and Cambridge dons opposed the full-fledged introduction of science and other "unsettled" studies, including history, economics, and English literature. And since mental discipline was by itself insufficient—indeed dangerous—moral discipline had to be cultivated at least as assiduously. By insisting that college tutors acting in loco parentis to their students could best promote moral discipline, these dons justified the truly distinctive feature of the ancient universities, their residential-cum-tutorial colleges. In fact, these colleges dominated life at Oxford and Cambridge. And though university professorships did exist, even in some branches of science, few professors taught regularly and even fewer could attract a significant number of students to courses that had little connection with the degree examinations in classics or mathematics. To the collegiate forces, whose convictions can only have been reinforced by their vested interests, a strong research-oriented university professoriate was suspect, for to join in the search for new knowledge was to neglect the more important and eminently collegiate task of developing student character.[131]

Quite apart from the ideological barrier that the "Oxbridge" collegiate system thus raised for science and modern studies in general, it also posed a more immediately practical problem. In the mid-nineteenth century, the Oxford student body of 1,300 to 1,400 students was distributed unequally among twenty-four residential colleges and halls, while at Cambridge a roughly comparable number of students was distributed among seventeen colleges. Since no college harbored more than a fairly small portion of the total student population, only

[130] W. R. Ward, *Victorian Oxford* (London, 1965), p. 59. More generally, see J. P. C. Roach, "Victorian Universities and the National Intelligentsia," *Victorian Studies,* 2 (1959), 131-150; and C. C. Gillispie, "English Ideas of the University in the Nineteenth Century," in *The Modern University,* ed. M. Clapp (Ithaca, N.Y., 1950), pp. 27-55.

[131] For a spirited defense of the collegiate system, see E. B. Pusey, *Collegiate and Professorial Teaching and Discipline, in Answer to Professor Vaughan's Strictures, Chiefly as to the Charges against the Colleges of France and Germany* (Oxford, 1854). More generally, see Chapter 4 below.

the very largest of them had any reason to engage tutors outside of the degree subjects of classics or mathematics. There was even less incentive for an individual college to build or contribute toward expensive laboratories that only a handful of its own students would need. Only on a university-wide basis did it make sense to provide for modern studies and especially for the sciences as they came to be increasingly dependent on large and expensive laboratories.

For their neglect or subordination of science and other modern studies, Oxford and Cambridge faced the slings and arrows of outside critics, notably the *Edinburgh Review* and geologist Charles Lyell, himself an Oxford graduate.[132] As preserves of Tory Anglicanism, they also drew fire from Liberals and dissenters, whose political power was on the ascendant by the fourth decade of the nineteenth century. Often coalescing with and reinforcing each other, these critics had gained sufficient strength by 1850 to goad Parliament into appointing a Royal Commission to investigate the two ancient universities. As we shall see in Chapter 4 below, the results of this commission were far from trivial, but it had little immediate impact on either university's conception of its purpose, and it left the relationship between the colleges and the universities fundamentally unchanged. Not until the late 1860's and early 1870's, when the rising political, industrial, and military power of the German states intruded on Victorian complacency, did critics win a truly receptive hearing within the colleges at Oxford and Cambridge. Only then did a substantial number of dons begin to doubt that their traditional mode of education fully met the needs of the rapidly changing world around them. At Cambridge, especially, the doubt gave rise to an expanded conception of university purpose and a new sympathy toward the claims of science and other modern studies.

Meanwhile, in keeping with his origins in a middle-class Nonconformist family, Michael Foster had attended University College London and had joined his father in the private practice of medicine. Just one decade after he took his M.D. in 1859, he was to become a major beneficiary of events that transformed the Cambridge scene. Indeed, had Cambridge remained wedded to the mid-century conception of its function in English society and culture, Foster's achievement there would have been impossible, and the face of late Victorian English physiology would have been drastically different.

[132] Charles Lyell, "State of the Universities," *Quarterly Review, 36* (1827), 216-268; Lyell, *Travels in North America in the Years 1841-2, with Geological Observations on the United States, Canada, and Nova Scotia,* 2 vols. (New York, 1845), I, pp. 215-251.

3. Foster on His Way to Cambridge

It is hardly more than the other day since I wrote to you, I remember, from Huntingdon, wondering whether I should ever get away from that treadmill practice work—and now this strange tangled web of events has given me all the outward things I could wish for—and at the same time taken away the mainstay of my life, so that I live now in un unreal world, beating the air and grasping shadows—has given with the one hand and taken away with the other, and in the strangest way has made the giving dependent on the taking—for I suppose if my wife had never been ill, I had never left Huntingdon. If the parsons could leave their bad metaphysics and worse science and grasp hold of this wonderful Providence which tosses us about, and tossing, guides us—they might have something to say worth hearing. Would I could go back to the old patients and pills and have my wife back by my side were it only for an hour a week—but it is not to be, and I suppose my past means that there is something in the future for me to do.

Foster to Huxley (May 1870)[1]

MICHAEL FOSTER sprang from yeoman stock. For generations his ancestors had farmed in Hertfordshire and Bedfordshire, between London and Cambridge. The Fosters were prominent in the cause of religious Nonconformity and liberty, and at one point several Foster brothers guarded the dell at Preston where the seventeenth-century Baptist leader John Bunyan preached to his persecuted sect. John Foster, Michael's grandfather, settled in Holywell, a small Bedfordshire village near Hitchin, where he carried on the family tradition of religious activism and cultivated antiquarian tastes as well as crops. It was at Holywell, on 22 April 1810, that Michael Foster's father (also named Michael) was born.[2]

The elder Michael Foster broke the yeoman tradition and prepared himself instead for a career in medical practice. Excluded as a Nonconformist from the elite physician class, he sought licenses from the Royal College of Surgeons and the Worshipful Society of Apothecaries. Both required five years of apprenticeship, so he attached himself in 1826 to a surgeon, Mr. Peck of Kimbolton. In 1831, his apprenticeship completed, he entered newly established University College London,

1 Foster to Huxley [May 1870], Huxley Papers, IV. 173.

2 Victor Gustave Plarr, *Plarr's Lives of the Fellows of the Royal College of Surgeons of England*, revised by D'Arcy Power, with the assistance of W. G. Spencer and G. E. Gask, 2 vols. (London, 1930), I, p. 413; *Lancet*, 1880 (2), 111; [*Huntingdon*] *County Quarterly*, April 1880; and "The Michael Fosters at U.C.L.," *University College Hospital Magazine*, January 1930, pp. 18-22.

where during the academic session 1832-1833 he won gold medals in midwifery and medical jurisprudence as well as silver medals in anatomy, practical anatomy, and the principles and practice of medicine.[3] He became licentiate of the Society of Apothecaries and member of the College of Surgeons in 1833. After further clinical training at one of the London hospitals, he settled into practice with Mr. Josiah Wilson at Huntingdon, an outwardly "quiet little town on the sedgy Ouse" and birthplace of Oliver Cromwell.[4] The younger Michael Foster was born there on 8 March 1836.

With seven siblings, young Michael grew up in a home where the demands of their father's growing medical practice competed with his deep involvement in civic and religious activities. Under the terms of the Medical Act of 1858, he was appointed medical officer to the Huntingdonshire hospital and surgeon to the county gaol. Almost simultaneously, he sat as mayor of Huntingdon. A prominent figure in the Baptist community, he served as deacon of the chapel, president of the Young Men's Society, and (for forty years) as leader of the Working Men's Bible Class. But he subordinated any narrowly denominational preferences to a more general commitment to evangelical dissent. When the American revivalist Charles G. Finney visited England in 1859-1860, Foster invited him to stay at Huntingdon, at least partly in the hope that Finney could convert a skeptical young Michael to the faith.[5] Like many evangelical dissenters of that period, the elder Foster leaned to the political left. He was, by all accounts, a remarkable man of deep religious faith, independent cast of mind, and enormous strength of character. When he died of Parkinson's disease in 1880, only three of his children survived. In ways that must sometimes have perplexed him, he left his strongest mark on the son who bore both his names.

Until the age of thirteen, the younger Michael Foster was educated at the Huntingdon Grammar School. As described none-too-affectionately by Michael's schoolmate Bateman Brown, the school then housed about one hundred boys ranging in age from six to seventeen. The younger boys faced harassment and cruelty from their elder schoolmates as well as severe corporal punishment from their teachers. A poor performance in the daily recitations brought forth the familiar cane. According to Brown, however, young Michael Foster quickly

[3] University of London, *Distribution of the Prizes and Certificates of Honour Session 1832-1833* (London, 1833, 16 pp.).

[4] J. George Adami, "A Great Teacher (Sir Michael Foster) and His Influence," *Publication no. 7, Medical Faculty, Queen's University (Kingston, Ontario, Canada)*, June 1913, pp. 1-17, on 2.

[5] Michael Foster (the elder) to Charles Finney, 9 March 1859; original in the possession of Sir Robert Foster.

found a way to ease his punishment. After once being caned until he fainted, Michael thereafter stayed the hand of any would-be vicious flogger with the words, "I feel faint; I must complain to my father again."[6] He escaped the Huntingdon cane forever upon his transfer to University College School, a preparatory school associated not only constitutionally but also in aim and spirit with University College itself. In 1852, after three years at the school, where he was captain of the cricket team, young Foster entered University College, from which he graduated B.A. in 1854. Up until this time, he had devoted himself largely to classical studies, a background that may help to explain the highly literary style of much of his later writing on scientific subjects. He was, moreover, no ordinary classics student. By combining his natural aptitude with hard work, he won the classics scholarship on the University College B.A. examination.[7]

Indeed, Foster had such obvious talent in the classics that it is believed he might have pursued a career in that field had not Cambridge restricted its fellowship competition to those who subscribed to the Thirty-nine Articles of the Church of England.[8] Instead, he entered University College Medical School in 1854, immediately after graduation from its undergraduate division, and quickly displayed an equally strong aptitude for medical and scientific studies. In 1856 he carried off two gold medals for his performance on the examinations in chemistry and in anatomy and physiology. He graduated M.B. from University College in 1858 and proceeded to the M.D. there the next year. All in all, Foster's student career evokes the image of a well-rounded and popular young man of impressive attainments in a variety of pursuits. A similar versatility and breadth of achievement marked his later career.

William Sharpey, Foster's mentor in physiology

It was at University College that Foster learned to love physiology. William Sharpey (1802-1880), professor of anatomy and physiology there, instigated and promoted the courtship. The posthumous son of an English shipowner from Folkestone, Sharpey grew up in the Scottish city of Arbroath, where his parents had migrated and where his widowed mother married Dr. William Arrot. As a student at the Uni-

6 Bateman Brown, *Reminiscences* (London, 1905), pp. 12-17, quote on 17. I owe this reference to Richard Dupuis.

7 For official verification of this last point, and of Foster's degrees, see *University of London General Register*, Part II (31 March 1896), p. 222.

8 See, e.g., W. H. Gaskell, "Sir Michael Foster, 1836-1907," *Proc. Roy. Soc., B80* (1908), lxxi-lxxxi, on lxxi.

versity of Edinburgh from 1817 to 1823, Sharpey may have been especially attracted to the anatomist Robert Knox, who later became embroiled in the sensational Burke and Hare grave-robbing and mass-murder scandal. This scandal, which inspired Dylan Thomas's screen-play *The Doctor and the Devils* (1953), produced a public outcry against Knox, and it has been alleged that Sharpey not only bowed to public pressure by abandoning a plan to enter into partnership with Knox, but actually joined in the attack against the man to whom he had earlier co-dedicated his M.D. thesis. In truth, one has little solid basis either for accepting or rejecting this allegation, for a mist enshrouds nearly all of Sharpey's activities before he came to University College in 1836.[9]

What we do know of Sharpey's career at Edinburgh does little to explain how he gained his appreciation for physiology as an independent science or his awareness that laboratories were essential to its advance. Whatever his actual relationship with Knox, the latter seems an unlikely source of inspiration for either insight. The most plausible suggestion is that both derived from his extensive first-hand experience of Continental science and medicine. By the time he graduated M.D. from Edinburgh in 1823, Sharpey had already spent nearly a year in Paris studying medicine and surgery with Dupuytren and Lisfranc. After graduation, he spent another year there before briefly joining his stepfather in the private practice of medicine. By 1827, he had apparently decided to abandon practice and to pursue anatomical and physiological studies. Returning to the Continent in that year, "staff in hand and knapsack on his back," he walked hundreds of miles, stopping for a few months each at a variety of Continental medical centers—at Pavia in Italy, at Vienna in Austria, and at a number of cities in the German principalities. At Heidelberg he worked under the physiologist Friedrich Tiedemann, but he stayed longest in Berlin, studying for nine months under Karl Rudolphi, professor of anatomy and physiology there and teacher of Johannes Müller, among others. Not until the autumn of 1829 did Sharpey return to Scotland, where he settled in Edinburgh and entered into microscopical and anatomical research.

This extensive Continental tour did not make a full-fledged practicing physiologist of Sharpey. He published very little, and nothing whatever in experimental physiology. His only noteworthy original contribution was a microscopic investigation of cilia and ciliary mo-

[9] For the most complete available account of Sharpey's life and work, see D. W. Taylor, "The Life and Teaching of William Sharpey (1802-1880): 'Father of Modern Physiology' in Britain," *Med. Hist.*, *15* (1971), 126-153, 241-259. On his Edinburgh period, see ibid., pp. 130-131.

tion in animals.[10] Nonetheless, he had made the decision, then rare if not unique in Britain, to forgo the demands and rewards of medical practice in order to devote his life to the pursuit and teaching of medical science. His exposure to the Paris of Magendie's era and to the ascendant German universities must have made him aware that on the Continent physiology was moving toward independent recognition and laboratory status at a time when in Britain it was still submerged in surgical anatomy. The roots of Sharpey's role in the reform of British physiology almost certainly lay abroad.

For the time being, though, Sharpey was content to pursue his anatomical and histological investigations and to aim for Fellowship in the Royal College of Surgeons at Edinburgh, which he needed in order to teach the traditional medical courses in anatomy and physiology. Admitted to Fellowship in 1830, he spent the summer of that year in Berlin and then began his teaching career at the extramural (or non-University) school in Edinburgh. He taught anatomy and "practical anatomy" (dissection), while the lectures in physiology were entrusted to Dr. Allen Thomson, whose central interests were also anatomical and who later held the chair in anatomy at the University of Glasgow for almost thirty years.[11] Sharpey continued to teach at the extramural school until 1836, by which time his lecture class had grown from 22 to 71 students and his class in practical anatomy from 39 to 88.[12] Already, it seems, Sharpey had given evidence of his talent as a teacher.

The event that revolutionized Sharpey's life and gave new scope to his talents was his appointment to the chair of anatomy and physiology at University College London in 1836. The appointment occasioned surprise in some quarters and total outrage in the *Lancet*, which called his selection an example of "hypocrisy, treachery, envy and fraud, superadded to the one ancient evil, *love of pelf*." Insisting that "the community, the profession and the students of the institution HAVE BEEN BETRAYED," the editorial went on to ask, "Who is Doctor SHARPEY? Where is he known as a discoverer—as a physiolo-

[10] See William Sharpey, "On a Peculiar Motion Excited in Fluids by the Surfaces of Certain Animals," *Edinburgh Medical and Surgical Journal, 34* (1830), 113-122. Cf. also Sharpey, "Account of the Discovery of Purkyne and Valentin of Ciliary Motions in Reptiles and Warm-blooded Animals: With Remarks and Additional Experiments," *Edinburgh New Philosophical Journal, 19* (1835), 114-118; and Sharpey, "Additional Observations and Experiments to Purkyne and Valentin's Paper on the Discovery of a Continued Vibratory Motion Produced by Cilia as a General Phenomenon in Reptiles, Birds, and Mammiferous Animals," ibid., pp. 125-128.

[11] On Thomson, see *DNB, 19*, 713-714.

[12] See Sharpey's letter of application to University College, 7 July 1836. Sharpey-Schafer Collection, Wellcome Institute of the History of Medicine, B. 10. 1. 1.

gist? . . . Yet a majority of [the University College] Council votes go 'in favour of the UNKNOWN MAN OF THE NORTH.' "[13] Sharpey's appointment certainly took place in the midst of one of the bitter internecine quarrels that rocked University College Medical School during its first three decades, and the *Lancet*'s allegations of intrigue carry some persuasion. But nothing suggests that Sharpey himself participated in these academic politics, and his credentials seem at least as impressive as those of the other declared candidates for the chair.[14] From his subsequent stature in the history of British physiology, it is tempting to conclude that the council had acted responsibly and wisely.

By creating the position that Sharpey eventually held, University College apparently sought to promote the teaching of physiology.[15] But the council almost certainly had in mind the anatomically oriented British conception of physiology, and most of Sharpey's teaching was in fact anatomical or histological in character.[16] Only gradually and partially did he seek or achieve recognition for physiology as an independent and essentially experimental science. Nonetheless, he did resist the British tendency to make physiology a mere handmaiden of surgical anatomy, and his lectures from the beginning displayed an uncommon breadth of vision. In the academic year 1837-1838, when he gave 154 lectures on physiology, he distributed them as shown.[17]

Topic	Number of lectures
Introduction and properties of animal bodies	5
Blood	7
General or microscopical anatomy	26
Digestion	23
Circulation	11
Respiration	11
Absorption, nutrition and secretion	11
Nervous system	40
—structure	(13)
—functions	(13)
—external senses	(14)
Sleep	1
Embryology	19

[13] *Lancet* (1835-1836), pp. 789-790. Cf. Taylor, "Sharpey," pp. 132-133.

[14] Taylor, "Sharpey," 131-137.

[15] Ibid., 132-134, esp. the quote from the *Lancet*, 13 August 1836.

[16] See ibid.; and H. Hale Bellot, *University College London, 1826-1926* (London, 1929), pp. 166-167.

[17] Richard Quain, *Lectures on Physiology by Professor Sharpey (1837-1838)*, 2 vols., holographic MS. deposited in the Yale University Medical Historical Library.

The basic content and character of Sharpey's lectures can be reconstructed from several student notebooks which have happily survived.[18] From these notebooks, ranging in date from 1836 to 1867, it is obvious that Sharpey invested enormous time and energy in his continuously updated lectures. If he came to a point in dispute, he weighed the evidence on both sides, naming his authorities, and then tried to reach an independent judgment. Although often giving sympathetic attention to works of the seventeenth century and even earlier, he more commonly emphasized the latest (and predominantly Continental) literature. He gave credit where he believed it was due, but avoided dogmatism so that physiology became a living and continuous enterprise. Simple experimental demonstrations occasionally accompanied the lectures, and in a few cases Sharpey reported the results of his own experimental attempts to settle controversial problems. On at least one topic, the action of the mucosa in digestion, he informed only his auditors of suggestive experiments that probably deserved independent publication.[19]

Such efforts did not go unappreciated. In Joseph Lister, for one, Sharpey's lectures awakened "a love of physiology that has never left me."[20] At times as many as three hundred and fifty students came to hear him, and Sharpey only enhanced his popularity by taking pains to remember the name and circumstances of each one. Many were especially impressed by the apparent spontaneity and extemporaneous quality of his lectures, which he always delivered from nothing more than a few scribbled notes.[21] Foster, who later adopted the same style in his lectures, attended Sharpey's course even before entering medical school, "breaking into my ordinary studies," he says, "in order to do that."[22]

Almost from the beginning Foster attracted special attention. He later recalled that Sharpey very quickly introduced him to the advancing edge of physiology. One day, probably in 1855, Foster went to Sharpey after a lecture on the functions of the liver. Sharpey told him that he had just received a paper from Claude Bernard announcing his celebrated discovery of glycogen in the liver. Students who took the lectures next year would be told of the discovery then, Sharpey continued, but he thought Foster would want to know about

[18] See Taylor, "Sharpey," p. 139.

[19] Ibid., pp. 139-141.

[20] See E. A. Sharpey-Schafer, *History of the Physiological Society during its First Fifty Years, 1876-1926* (Cambridge, 1927), p. 18, n. 1.

[21] Ibid.; *Proc. Roy. Soc., 31* (1881), x-xix, on p. xiv; Bellot, *University College*, p. 129; and Taylor, "Sharpey."

[22] Michael Foster, "Reminiscences of a Physiologist," *Colorado Medical Journal, 6* (1900), 419-429, on 419.

it immediately.[23] By winning the gold medal in anatomy and physiology in 1856, Foster became in effect "Sharpey's medallist," and more than ever the special beneficiary of his experience, insight, and influence.

Sharpey's training in laboratory procedures seems to have been limited to the essentially histological techniques he learned on the Continent. One of the first in England to use the microscope as a teaching instrument, he constructed for this purpose a round table with a hole at its center. Into this hole an iron bar was inserted vertically, and from the iron bar extended a metal arm bearing a microscope at its free end. This free end was then attached to a brass strip inlaid around the periphery of the table in such a way that the microscope could be revolved from student to student around the table.[24] This ingenious device allowed Sharpey to extend the benefits of a single microscope to several students and to present microscopical demonstrations simultaneously with his lectures on histology. Such demonstrations apparently became a regular and prominent feature of Sharpey's lectures soon after his appointment, and by the time Foster entered the medical school at University College, Sharpey had already instituted a separate course in "practical" histology.[25]

During Foster's second year in the medical school (1855-1856), a new departure was made by the founding of an independent lectureship in practical physiology and histology. The appointment went to George Harley, a young man of impressive background. After graduating M.D. from Edinburgh in 1850, he had spent more than four years at a number of Continental universities and had worked in physiology with Claude Bernard, in histology with Albert von Koelliker, in pathology with Rudolf Virchow, and in chemistry with Charles Wurtz. Only twenty-six years old, he had already shown great promise in five published papers that had resulted from his own research work.[26]

At University College, however, Harley did not do for physiology all that he might have done. Whether in search of more work or a larger income, he combined his lectureship with an expanding practice as a consulting physician and, after 1859, with a second appointment at University College as professor of medical jurisprudence.[27] Not surprisingly, his course in practical physiology never really developed beyond its original narrow bounds. Instituted "with the view

[23] Ibid., p. 421.
[24] Bellot, *University College*, p. 167.
[25] Ibid., p. 313.
[26] Mrs. Alec Tweedie, ed., *George Harley, F.R.S., The Life of a London Physician* (London, 1899), esp. pp. 39-41, 101, 113. This lively but romantic account of Harley's career should be read with a skeptical eye.
[27] Ibid., pp. 124-127.

of supplying the Medical Students with instruction in the use of the Microscope in examining the textures and fluids of the body," this course was at first given only in the summer term, in a portion of the dissecting room temporarily partitioned off and arranged for the purpose. By 1859, a separate room had been allocated, but the course still "consisted in little more than distributing sections made by the [instructor]."[28] In effect, the new course in "practical physiology" amounted to an introductory course in microscopy, and experimental physiology remained untaught at University College.

This was not the result of any antagonism toward experiment on Sharpey's part. In an address of 1862 to the British Medical Association, he deplored "the indignant but misdirected declamation . . . against experiments on animals," insisting that the recent spectacular progress in physiology was due precisely to such experiments and to "the establishment of schools of Practical Physiology in various parts of Europe."[29] And while Sharpey's lectures only rarely discussed an experiment of his own making, they did refer frequently to the experiments of others and conveyed a clear appreciation of the principles on which they were conducted. In the absence of experimental apparatus to use in illustration of his lectures, he displayed a charming resourcefulness. Foster and others recalled that when Sharpey lectured on blood pressure, and needed to explain the principles of Ludwig's new kymograph, "all he had to help him was his cylinder hat, which he put upon the lecture table before him and with his finger traced upon the hat the course of the curve."[30] Fully aware of his own inadequacies as an experimental physiologist, Sharpey "was none the less ready to encourage others to undertake research and was prepared to afford opportunities for carrying out such work at University College."[31]

Foster was among the very few who took real advantage of the encouragement and opportunities Sharpey had to offer. In 1859, the year he graduated M.D. from University College, Foster carried out there a series of experiments on the excised hearts of snails. Having found that even tiny pieces cut from the beating heart continued for some time to beat rhythmically, he concluded that the heartbeat could not be the result of any localized nervous mechanisms, but must instead be viewed as a peculiar property inherent in the general cardiac tissue. That September, when the British Association for the Advance-

28 Bellot, *University College*, p. 313.

29 W. Sharpey, "Address in Physiology," *Brit. Med. J.*, 1862 (2), 162-171, on 163.

30 Foster, "Reminiscences," p. 420.

31 Sharpey-Schafer, *Physiological Society*, p. 1.

ment of Science met at Aberdeen, Foster described his experiments and his conclusion before the section on botany and zoology.[32] This work created no discernible excitement at the time, but it introduced Foster to the world of scientific research and awakened in him an interest in the problem of the heartbeat, an interest that never died and that was to become of crucial importance in the formation of the Cambridge School of Physiology.

Voyage to the Red Sea and medical practice in Huntingdon, 1859-1867

In the very week that Foster was at Aberdeen describing these experiments, his father wrote to revivalist Charles Finney about Michael's progress toward a very different goal—religious conversion. Six months before, in March 1859, the elder Foster had invited Finney to Huntingdon, noting that "my son remains, apparently, 'in statu quo,' but my confidence is unshaken that he will be restored." He prayed that the restoration might occur "through your visit to this part."[33] In his *Memoirs* Finney described the invitation and its results in the following fashion:

> Such was the state of my health that I thought I must return home. But Dr. F[oster], an excellent Christian man living in Huntington [*sic*], urged us very much to go to his house and finish our rest, and let him do what he could for me as a physician. We accepted his invitation and went to his house. He had a family of eight children, all unconverted. The oldest son was also a physician. He was a young man of remarkable talents, but a thorough sceptic. He had embraced Comte's philosophy, and had settled down in extreme views of atheism, or I should say, of nihilism. He seemed not to believe anything. He was a very affectionate son but his scepticism had deeply wounded his father, and for his conversion he had come to feel an unutterable longing.
>
> After remaining at the doctor's two or three weeks, without medicine, my health became such that I began to preach. . . . I soon found opportunity to converse with young Dr. F[oster]. I drew him out into some long walks, and entered fully into an investigation of his views; and finally, under God, succeeded in bringing him to a perfect stand-still. He saw that all his philosophy was vain. At this time I preached one Sabbath evening on the text: "The hail shall sweep

[32] An abstract four sentences in length appeared in the published report of the meeting. See *Brit. Assoc. Rep.*, *29* (1859), Transactions, p. 160.

[33] M. Foster (the elder) to Finney, 9 March 1859; original in the possession of Sir Robert Foster.

away the refuges of lies, and the waters shall overflow the hiding places. Your covenant with death shall be disannulled, and your agreement with hell shall not stand." I spent my strength in searching out the refuges of lies, and exposing them; and concluded with a picture of the hail-storm, and the descending torrent of rain that swept away what the hail had not demolished. The impression on the congregation was at the time very deep. That night young Dr. F[oster] could not sleep. His father went to his room, and found him in the greatest consternation and agony of mind. At length he became calm, and to all appearance passed from death unto life. The prayers of the father and the mother for their children were heard. The revival went through their family, and converted every one of them. It was a joyful house, and one of the most lovely families that I ever had the privilege of residing in.[34]

Not atypically, Finney's account at the very least exaggerates the speed and totality of his triumph. In the letter to Finney of September 1859, the elder Foster wrote of his son then at Aberdeen:

> I cannot say there is any tangible thing that I can name to you as indicating any change in his state. I find *my* strength in the love of God, & upon His love I hold by prayer, without fainting. Why is it I should have doubt of his conversion removed from me, if the Lord intends not to fulfil this desire of my soul? I feel thankful he had your ministrations. I know *now* he has a foundation if he *will* build upon it.[35]

He hoped young Michael would find time on his way home to pass through Edinburgh, where Finney was then staying, and "thus have another opportunity of seeing you." If this second meeting took place, its impact must have been less than monumental, for Michael's conversion remained in doubt that October, when he embarked on a voyage to the Red Sea. There he was to serve as ship's surgeon to the S.S. *Union* and thus to join a company that had as its charge the building of a lighthouse on the Asaruji Rock opposite Mt. Sinai.

Foster had signed on after an otherwise inexplicable decline in his health led to fears of consumption. The favorite Victorian remedy for this affliction being a protracted sea voyage to warmer climes, Michael applied for duty as surgeon partly to cover the expenses of the prescrip-

[34] Charles G. Finney, *Memoirs* (New York, 1876), pp. 452-453. My thanks to Richard Dupuis for this reference and for the identification of "Dr. F" as the elder Michael Foster.

[35] M. Foster (the elder) to Finney, 11 September 1859; original in the possession of Sir Robert Foster.

tion. But he had also been led to believe that he would act as part-time naturalist to the expedition, and he did in fact make some observations on marine fauna, perhaps in emulation of T. H. Huxley's activities during his four-year voyage as ship's surgeon to H.M.S. *Rattlesnake*.[36] From a scientific point of view, however, Foster's voyage pales into insignificance beside Huxley's or those of other Victorian naturalists—including Joseph Hooker, Alfred Russel Wallace, and Charles Darwin himself—whose work and thought had been profoundly shaped by spending several years of their early manhood at sea. If Foster seriously aspired to join this illustrious circle, it seems odd that he signed on for a mere six-months' tour. Insofar as he did hope to accomplish serious scientific research within that time, he found the way blocked by a series of frustrating obstacles. He recorded these frustrations, along with his other experiences and occasional adventures, in a score of long and vivid letters to his family back in Huntingdon.[37]

Sailing out of Southampton in late October 1859, Foster reached the Bay of Suez nearly a month later after stopovers in Malta, Alexandria, and Cairo. On the voyage out, aboard the *Delta*, he complained of little except enforced idleness. His remarkably colorful accounts of Alexandria and Cairo did not conceal his eagerness to get on to Suez, where he was to join the *Union* and where he hoped to begin his research in earnest. But when he finally did board the *Union*, in late November, no magic transformation took place. For a week or two, he seemed content enough, though his living quarters gave him "a vivid conception of what Shakespeare meant when he spoke of cabined, cribbed, confined," and though he found it difficult to accomplish much dissection:

> Dissecting on board ship requires a very great deal of patience indeed. First whenever I pull out my microscope, everybody comes bothering me with questions about what I am looking at and requests to have a peep. Sailors are certainly more obtuse to hints than anybody else—and as they never enjoy any privacy themselves, they slip into the idea that nobody else ever desires such a thing. Then of course light is precarious. If I attempt to dissect on deck, I am in danger of myself and my microscope being swept away by a howser,

36 See Henry Dale, "Sir Michael Foster, K.C.B., F.R.S., A Secretary of the Royal Society," *Notes and Records of the Royal Society (London), 19* (1964), 10-32, on 14.

37 These letters remain in the possession of Foster's grandson, Sir Robert Foster, who kindly allowed me to photocopy the entire set. In doing so, I arranged the letters chronologically, made one photocopy page for each foolscap sheet Foster sent home, and paginated (1-83) the accumulated material. In the following footnotes, my page references are to that photocopied version of the letters, which is deposited in my personal files.

> and in certainty of being surrounded by a large group of inquisitive seamen and natives. If I dissect in my berth, . . . my light only comes from a porthole 7 inches in diameter, and as soon as I set my instruments out, a boat comes alongside and deprives me even of that small pittance. . . . [D]ifficulties, people say, ensure success, and perhaps they are right, though it does not look much like it at first sight.[38]

Nonetheless, Foster had begun to collect and study small marine animals and his spirits remained bright in anticipation of the *Union's* scheduled trip to Ushruffee Island, where he obviously expected to find more interesting fauna. His hopes gradually sagged, however, as the *Union* languished in the Bay of Suez for more than a month and as the lighthouse expedition ran into repeated delays and difficulties. By Christmas, Foster was beginning to complain of idleness again, and though he dutifully performed his minor medical tasks, he did so with little sense of excitement. In fact, he was getting tired of being constantly called "Doctor":

> There is one thing among many, many others that I shall be glad in when I get home and that is that I shall be called by my proper. I have not been called Foster above 3 or 4 times since I left home. My name is "Doctor" and it is dinged into my ears all day long. "Good morning, Doctor. Fine evening, Doctor. Oh, Doctor, there is a man wants to see you please, Doctor. Well, Doctor, have you made any discoveries today, Doctor. Oh Doctor, please Doctor, will you give me a dose of salts, Doctor. Thank you, Doctor. Good night, Doctor." And so it is from the time I get up (or rather before that, for the first thing I hear—at about 5:30—is "Here's your coffee, Doctor. Half past five, Doctor") to the time I am asleep.[39]

Foster's humorous lament was symbolic of a more serious problem of which he had not yet become fully aware. Until the end of January 1861, he seemed to believe that his efforts to do natural history had been hampered only by the general difficulties confronting the lighthouse expedition and by the "want of boats" available to him for expeditions in search of specimens.[40] On 31 January 1861, however, he wrote of the Captain's message that "he considers me as Surgeon and not officially naturalist in any way and that all the opportunities of working that he gives me are to be considered as personal favours. This is of course a very different understanding from that on which I thought I came out and so the sooner I come home the better." A week later, from the Bay of Suez, Foster told the story in somewhat greater detail:

[38] Ibid., pp. 28, 34-35. [39] Ibid., p. 52. [40] Ibid., p. 61.

> I hope I sufficiently explained to you why I had better come home. I am quite well and as comfortable as one can be on board ship but I cannot do that amount of work which I expected and which would justify me in remaining at my inconvenience and (what is more) at *yours*, especially during the spring months when I hope to be of use to you. I can't dredge because I have no boat at my disposal which Parkes said I should have. Again there are few animals on the reef and I can have no boat to go to the island or elsewhere. I am limited to fishing with my townets off the side of the vessel. The first three weeks I worked very hard, but at the end of that time Capt. Baker spoke to me in terms of complaint about my basin and instruments, etc. in a way which I think was uncalled for. This upset me rather and I did no work for 2 or 3 days, then the gale of wind came on and when that ceased we came up here.
>
> The fact is this expedition is a queerly and badly managed affair. The ship is not in any way under Mr. Parkes' direction but quite independent of him. He cannot alter a single thing in her. She is employed by the P. & O. Company to assist in building the lighthouse —that is, in affording a home to them during its erection and in bringing and taking them . . . to and from the reef. Now I am surgeon to the ship, and although Mr. Parkes spoke of my appointment as that of Surgeon-Naturalist, Capt. Baker maintains that in the eyes of the Company, I am simply surgeon and that I have no *right* to ask for any assistance whatever in my natural history work. He says that Mr. Parkes had no authority in promising me boats or indeed had anything to do with my appointment except the finding of me. This being the case, it was of course high time I should ask for my dismissal.[41]

Although his six-months' tour of duty did not officially end until 21 May, Foster apparently had little difficulty securing an early release and sailed for home at the end of March. What little natural history he had been able to do resulted in no published work, and the voyage had no discernible impact on his scientific career.

Nor, it seems, did the voyage do much to deliver Foster from the shadow of religious doubt. His letters home do contain accounts of the church services aboard ship and frequent Biblical allusions. Indeed, at first blush, the letters seem to support Finney's claim as to Michael's conversion, for they record his dutiful reading of the Bible and beseech his father's prayers "that I may grow more and more in grace." That he felt more religious "the more that mere externals are lacking" and

[41] Ibid., pp. 64-66, 69.

that he did not "suffer much from want of sermons" merely indicate his conformity with the evangelical preference for private over public worship, for personal study of the Bible over institutionalized ceremony. Like "any true religion whatever," Foster wrote, his was "not dependent on any outward means . . . for I am thrown all the more on my Bible and my own thoughts."[42] His geographical situation, his proximity to Biblical scenes, obviously enhanced the appeal of the testaments for him:

> After breakfast I sat upon the quarterdeck reading my Bible and everything seemed to join together to enable me to enjoy it. On deck all was quiet, no work scarcely being done, no shrieking of natives, no loud commands of officers—everybody sitting reading or walking quietly about. And the sea was so calm and the sun so bright and the air so clear. I read about Moses and the children of Israel and around me was the Red Sea, before me lay the long yellow range of desert over which they trod, and in the horizon the solitary silent hills that barred their way to the promised land.[43]

Nonetheless, in a letter of 6 February 1861, Foster confessed that his odyssey toward conversion remained unfinished and revealed his fear that the progress he had made might be merely temporary and circumstantial. Having gone ashore at Suez the Sunday before, he took a room in a hotel, locked his door and "spent the afternoon over my Bible," from which he strayed only to think of his family in Huntingdon:

> About 4 o'clock there was a very heavy shower of rain, and when it ceased I got up and went out for a walk in the desert. I walked . . . toward Jerusalem, and kept thinking and hoping that such a walk might typify my course on the spiritual desert of life. . . . I often fear that I may not feel so religious when I am at home again as I do in my loneliness and dependency here. Pray for me that in the luxuries and comforts of home I may be even stronger than I am now. It has always been an unhappy fact in reference to myself that the farther I am from preachers and even general worship, the more I feel the need of a religion, but I have never been so thoroughly removed from such things as now and I hope that I shall be cured of what is after all a mental weakness.[44]

From Foster's later indifference to religious matters, from his ready acceptance of Darwinian evolutionary theory, and from his active participation in the scientific naturalism that appalled many Victorians

[42] See ibid., pp. 5, 23, 32-33, 40, 48, 57-59.
[43] Ibid., p. 32.
[44] Ibid., pp. 67-68.

and against which even some scientists rebelled—from all of this, it seems clear that his course on "the spiritual desert of life" never got beyond that incomplete journey toward Jerusalem.[45] Despite Finney's ministrations, and despite his obvious desire to do as his father wished, Foster's ultimate religious position probably resembled that of the skeptical Huxley, who invented the word "agnostic" to define his own position.

Less aggressive than Huxley in his agnosticism as in other matters, Foster enjoyed a reputation for perfect rectitude, except perhaps in the eyes of those fervent antivivisectionists who considered all experimental physiologists morally depraved. From the photograph which serves as the frontispiece for this book, one can imagine why a contemporary described him as "the gentlest man that ever wore a red tie"—a sartorial aberration that puzzled his Victorian friends, but which the same writer traced to his love of vividly colored flowers, those "smiles of nature."[46] In the careful propriety of his moral conduct, Foster in fact differed little from Charles Darwin or Huxley himself. For none of them did agnosticism imply a loosening of the basic mores to which respectable Victorian society paid public allegiance.

If, then, Foster's voyage to the Red Sea had no permanent impact on his scientific or religious life, what of its effect on his health? His letters home refer to his health just often enough to suggest that it was a matter of concern, at least to his family, but they reveal no dramatic improvement or decline. Since his few brief allusions to his physical condition are in fact invariably favorable, one may doubt the story that he headed for home convinced of his imminent death and so, during a stopover in Paris, carried his name and address with him as he paced a square near the Louvre, expecting the worst.[47] One may even doubt that he had any serious pulmonary disease at all, though fear of "consumption" was entirely natural in view of the uncertain state of its diagnosis and its high incidence in Victorian England. To those of us reared in an age of antibiotics, Foster's frequent later complaints about his health, and especially his dyspepsia, may seem hypochondriacal.

[45] Late in life, it is said, Lady Foster (Sir Michael's wife) was told that the residents on her hill (presumably including her husband and fellow-physiologist, W. H. Gaskell) had a reputation for being atheists. " 'That's too bad,' she said. 'I often go to the village church, and once, yes once, I got Michael to go too.' " William Cecil Dampier, *Cambridge and Elsewhere* (London, 1950), pp. 29-30. On "scientific naturalism" and the reaction against it, see Frank Miller Turner, *Between Science and Religion: The Reaction to Scientific Naturalism in Late Victorian England* (New Haven, 1974).

[46] See *The Hunts County News*, 9 February 1907, p. 5.

[47] W. T. Thiselton-Dyer, "Michael Foster—A Recollection," *Cambridge Review*, *28* (1907), 439-440, on 439.

Among his contemporaries, however, such complaints were far from unusual, and dyspepsia in particular was yet another bond uniting him with Darwin and Huxley. At one point, comparing his infirmities with those of Huxley, Foster thought their "coincidence . . . really worth communicating to the Psychical Society" as an example of the sympathetic "transference of maladies."[48]

In any case, the concern over Foster's health remained so serious that it definitely influenced his later decision to go to Cambridge and probably influenced his choice of residence there.[49] It also induced him to take up gardening as a form of relaxation and recreation.[50] Before long, however, he had transformed this relaxing "hobby" into another scientific pursuit. He hybridized several new varieties of iris and was for a time the acknowledged expert on the genus. As a matter of fact, he eventually published more papers in horticulture than in physiology.[51]

However meager the results of Foster's frustrating voyage to the Red Sea, it failed to extinguish his interest in scientific research. In June 1860, two months after his return to England, he presented to the Royal Society a paper on the physiological effects of freezing isolated frog muscles. In this paper he reported his unsurprising observation that muscles and nerves lose their physiological attributes when frozen, and if the freezing is not too prolonged, regain them upon being thawed. Muscles frozen more than ten minutes, he reported, "never regain their irritability."[52] He offered no explanation for the failure of a heart to resume its beat even when frozen only very briefly. This paper in fact contains little of general interest, though it does suggest Foster's early concern with the conditions affecting muscular irritability.

Before he left on his voyage to the Red Sea, and then again upon his return, Foster pursued postgraduate clinical training, partly at the hospital associated with University College London and partly in Paris. Better known and more accessible than the German universities, the Paris hospitals remained at this period the chief Continental at-

48 Foster to Huxley, 2 November 1894, Huxley Papers, IV. 384.

49 That health was a major factor in Foster's decision to move from London to Cambridge is clear from his father's letters to Huxley of 3 May 1870 and 5 May 1870. Huxley Papers, IV. 176-179. That it influenced his choice of residence at Cambridge is the claim of Gaskell, "Foster," pp. lxxi-lxxii.

50 Gaskell, "Foster," p. lxxii; Thiselton-Dyer, "Foster," p. 439; and Foster to Huxley [undated, c. 1890], Huxley Papers, IV. 361.

51 See Dale, "Foster," pp. 24-25 and the bibliography, 31-32; and William R. Dykes, *The Genus Iris* (London, 1913), recently reprinted in clothbound edition (New York, 1975).

52 Michael Foster, "On the Effects Produced by Freezing on the Physiological Properties of Muscles," *Proc. Roy. Soc., 10* (1860), 523-528, quote on 525.

traction for English medical students. For one interested in physiological research, Paris offered the special attraction of Claude Bernard, the leading name in physiology. Bernard filled his lectures at the Collège de France with the results of his own research, including that on the glycogenic function of the liver, about which Foster had learned long before from Sharpey. Of all the Continental physiologists, Bernard was certainly the one Foster most admired, and he later gave tangible expression to his admiration by writing a biography of Bernard. In that biography, however, Foster described himself as one who had never even looked upon the face of the great French physiologist,[53] an opportunity perhaps denied him because Bernard suffered a breakdown in his own health just as Foster was returning from the Red Sea.[54]

In 1861 Foster settled at home in Huntingdon and joined his father in the private practice of medicine. By 1863 he apparently felt secure enough, physically and financially, to marry a young woman from his native town. He continued to practice at Huntingdon until 1867, wishing all the while that he could pursue instead his interest in scientific research and teaching. By his own account, he worked at physiology "whenever my patients would leave me alone" and felt thankful that during those years "I did no more mischief than I did."[55] From 1861 to 1867, his "Huntingdon years," Foster published perhaps a dozen popular articles in the *Christian Spectator* on a wide variety of scientific topics, including one on the snail's heart; a review of Karl Ernst von Baer's autobiography; an unspectacular review article on the coagulation of blood; and a brief paper reporting his discovery of a surprisingly large amount of glycogen in the tissues of *Ascaris lumbricoides*, a species of parasitic roundworm.[56] Despite the frequent literary merit of the other contributions, only this last paper even approaches scientific distinction.

Mainly because of Claude Bernard's work on the glycogenic function of the liver and his discovery of glycogen in fetal muscle tissues, glycogen had lately much attracted the attention of physiologists. Bernard's work on vertebrate species had been extended by several workers to invertebrates, in many species of which large quantities of glycogen had also been found. What puzzled Foster about the presence of glycogen in the roundworm was that such organisms would seem to have little need for it under the prevailing view of the role of glycogen in

53 Michael Foster, *Claude Bernard* (London, 1899), p. vii.

54 See J. M. D. and E. Harris Olmsted, *Claude Bernard and the Experimental Method in Medicine* (New York, 1952), p. 114.

55 Foster, "Reminiscences," p. 421.

56 M. Foster, "On the Existence of Glycogen in the Tissues of Certain Entozoa," *Proc. Roy. Soc., 14* (1865), 543-546.

the animal economy. According to this standard theory, glycogen served in the adult animal as a reserve fund of carbohydrate material that could be drawn upon as a source of "animal heat" during periods when food was not taken. Its concentration in fetal muscles was believed to accord with their need for a large amount of carbohydrate material to serve as a source of the energy required for their future differentiation.

But since intestinal worms already had "a constant temperature secured to themselves" through the warmth of their hosts, they seemed to Foster "the very last of creatures to need . . . 'calorifacient material.' " He therefore rejected the possibility that glycogen had any respiratory use in the roundworm. And concentrated though it was in the muscular parietes, he could not believe that this glycogen had any "muscular future" since the worms under investigation were adult and ova-producing animals. Having rejected these two possibilities, Foster proposed instead the hypothesis that the glycogen being stored in *Ascaris* was intended for the ova and embryos. Faced with the apparently contradictory fact that very little glycogen could be obtained from the ascarid reproductive system, he suggested that glycogen might later "migrate" to the reproductive system from glycogen-storing tissues, somewhat as starch was known to migrate from starch-storing tissues in plants. He closed with a question: "May not 'migration,' which plays so important a part in vegetable physiology, occur in the animal economy in reference to other substances besides fat?"[57]

Although not of great intrinsic importance, this paper is interesting for what it reveals about Foster's attitude toward physiological investigation. It suggests that he was eager to place his rather minor observation in a broader context and to draw from it a principle of potentially wide applicability. Beginning with an apparently isolated fact, he managed to direct attention to a significant problem of active concern; and by relating his particular hypothesis for the function of glycogen in *Ascaris* to the known migration of starch in plants and of fat in animals, he revealed an almost instinctive interest in the processes basic to all living organisms.

There is, as Sir Henry Dale first recognized,[58] a subtle kinship between the impression conveyed by this paper and that conveyed by Foster's review of the work that had been done on the coagulation of blood.[59] In that review, too, Foster had found an opportunity to transcend the apparent constraints of his particular topic and to pro-

[57] Ibid., p. 546.

[58] Dale, "Foster," pp. 14-15.

[59] M. Foster, "The Coagulation of Blood," *Natural History Review*, *4* (1864), 157-187.

pose a general theory, this time about the progress of knowledge. Suggesting that knowledge moved in spirals rather than in circles or straight lines, he envisioned successive theories as being directed "towards the same point of the compass," but "on a higher level, with a wider horizon." This idea, like Foster's hypothesis for the function of glycogen in *Ascaris*, was less than fully demonstrated by the evidence at hand; and neither idea is as remarkable for its own merit as it is for the strength of the impulse that brought it forth. It is hard to know by what name this impulse should be called. Dale called it Foster's "lively philosophical attitude,"[60] and perhaps that is as good a name as any. It was in any case an impulse essential to Foster's intellectual makeup and an impulse from which much of his later influence may well have sprung.

The paper on glycogen in *Ascaris* was communicated to the Royal Society in November of 1865 by Thomas Henry Huxley. Foster had first met Huxley sometime in the 1850's[61] and from the beginning looked up to Huxley, eleven years his senior, for guidance, inspiration, and encouragement. The bond between them had become so strong by 1865 that Foster was confiding his innermost thoughts. In a letter sent to Huxley in October of that year, Foster wrote: "Between yourself and me . . . I fancy I have one poor talent, that of teaching—and that is at present wrapt up very tight in a dirty napkin. I only wish I were obliged to throw the napkin away."[62] That Foster was able to recognize so early wherein his chief talent lay is all the more remarkable since there exists no evidence that he had as yet acquired any teaching experience.[63]

Like Huxley, Foster had strong convictions about education and by 1867 had already contributed a popular article in support of Huxley's campaign to bring science into the English school curriculum.[64] Whether or not Foster had come to his convictions independently, this article is pure Huxley, right down to its vigorous and witty style. It inveighs against the belief among English schoolmasters that only classics and mathematics are capable of training the mind, "as if . . .

60 Dale, "Foster," p. 15.

61 Probably in 1856, when Huxley examined Foster in anatomy and physiology for the University of London. See Michael Foster, "T. H. Huxley," *Proc. Roy. Soc.*, *59* (1897), xlvi-lxvi, on liii.

62 Foster to Huxley, 22 October 1865, Huxley Papers, IV. 153-154.

63 Though the possibility remains, of course, that he had on an informal basis assisted Sharpey in his courses at University College.

64 [Michael Foster], "Art. VI. Science in the Schools," *Quarterly Review*, *123* (1867), 244-258. Attributed to Foster in *Wellesley Index to Victorian Periodicals*, ed. Walter E. Houghton (Toronto, 1966), p. 750, item 1578. For Huxley's views on education, see Cyril Bibby, *T. H. Huxley, Scientist, Humanist and Educator* (London, 1959).

to lean on [science] as a chief means of general culture, would be to throw away all hopes of bringing up stout masculine minds, and of training the young English intellect to sound and vigorous thought." Part of the reason science suffered so badly was that it had "unhappily in former days been stigmatised as 'useful knowledge,' " while the schoolmasters were convinced that the value of classic studies "lies precisely in their being what the busy world calls useless."[65]

Besides being useful, Foster insisted, science was at least as capable as any other study of instilling intellectual and even moral culture, but its full benefits could be reaped "by those only who have had the opportunity of spending some time in actual original research." The method of teaching was "all in all." It was absolutely essential that the teacher introduce the student to the problems of nature "as if they had never been solved before" and that he "recognize and utilize the actual condition of the pupil's mind." Taught as it now was in England, as mere useful knowledge, "or as flimsy material for ornamental fringes," science might best be left untaught.[66]

Foster closed his article by linking the sorry state of science teaching in England with her poor showing at the recent International Exhibition in Paris and with her precarious position in the general world of science. By far the greatest amount of solid experimental science was contributed by men "possessing none but a German degree," while in England "the learned societies groan under the burden of memoirs presented by gentlemen who would fain fly in science before they have learned to creep." This situation would continue in England "so long as physics and chemistry play hide and seek with dancing and gymnastics, [and] so long as science is offered to boys on half-holidays as an obstacle to cricket and foot-ball." There was a practical aspect to the problem. All the emoluments and prestige belonged to the classics and mathematics, while very few posts of any kind were open to scientific men. Quite apart from the inadequacy of their training, Englishmen were deterred from a career in science by the "prospect of speedy starvation."[67]

University College and the call to Cambridge, 1867-1870

Almost certainly, this prospect had deterred Foster himself from entering immediately into a career in science and teaching. As it happened, however, he was not obliged to accept the usual fate and to spend the rest of his days in an occupation that, however useful and

65 [Foster], "Science in the Schools," p. 245.
66 Ibid., pp. 250-251, 258.
67 Ibid., pp. 256-258.

important, would have been wasteful of his rarer talents and uncongenial to his natural interests. Foster threw his "dirty napkin" away when Sharpey offered him the instructorship in practical physiology and histology at his alma mater, replacing George Harley, who had fallen victim to a debilitating eye affliction. Foster accepted in January of 1867, moved to London, and never again practiced medicine.[68]

H. Hale Bellot has described Foster's appointment as the first of the revolutionary changes that transformed the study of physiology at University College in the late 1860's.[69] The transformation had its origin in a course Foster instituted soon after his appointment; its nature has been described in admirable detail by Sharpey-Schafer, one of the first of Foster's students:

> Having accumulated certain physiological apparatus, including microscopes, [Foster] organised a course of practical instruction in a room which was allotted to him in the College, and which received the title of Physiological Laboratory. The course was not compulsory and was attended by students who displayed a special interest in the subject and were willing to pay an extra fee. It consisted, as regards histology, of the examination of teased preparations of fresh tissues and of sections of organs made with a razor by chopping them on a glass slide, or, in the case of firmer tissues, by cutting them while held in a split cork. The chemical part embraced a study of the constituents of blood and serum, the spectroscopic appearances of haemoglobin and its derivatives, the components of bile and urine, the phenomena of gastric and pancreatic digestion, the general properties of albumins, carbohydrates and fats. The experimental part was less complete, but the phenomena of nerve- and muscle-physiology were investigated in the frog, as was also the action of the heart and the circulation—the latter both in the frog and mammal.[70]

Foster's own description of this course, in his reminiscenses, agrees very nearly with that given by Sharpey-Schafer:

[68] See Bellot, *University College*, p. 314. Confusion exists, however, about the exact date and manner of Foster's appointment and about its relation to Harley's illness. From Tweedie, *Harley*, pp. 121, 173-179, 187, 192-193, it would appear that Harley had been promoted to professor and did not resign until 1869, though Foster had been largely performing his duties unofficially since autumn 1866. J. N. Langley, "Sir Michael Foster—In Memoriam," *J. Physiol.*, *35* (1907), 233-246, leaves the impression (p. 235) that Foster held an official position from 1867 on as "Teacher of Practical Physiology" concurrently with Harley's retaining his professorship. For utter confusion, see Foster, "Reminiscences," p. 421.

[69] Bellot, *University College*, pp. 313-314.

[70] Sharpey-Schafer, *Physiological Society*, p. 2.

> . . . what could be done then was very little. I had a very small room. I had a few microscopes. But I began to carry out the instruction in a more systematic manner than had been done before. For instance, I made the men prepare the tissues for themselves. That was a new thing then in histology. And I also made them do for themselves simple experiments on muscle and nerve and other tissues on live animals. That, I may say, was the beginning of the teaching of practical physiology in England. . . .[71]

From both descriptions, it is clear that Foster was at this point rather less concerned with teaching experimental physiology itself than with giving his students the chemical and histological preparation they would need in order later to pursue experimental physiology on their own.

Foster's course was probably most complete in the histological training it gave. Like Sharpey before him, Foster did not immediately or totally dismiss the preeminently British tendency to approach physiology by way of anatomy. But by focusing on histology instead of gross anatomy, he preserved Sharpey's earlier innovation; and by making his students prepare their own sections, he went beyond it. As might be expected from his training under Sharpey, Foster possessed a thorough knowledge of histological techniques. In the year of his appointment at University College he published a valuable report on modern microscopes, and three years later described his own method of embedding tissues in paraffin.[72]

It is doubtful that Foster could have done more for English physiology by trying to emancipate it more quickly and more completely from microscopical anatomy. Between the Magendian period of physiology and the physicochemical or German period, histology had established itself as a field of great relevance to physiology. And just as Magendie had promoted the emancipation of physiology from anatomy only after being thoroughly trained in anatomy, so the German physiologists of the later period promoted the emancipation of physiology from histology only after they were themselves thoroughly trained in the latter. During the middle years of the nineteenth century, when most of the great leaders of the physicochemical school were being trained, the leading feature of German physiology was precisely its

71 Foster, "Reminiscences," p. 421.

72 M. Foster, *Report on Modern Microscopes, and Recent Improvements in Microscopical Apparatus* (London, 1867), 37pp.; M. Foster, "On Imbedding Substances for Microscopic Section," *Quarterly Journal of the Microscopical Sciences, 10* (1870), 124-126.

attention to minute morphology.[73] When the more famous emphasis on experimental and physicochemical techniques did become the dominant feature of physiology in the German universities, the movement was led for the most part by men already well prepared in histology. Nor did they all forget even then the importance of histology. When his new institute for physiology was built at Leipzig about 1870, Carl Ludwig made extensive arrangements for the teaching of histology and included at least one separate room for histological research.[74]

In England, meanwhile, very few students had acquired an adequate preparation in histology even by the late 1860's. If the German scenario could be trusted, the students who came to Foster's course poorly trained in histology were also poorly prepared to plunge immediately into a full course in experimental physiology. So it was in keeping with the German experience, as well as with his own capacities, that Foster made a special effort to fortify whatever preparation his students may have received in histology.

By introducing also the practical teaching of what might be called physiological chemistry, Foster added an element apparently absent from any of Sharpey's courses at University College, despite his appreciation of its importance.[75] When Foster was a medical student there, the leading force in the chemistry department was A. W. Williamson, a man of extensive Continental experience and a graduate of Liebig's training school at Giessen.[76] It was very probably through Williamson that Foster acquired his sound training in laboratory chemistry. The capacity that he displayed by winning the department's gold medal was later confirmed in published papers. In his article of 1865 on glycogen in *Ascaris,* Foster had already demonstrated his ability to enlist the aid of chemical methods in physiological research. In 1867, shortly after his appointment to the faculty of University College, he published

73 Cf. Theodore Billroth, *The Medical Sciences in the German Universities: A Study in the History of Civilization,* trans. from the German with an introduction by William H. Welch (New York, 1924), p. 39. Of the four most famous leaders of the German school of "organic physics," three were trained under Johannes Müller, who made important contributions of his own to microscopy and who was among the first to realize the value of chemical reagents in microscopic work. Cf. E. S. Russell, *Form and Function* (London, 1916), esp. pp. 173, 188, 209n.

74 See H. P. Bowditch, "The Physiological Laboratory at Leipzig," *Nature, 3* (1870), 142-143; and Charles Wurtz, *Les hautes études pratiques dans les universités allemandes, rapport présenté à son exc. Le Ministre de l'Instruction Publique* (Paris, 1870), pp. 59, 67.

75 In the spring of 1852, Sharpey was working "all day in the [chemistry] laboratory with Dr. Williamson telling him how to do the things." See Bellot, *University College,* p. 168; and Tweedie, *Harley,* p. 117.

76 See Bellot, *University College,* pp. 284-286.

a more purely chemical paper on the properties of certain of the "ferments" (i.e., enzymes) that convert starch into sugar.[77] Foster was, then, qualified to teach at least the rudiments of physiological chemistry, and he chose to do so at a time when the subject was rarely if ever taught even to chemistry students in England. The chemical portion of his course must have been a revelation to many of his students.

So even though the experimental part of Foster's new course at University College was, in Schafer's words, "less complete" than its histological and chemical portions, this course was nonetheless the most complete preparation then available in England for students of physiology. In terms of the limited objectives sought, this course and Foster's whole brief career at University College were markedly successful. By the time his appointment was two years old, Foster had already greatly expanded preparation for and laboratory teaching in physiology, had secured a physiological laboratory of sorts, and—even though his course was an elective—had succeeded in gathering around him a small but enthusiastic band of students, most notably Edward Schäfer (later Sharpey-Schafer) and Henry Newell Martin. These rapid accomplishments did not go unnoticed, and Foster was elevated to the professorship in practical physiology and histology as early as 1869.[78]

The year following this promotion was an eventful and crucial one in Foster's life. In a way it was the best of times and the worst of times. It was the year he was called to begin his great career at Cambridge, but it was also the year he had to break his old and special ties with University College. It was a year of substantial personal growth and achievement, but it was also the year of his greatest personal tragedy, the death of his first wife.

As the year 1869 began, Foster was working up for publication a histological paper on the epithelium of the frog's throat[79] and a brief note on the effects of the interrupted current on the frog's ventricle. In the latter he attempted to extend to the frog's heart the conclusion drawn from his earlier work on the snail's heart—namely, that the rhythmic beat is an inherent property of the cardiac musculature.[80] Far deeper insight into Foster's position on the problem of the heartbeat can be gained from a series of three lectures on the involuntary move-

[77] M. Foster, "Notes on Amylolytic Ferments," *J. Anat. Physiol.*, *1* (1867), 107-113.

[78] Bellot, *University College*, p. 315.

[79] M. Foster, "On Some Points of the Epithelium of the Frog's Throat," *J. Anat. Physiol.*, *3* (1869), 394-400.

[80] M. Foster, "Note on the Action of the Interrupted Current on the Ventricle of the Frog's Heart," ibid., *3* (1869), 400-401.

ments of animals, delivered before the Royal Institution in February 1869. Like the rest of Foster's work on the problem of the heartbeat, these lectures receive more detailed analysis in Chapter 5 below.

In April 1869, Foster was appointed Fullerian Professor of Physiology at the Royal Institution, succeeding Huxley, who had told Foster the year before of his intention to resign and had asked his protégé to take his place.[81] This appointment did not require Foster to give up his position at University College. Largely because of the inadequate stipends attached to the professorships at the Royal Institution, it had become standard for the occupants of the Fullerian Chairs to hold them in conjunction with another position.[82] There was no obligation to reside at the Royal Institution, and the duties were such that Foster's "professorship" really amounted to a visiting lectureship. When the appointment became official, Huxley wrote him a congratulatory note, but then, just a week later, found himself regretting that Foster had accepted the position (and Huxley's own advice) because the pay was so inadequate and because he now believed that Foster should lecture less and write more.[83]

By autumn, Foster's attention was diverted from these professional matters by a domestic problem of far greater urgency. His wife's health, long a matter of concern, was deteriorating drastically, and in late October he informed Huxley that she was dying. "She cannot last much longer . . . My Father and Mother are here and we have all we need except—the life." On 3 November, he added "just a line to say my poor darling died at 3:20 this morning . . . My whole life seems broken."[84] Foster was left the task of raising their two young children. Distraught and disoriented, he gave up his professorship at the Royal Institution[85] and moved from London to his father's house at Huntingdon, in which familiar and peaceful surroundings he might hope to find himself again. A few weeks later he embarked on a foreign tour, from which he returned in January, earlier than planned, feeling ill and homesick, his sense of well-being not yet fully restored. For the next few months, he cloistered himself in his father's house, vacillating,

81 Huxley to Foster, [Nov. 1867], Huxley Papers, IV. 3.

82 See *Report on the Past, Present, and Future of the Royal Institution, Chiefly in Regard to Its Encouragement of Scientific Research*, by the Honorary Secretary (London, 1862), 16pp. pamphlet, pp. 9-11.

83 Huxley to Foster, 20 April and 28 April 1869, Huxley Papers, IV. 15-16. According to Huxley, the pay was ten guineas a lecture. Huxley to Foster, 10 April 1869, ibid., IV. 13.

84 See Foster to Huxley, letters of 20 August 1866, 29 October 1869, and 3 November 1869, ibid., IV. 160, 166-167.

85 See *Gardener's Chronicle, 41* (1907), 78-79, on 79.

"lazing," unable to concentrate, uncertain of his plans—suffering, in short, from an indisposition that he himself diagnosed as "more mental than physical."[86]

Meanwhile, in Cambridge, the first steps had been taken toward establishing the praelectorship that Foster was eventually to hold. On 9 November 1869, the Master and Senior Fellows of Trinity College held a meeting to discuss the appointment of a praelector under the terms of a statute that had been worked out as a compromise between the college and the Statutory Commission on the university. The commission had proposed in 1857 that Trinity should appoint at least six praelectors to "give lectures in mathematics, languages, philology, history, and in such special departments of natural and moral science as shall appear to the Master and Seniors best suited to advance the interests of the College and to place its system of instruction in harmony with the extended range of studies encouraged and cultivated in the University." The commissioners expected that these praelectors would ordinarily be chosen from among the Fellows of Trinity College, but they specified that "if the efficiency of the College requires it, any person, who has formerly been a Fellow of the College, or even if necessary, any other person" could be appointed.[87]

Reluctant to sacrifice the amount of money such a scheme would require, the college had opposed the commission proposal and managed to secure instead a compromise measure that pledged them to appoint only half the number of praelectors originally recommended. And instead of specifying in what subjects the praelectors might be expected to lecture, the compromise statute merely stated that they were to "give Lectures in such subjects as the Master and the Seniors shall from time to time determine."[88] Even then, nothing was done for many years to give life to this statute. At its general meeting in 1866, the college voted down a proposal "that considering the great need of providing for the direction of studies in the College, especially the study of Physical Science, at least one Praelector be appointed without delay in accordance with Statute XIV." Not until after the summer of 1869, when certain minor changes were made in the statute, did serious deliberations begin.[89]

[86] Foster to Huxley, letters of 3 November 1869, 8 November 1869, 20 January 1870, 28 January 1870, 10 February 1870, and 1 March 1870. Huxley Papers, IV. 167-170; XVI. 202-207.

[87] See Robert Robson, "Michael Foster and Trinity College," *Trinity Review*, Easter 1965, pp. 9-11; and D. A. Winstanley, *Early Victorian Cambridge* (Cambridge, 1940), p. 354.

[88] Winstanley, ibid., p. 364.

[89] Robson, "Foster and Trinity," pp. 9-10.

At the meeting of 9 November, Adam Sedgwick, professor of geology, "advanced the claims of the higher parts of experimental physics," and it was determined that the Master should extend an offer to Sir William Thomson (later Lord Kelvin). After Kelvin declined the praelectorship on the grounds of his wife's health, the matter was discussed again by the Master and Seniors in March 1870, and Foster was offered the position two months later.[90]

To explain why, in their second attempt, the Trinity Seniority chose physiology as the subject of the new praelectorship, a variety of more or less elaborate suggestions have been made. Most of these suggestions seem to follow the account given by Walter Holbrook Gaskell, who was not only a student in Trinity at the time but who also became one of Foster's very favorite students. According to Gaskell, physiology was chosen under the following circumstance:

> The original suggestion came apparently from George Eliot and George Henry Lewes, who were great friends of W. G. Clark [philologist and Fellow of Trinity College]. He and Coutts Trotter felt, and persuaded the College, that the time had come when it would be of advantage to the University for separate teaching in Physiology to be given.[91]

Edward Sharpey-Schafer, another of Foster's favorites, later embellished this account by including among the influential group Dr. George Paget, then on the staff of Addenbrooke's Hospital in Cambridge and later Regius Professor of Physic at the university; and more especially George Murray Humphry, then professor of human anatomy, a man of broad and decisive views on medical education, who was "anxious for its development in his own University" and who "recognized that the first necessity for this was the foundation of a chair of Physiology." Sharpey-Schafer repeats the claim that the original suggestion came from Lewes and Eliot.[92]

In support of the authority of Gaskell and Sharpey-Schafer, it may

[90] Ibid., p. 10. Cf. Trinity College Minute Books (Nov. 1868-May 1870), pp. 119-121, 172, 178, 192, 197. I am most grateful to the Master and Seniority of Trinity College for permission to examine and to quote from these minute books.

[91] Gaskell, "Foster," p. lxxiii.

[92] Sharpey-Schafer, *Physiological Society*, p. 3 and n. 2. Cf. Humphry Davy Rolleston, *The Cambridge Medical School: A Biographical History* (Cambridge, 1932), p. 79; J. J. Thomson, *Recollections and Reflections* (London, 1936), p. 282; and Walter Langdon-Brown, *Some Chapters in Cambridge Medical History* (Cambridge, 1946), p. 89. Abraham Flexner's claim that physiology was chosen because "the Master of Trinity at that time [W. H. Thompson, a classicist] chanced to be interested in the subject" has no visible means of support, direct or indirect. Flexner, *Universities: American, English, German* (London, 1930), p. 286, n. 27.

be added that all of these people played a more or less important part in the career of the new praelector once he had been chosen. But there exists no direct evidence that they were, either singly or in combination, actually responsible for creating the position he held.[93] The specific claim for Lewes and Eliot receives no confirmation in Gordon Haight's monumental edition of their letters, nor in his recent biography of Eliot.[94]

Impressive evidence exists, on the other hand, that the choice of physiology as the subject of the praelectorship was directly related to the choice of Foster as the man to fill it, and that both suggestions came first and foremost from Foster's old guide, philosopher, and friend, T. H. Huxley. As early as 1866 Trinity College was in touch with Huxley about their proposed praelectorship in natural science,[95] and in April 1870 W. G. Clark wrote him that his recommendations had now gained support:

> I read your letter to the Seniority yesterday. Your suggestion as to the Praelectorship of Physiology and the man to fill it, was most favourably received.
>
> If Dr. Carpenter, whom we have also consulted, gives the same advice, I have no doubt it will be followed. We shall not have another meeting till after the vacation. I am glad to see that Tests Bill is to come on before Easter.
>
> If Mr. Foster will come and spend a day with me, I can explain everything to him. . . .[96]

Because the Seniority solicited the advice of Huxley and Carpenter, it may appear that they had already decided to establish a praelectorship in physiology. But even this decision may not have crystallized until after Huxley's recommendations had been received. Before then, some testimonials supported instead the establishment of a praelectorship in chemistry.[97] In any case, Carpenter's advice confirmed Huxley's,[98] and Foster was offered the position.

[93] Cf. Dale, "Foster," pp. 16-18; and Robson, "Foster and Trinity," p. 10.

[94] Gordon S. Haight, *The George Eliot Letters*, 7 vols. (New Haven, 1954-1955); Haight, *George Eliot, A Biography* (Oxford, 1968).

[95] See Bibby, *Huxley*, p. 183.

[96] Clark to Huxley, 2 April [1870], Huxley Papers, IV. 172. The Trinity College Minute Books (Nov. 1868-May 1870), p. 192, record for April 1st: "The Vice Master read a letter from Prof. Huxley advocating the appointment of a Praelector in pure Physiology, as distinguished from Medicine, and recommending Mr. Michael Foster."

[97] Ibid. Cf. Robson, "Foster and Trinity," p. 10.

[98] Trinity College Minute Books (Nov. 1868-May 1870), p. 197.

When the offer came, Foster was still ensconced in his father's house, suffering from mental depression. The decision to accept was made at least as much by his father as by Michael himself. For the elder Foster, the primary concern was the effect the appointment might have on his son's health; of secondary importance was a desire not to do anything that might cause rancor between University College and his son. Once again it was Huxley who mainly determined the final outcome. On 3 May 1870, the elder Foster wrote Huxley that he felt

> a 'pang' at the prospect of [Michael's] leaving Univ[ersity] Col[lege]. I had set my heart upon seeing him Prof. of Physiol. there . . . It is not to be; for looking at the state of his health and the *still* poor prospects pecuniously [?]—I cannot but feel that it is in every way [best?] for him to accept the Cambridge offer.
>
> Is this your judgement looking at his health especially? Michael looks to your judgement and banks upon it very confidently.[99]

Within two days Huxley had apparently removed whatever reservations the elder Foster may have had about the new appointment. On 5 May, Michael's father declared himself "quite overjoyed with your account of Cambridge. I was unwilling that Michael should do anything not right [?] for his old college." He continued:

> Your [remarks?] settle the matter; the opening there is marvellously adapted [?] to his present conditions—as it regards health, integrity to our home, light work and good pay and useful-honorable position.
>
> A thousand thanks for your efforts.[100]

After he had formally accepted the offer, the younger Foster wrote Huxley a remarkable letter, indicating, among other things, that he himself looked upon Huxley as the chief architect of his new opportunity:

> Reverend Sir,
>
> Most people every now and then take stock of what they possess, and during the last few weeks I have been taking stock of what I owe you. The quarter of a century you spoke of on Wednesday reminded me of the more than half a quarter century which has gone by since I first read an Oct. no. of Brit. Chir. Rev.[101]—and became aware of the existence of a certain T. H. H., holding broad views on

[99] M. Foster (the elder) to Huxley, 3 May 1870, Huxley Papers, IV. 186.

[100] M. Foster (the elder) to Huxley, 5 May 1870, ibid., IV. 178.

[101] See T. H. Huxley, "The Cell Theory," *British and Foreign Medico-Chirurgical Review, 12* (1853), 285-314.

> physiology which in the sweet ignorance of nineteen, I then thought were confined in England to myself.
>
> From the time when I trembled before you at a corner of the Royal Society's old tearoom, or when you made me blush at Aberdeen in the reddest and hottest manner by patting me on the back after my first little shot,[102] up to last week I have had nothing but help and sympathy from you. All that time I have been dreading for the hour to come when you and others should *find me out* and kick me out as an impostor. Now a days I begin to fancy that perhaps after all I shall sneak through life with my feebleness undetected. However the only return I can make is to send you a poor paper IOU, and to promise to pass the help on—to help others as far as I can as you have helped me. I can never of course expect to give any one anything like this last lift—the importance of which I as yet can hardly realize.[103]

In the moving passage with which this letter ends (quoted in full at the head of this chapter), Foster wrote of "this wonderful Providence which tosses us about, and tossing, guides us," and of his belief that "my past means that there is something in the future for me to do." The great turning point in his life had been reached. From this point on, he began to recover his powers. In the summer of 1870 he toured several of the German universities with Sharpey, inspecting their physiological laboratories and asking about their methods of teaching.[104] In November he gave his first lecture at Cambridge and began very quickly to exert that special influence which was to alter the course of Cambridge biology and English physiology.

[102] That is, when he read his paper on the snail's heart before the British Association in September 1859.

[103] Foster to Huxley, [May 1870], Huxley Papers, IV. 173.

[104] Cf. Langley, "Foster," p. 236; Dale, "Foster," p. 16; and Bellot, *University College*, p. 315.

PART TWO

The Institutional Framework for Foster's Achievement

4. Foster Meets Cambridge: Trinity College, University Reform, and the Rise of Laboratory Science

Of the many environmental forces which injected scientific thought into British universities, two may be singled out as of special importance. Both came from outside England. One was an ideal. The other was an anxiety. The ideal was the German concept of *Wissenschaft*: the university as a centre for research. The anxiety was that continental countries by their vigorous application of science to industry would overtake Britain's industrial supremacy.

Eric Ashby (1963)[1]

A traditional institution like Cambridge, under public but not authoritarian pressure, may draw upon its own history, heritage and ideals to interpret the demands upon it in a unique and unexpected way. A university which is being asked to reform, but is still allowed a high degree of internal freedom, may restructure itself to acquire an identity and function which few expected. When Peer Gynt, searching for his true self, left the kingdom of trolls, he found the vague, formless and invisible Boyg blocking his path and was advised by a voice from the blackness to continue his quest by going "round about." It is the "round about," . . . so frequently neglected in histories of universities, which needs special attention.

Sheldon Rothblatt (1968)[2]

When Foster came to Cambridge in 1870, he met a university in flux and transition, a university at which a growing spirit of reform was challenging institutional aims and practices taken for granted a generation before. Instead of viewing their positions as temporary way stations, usually on the road to a career in the church, college tutors increasingly looked forward to a full-time career in teaching. Instead of so often leaving Cambridge entirely, and so often using their awards as sinecures, the holders of college fellowships increasingly taught or performed other duties in college. Instead of accepting their hitherto peripheral role in Cambridge life, university professors took increasing responsibility for the production and transmission of knowledge. Instead of dismissing the natural sciences and other modern studies as

1 Eric Ashby, *Technology and the Academics: An Essay on Universities and the Scientific Revolution* (New York, 1963), p. 40.

2 Sheldon Rothblatt, *The Revolution of the Dons: Cambridge and Society in Victorian England* (New York, 1968), p. 26.

dangerous distractions from the collegiate aim of cultivating discipline, members of the Cambridge Senate gave an increasingly sympathetic hearing to pleas for an expanded curriculum. In these and other ways, many Cambridge dons had become converts to a new conception of their role, and emissaries of an attitude so different from the traditional that Sheldon Rothblatt has called its emergence "the revolution of the dons."[3]

The causes of this revolution remain somewhat obscure. Almost certainly, it would not have occurred when and as it did in the absence of earlier governmental intervention. By establishing the Royal Commission on the Universities in 1850, Parliament had served notice that it was dissatisfied with the pace and direction of efforts to reform Oxford and Cambridge from within. These efforts at internal reform had resulted in the creation and codification of a system of examinations, including the rigorous Mathematical Tripos at Cambridge, that helped to reduce the once conspicuous indolence of many dons and students. But until the Royal Commission loomed on the horizon, nothing had been done to extend these examinations beyond the traditional subjects of mathematics or classics, which were taught by college tutors or private "coaches" rather than university professors. Nor had anything been done to reduce the social and religious exclusiveness of the ancient universities. Indeed, Oxford had never been more expensive or socially homogeneous than it was in the three decades before 1850,[4] and that probably applies to Cambridge as well. In a period when science was growing in esteem and when the political power of religious dissenters and social reformers was on the ascendant, such a situation naturally invited scrutiny and criticism. Since, by 1850, Parliament was under the control of Liberals and Nonconformists, the Royal Commission was not an entirely unpredictable event.[5]

Disappointed, not to say outraged, that the Royal Commission was named despite their efforts at internal reform, most Cambridge dons devoutly wished to avoid further government interference. Partly to prevent it and to retain maximum autonomy, they went a long way toward meeting the recommendations set forth in the Cambridge commissioners' report of 1852. Chief among those recommendations were

[3] Ibid.

[4] See Lawrence Stone, "The Size and Composition of the Oxford Student Body, 1580-1909," in *The University in Society. Volume* 1. *Oxford and Cambridge from the 14th to the Early 19th Century*, ed. Stone (Princeton, 1974), pp. 3-110, esp. 60-64, 73.

[5] See, e.g., D. A. Winstanley, *Early Victorian Cambridge* (Cambridge, 1940), Chapter 9; W. R. Ward, *Victorian Oxford* (London, 1965), Chapters 5-7 et passim; and J. W. Adamson, *English Education, 1789-1902* (Cambridge, 1930), esp. pp. 171-202.

the introduction of free competition for all scholarships and fellowships, an extension and strengthening of the university professoriate, and an increase in the resources devoted to modern studies in general and the natural sciences in particular. It can hardly be accidental that the Cambridge Foster met two decades later conformed so closely to the sort of Cambridge envisioned by the commissioners.

And yet, without at all denying the crucial role of the Royal Commission of 1850, the revolution of the dons can scarcely be ascribed to it alone. As a matter of fact, the Cambridge commissioners' report of 1852 and the resulting Act of 1856 were in certain important respects surprisingly conservative documents. Thus, while the commissioners urged "the restoration . . . of the ancient supervision of the University over the Studies of its Members, by the enlargement of its Professorial system," they also defended the collegiate system as "one of the most striking and valuable characteristics of our English Universities . . . by which habits of order and moral control are most satisfactorily obtained."[6] On this ground, they rejected the idea that students be admitted to the university without collegiate affiliation. And though the Executive Commission established by the Act of 1856 did go so far as to recommend that the colleges contribute five percent of their gross income to university purposes, no mechanism for enforcement was provided and "the University never got its five per cent."[7] In short, the Cambridge Act of 1856, like the Oxford Act of 1854, left the relationship between the wealthy colleges and the impoverished university fundamentally unchanged.

Far more revolutionary in this respect was the Universities Act of 1877, which required the colleges at both ancient universities to devote a portion of their annual income to university purposes. For some, the rise of science and other modern studies can be attributed largely to this legislative action, which removed at last the obstacle of collegiate "selfishness."[8] Although appealing in its simplicity and un-

[6] *Report of the Commissioners Appointed to Inquire into the State, Discipline, Studies and Revenues of the University and Colleges of Cambridge* (London, 1852), pp. 202, 143.

[7] A. I. Tillyard, *A History of University Reform from 1800 A.D. to the Present Time, with Suggestions Towards a Complete Scheme for the University of London* (Cambridge, 1913), p. 160.

[8] To no small extent, this interpretation runs through Tillyard's book (ibid.) and through D. A. Winstanley, *Later Victorian Cambridge* (Cambridge, 1947). Historians of science, eager to see the object of their study arrive, are perhaps especially susceptible to it. See, e.g., Romualdus Sviedrys, "The Rise of Physical Science at Victorian Cambridge," *Historical Studies in the Physical Sciences, 2* (1970), 127-146, esp. 127-129, 135-138; and, for a sometimes blatant example, G. L. Geison, unpublished Ph.D. dissertation, "Sir Michael Foster and the Rise of the Cambridge School of Physiology" (Yale, 1970), Chapter 3.

doubtedly true to some extent, such an interpretation fails to explain the large increase in expenditures on new subjects before 1883, when the collegiate payments mandated by the Act of 1877 actually began. Like so much of the literature on Victorian Cambridge, it suffers from the "whiggish" tendency to speak of dons as though they were motivated entirely by partisan political considerations, or to speak of the university as though its constitution and laws determined entirely its spirit and direction.[9]

Almost as unpersuasive are attempts to trace the revolution of the dons to a change in the social composition of the Cambridge student body. During the second half of the nineteenth century, to be sure, Cambridge (like Oxford) did become more accessible to students of moderate means than it had been during the previous several decades, and (again like Oxford) it experienced an immense expansion in undergraduate admissions. Moreover, that expansion may well have taken place largely at the expense of the sons of clergy and the landed gentry. At Oxford, in any case, these groups were ultimately swamped by students from the new middle classes. Yet most of this demographic shift apparently occurred after 1870—at which point sixty percent of Oxford undergraduates still came from the clergy and landed gentry—and there is little reason to suppose that the social composition of the Cambridge student body was strikingly different.[10] In short, the revolution of the dons was already underway by the time Cambridge admitted large numbers of students from those emerging industrial classes which have been rather too casually linked with reform in general and educational reform in particular.

At first blush, another sort of change in the student body may seem more significant. Under the Cambridge Act of 1856, non-Anglicans won the right to proceed to the B.A. degree. Because religious dissent tended to be associated with reformist sentiment and scientific activity,[11] it is tempting to suppose that this provision contributed importantly to Cambridge reform and to the new sympathy for science. In fact, however, we do not yet know how quickly or in what numbers dissenters chose to exercise their new option, and the impact of any change in the religious profile of Cambridge students could only have been indirect at best. For despite their new opportunity to graduate from Cambridge, Nonconformists remained barred from fellow-

[9] For a critique of "whiggish" histories of university reform, see Rothblatt, *Revolution*, pp. 17-26.

[10] See Stone, "Oxford Student Body," esp. pp. 65-67. For Cambridge, see Rothblatt, *Revolution*, esp. pp. 86-88.

[11] See, e.g., George Haines, *German Influence upon English Education and Science, 1800-1866* (New London, Conn., 1957).

ships or the M.A. degree, a requirement for participation in the governing of the colleges or the university. By 1871, when these barriers also fell, many dons and university officials had already been captured by the spirit of reform. Thus, while the abolition of religious tests (like the admission of the middle classes) doubtless accelerated the pace of reform and scientific activity at Cambridge, religious dissenters cannot have played a major direct role in the early stages of the revolution of the dons.

In describing the emergence of the new Cambridge, Rothblatt seeks to transcend "whig" or "class conflict" interpretations by considering the role of influential individuals who acted out of intellectual and moral conviction, or perhaps out of personal quirk, as well as out of partisan political considerations. His account serves as a valuable corrective to the tendency of external critics to think of Oxford and Cambridge as monolithic political entities.[12] Almost paradoxically, it also helps to remind us that the "revolution" of the dons involved neither physical violence nor a wholesale rejection of the past. The distinctive system of residential-cum-tutorial colleges that Cambridge shared with Oxford remained intact. So did the commitment to undergraduate education and to the collegiate aim of building character. Even the leaders of the reform, despite their enhanced appreciation for science and scholarship, sought not so much to abrogate their traditional collegiate functions as to graft onto them the additional functions of advanced instruction and research—to expand the Cambridge conception of university purpose, not to overthrow it.[13]

The revolution of the dons, that is to say, had less to do with changes in the legal or social structure of the university than with a major shift in the attitudes of those who lived and worked there. As universities go, the shift was sufficiently profound and sufficiently widespread to be called a revolution. Perhaps, as Rothblatt suggests, no such attitudinal change could be expected to result entirely or even chiefly from legislative fiat, and any dramatic changes in the social or religious profile of Cambridge apparently came too late to account for the early stages of the shift. What factors, then, can explain the emergence of the new Cambridge? It is precisely here that Rothblatt leaves

[12] For contributions toward the same end, see Arthur J. Engel, "The Emerging Concept of the Academic Profession at Oxford, 1800-1854," in *University in Society* I, ed. Stone, pp. 305-352; and Engel, "From Clergyman to Don: The Rise of the Academic Profession in Nineteenth Century Oxford," unpublished Ph.D. dissertation (Princeton, 1975).

[13] See, e.g., the discussion of J. R. Seeley in Rothblatt, *Revolution*, esp. pp. 174-176.

us in the lurch. His account of the shift is almost aggressively local and parochial. If we gain some insight into the nature of the change as it was viewed from within Cambridge, we are left to guess at the forces that might have provoked it. In the end, Rothblatt reduces Victorian Cambridge to virtual chaos. And however important it is for us to recognize the particular, the idiosyncratic, and the serendipitous elements in past events, historical understanding also demands coherence. At least for now, that demand can best be met by considering the revolution of the dons as a locally conditioned response to powerful external forces.[14]

Governmental intervention, or the threat of it, was one of the most powerful of these external forces. By 1870, however, there was perhaps an even more important one. For some time evidence had been accumulating that science and scholarship flourished best under the German university system, with its emphasis on abstract research, its tradition of decentralized autonomy, and its open and competitive system. Joseph Ben-David has recently emphasized the comparative advantages for science and research of this system over the highly centralized French system and the largely noncompetitive and more subtly centralized English university system.[15] Fueled in large part by the rivalry between politically independent principalities, the German university system encouraged the proliferation of new specialties and the development on a large scale of the modern research laboratory, of which Liebig's chemistry laboratory at Giessen was the chief prototype.[16] The triumphs of German philological scholarship and scientific research greatly impressed some English observers, including

[14] Cf. Ashby, *Technology and the Academics*, esp. pp. 46-49.

[15] See Joseph Ben-David, "Scientific Productivity and Academic Organization in Nineteenth Century Medicine," *American Sociological Review, 25* (1960), 828-843, on 841-842. Charles Lyell was one who perceived early the benefits to science of decentralization and competition. See Lyell, "Art. VIII. Scientific Institutions," *Quarterly Review, 34* (1826), 153-179, esp. 174-179.

[16] This development still awaits its historian, seventy-five years after William H. Welch drew attention to the need for a monographic treatment of the general history of scientific laboratories. See Welch, "The Evolution of Modern Scientific Laboratories," *Johns Hopkins Hospital Bulletin, 7* (1896), 19-24. His own modest effort provides ample evidence that Germany was the birthplace of laboratory science. Informative and detailed descriptions of several nineteenth-century German laboratories can be found in Charles Wurtz, *Les hautes études pratiques dans les universités allemandes, rapport présenté à son exc. M. Le Ministre de l'Instruction Publique* (Paris, 1870); and Wurtz, *Les hautes études pratiques dans les universités d'Allemagne et d'Autriche-Hongrii, deuxième rapport présenté à M. Le Ministre de l'Instruction Publique* (Paris, 1882). On Liebig's laboratory as a model, see Wurtz's report of 1870, esp. pp. 15-16. Cf. Welch, p. 21; and John Theodore Merz, *A History of European Thought in the Nineteenth Century*, 4 vols. (Edinburgh, 1896-1914), I, p. 188.

Matthew Arnold,[17] and a few wished to import the German university model to English soil.

Cambridge had its share of admirers of Teutonic scholarship, and the revolution of the dons clearly owed something to the Germanic ideal of *Wissenschaft*. But the German idea of a university had arisen in a unique social and intellectual milieu—in a group of fragmented, poor, and politically weak principalities united only by cultural and linguistic bonds. Neither Oxford nor Cambridge could be expected to adopt that conception immediately or totally. On the whole, the two ancient universities, like the rest of English society, paid relatively little attention to the German states before the late 1860's.[18] Even those Englishmen who recognized the growing German domination of scholarship and science rarely felt threatened by the recognition. Like their counterparts in France,[19] they had not quite yet reached the conclusion that it was really important to excel in scholarship or basic scientific research.

Toward the end of the 1860's, however, there emerged in England an urgent new awareness of the German *Sprachgebiet*, as its constituent principalities demonstrated increasing military, industrial, and political strength. Three events in the late 1860's and early 1870's shook all of England forever out of its complacency about the Germanies: (1) the demonstration of German industrial superiority at the International Exhibition in Paris in 1867; (2) the victory of the Prussian army over the French in the Franco-Prussian War of 1870; and (3) the union of the German principalities into a nation state in 1871, under the impetus of Bismarck's "Blut und Eisen" brand of politics. These events made it abundantly clear to everyone that powerful forces had been at work in the German states during the previous several decades. Not surprisingly, many French and English intellectuals began to insist that the source of Germany's new strength and vitality was to be sought in its educational system, and especially in its emphasis at the university level on research and science. In the face of the new economic and political realities, English society became susceptible as never before to German influences, and critics of

[17] See Matthew Arnold, *Schools and Universities on the Continent* (London, 1868).

[18] This is the central thrust of Haines, *German Influence*.

[19] See, e.g., Joseph Ben-David, *The Scientist's Role in Society: A Comparative Study* (Englewood Cliffs, N.J., 1971), Chapter 6. In private conversation, Robert Fox has stressed to me the premium placed on oratory (and ornate lecture halls) at the expense of research and laboratories in mid-century France. Among leading French scientists, Bernard and Pasteur complained bitterly of their laboratory facilities compared to those in Germany.

Oxford and Cambridge, both internal and external, gained a vastly more sympathetic hearing for their appeal to the German university model.[20]

Even then, however, these external forces scarcely required any ineluctable response, let alone a revolutionary one. The quite different responses of Oxford and Cambridge illustrate forcefully the danger of trying to impose unitary interpretations of whatever sort on the two ancient universities. Both universities had been investigated by the Royal Commission of 1850, and the recommendations of the Oxford commissioners, though slightly more radical in tone, were really quite similar to those of the Cambridge commissioners. The provisions of the Oxford Act of 1854 resembled very closely those of the Cambridge Act of 1856. The two universities remained equally vulnerable to the threat of additional government interference, and the Universities Act of 1877 treated them very similarly. The German example, or threat, obviously bore equally upon them. Nor, it seems, can any sharp distinction be drawn between the social or religious composition of the two universities. And yet, under virtually identical external pressures, and in spite of their social homogeneity, Cambridge and Oxford adopted quite different postures vis-à-vis the issues of university reform in general and vis-à-vis the claims of natural science in particular.

The nature of this difference requires some specification, for it was decidedly not the case that Oxford remained static as Cambridge underwent its "revolution." In fact, in the sense in which Rothblatt seems to use the phrase, mid-Victorian Oxford experienced a "revolution of the dons" scarcely less profound than that taking place at Cambridge. At Oxford, as at Cambridge, dons sought to make a permanent career of teaching and scholarship; there too, as Arthur Engel has shown, the professional academic was on the rise.[21] If at Oxford the process of change moved at a more leisurely pace and evoked more resistance than at Cambridge, nonetheless the general direction and tangible symbols of the process were strikingly similar at the two universities. The differences, though real and significant, might easily have escaped the casual observer who compared university life on the Isis with that on the Cam.

[20] Cf. George Haines, "German Influence upon Scientific Instruction in England, 1867-1887," *Victorian Studies, 1* (1958), 215-244, esp. 216-222. For a specific example of the way the Cambridge reformers made use of these developments in Germany, see the account below (p. 158) of G. M. Humphry's article in the *University Reporter,* October 1870.

[21] Engel, "Clergyman to Don."

Science at Oxford and Cambridge c. 1870: the outward similarities

Even in the case of the natural sciences, university statutes or bricks and mortar provided little basis for drawing sharp distinctions between Oxford and Cambridge. On the surface, if only on the surface, Oxford might actually have seemed more receptive to natural science than Cambridge. Had Foster been offered a position at Oxford at the same time that Trinity College, Cambridge, offered him its praelectorship, and had his decision rested solely on statutory or material considerations, he might have found it difficult to make a choice.

Science had become an examination subject in the two universities almost simultaneously, the Cambridge Natural Sciences Tripos having been founded in 1848 and the Oxford Natural Sciences School in 1850, on the eve of the calling of the first Royal Commission on the Universities. And though the new Oxford examination had been created at least partly in a vain attempt to head off some such "Paul Pry Commission,"[22] neither had the new Cambridge tripos made much headway until dons there sensed "the whig wolf prowling around the academic door."[23] In fact, William Whewell, Master of Trinity College and one of the architects of the new tripos, had reversed his earlier opposition to an expanded curriculum precisely because of "tolerably plain indications that the old Universities are not to expect a continuance of the protection they have been accustomed to receive at the hands of Government."[24] He hoped that his plan for bringing the natural sciences into the Cambridge curriculum would remove "the alleged neglect of the Inductive Sciences in the University, without any great disturbance in our existing system."[25] At neither university, then, did this "internal reform" arise spontaneously or in a vacuum; at both, it depended importantly on the fear that a Parliament controlled by Liberals and Nonconformists would seize on the alleged neglect of science as a pretext for initiating a governmental investigation of two such Tory and Anglican strongholds.

By 1870, when Foster was pondering his offer from Trinity College, the examination statutes revealed no remarkable differences between

[22] See J. B. Atlay, *Sir Henry Wentworth Acland: A Memoir* (London, 1903), pp. 131-132. The Natural Sciences School won final approval only after "fresh rumours of state intervention" ran through Oxford. Ward, *Victorian Oxford*, pp. 147-151.

[23] Winstanley, *Early Victorian Cambridge*, p. 203; cf. pp. 209-211.

[24] Ibid., p. 198.

[25] William Whewell, *Of a Liberal Education in General and with Particular Reference to the Leading Studies of the University of Cambridge* (London, 1845), p. 224; see also pp. 121-127.

the legal status of science at Oxford and at Cambridge. Originally, in 1850, those statutes might have made Oxford seem a rather more attractive soil for science. At that point, every candidate for the Oxford B.A. had to pass an examination not only in the "final school" of *Literae Humaniores,* but also in one of three other final schools—natural science, mathematics, or law and modern history.[26] Thus, though he certainly need not do so, the Oxford student of 1850 could study natural science in the ordinary course of taking his degree. At Cambridge, by contrast, only those who had already qualified for the B.A. could sit for the Natural Sciences Tripos, and for this extra effort, the winning of honors was the sole possible reward.[27]

In 1860, the Cambridge regulations were altered to provide that a student could earn a degree, as well as honors, by studying science.[28] To all intents and purposes, these new statutes elevated the Natural Sciences Tripos to constitutional equality with the older Mathematical and Classical Triposes, and Charles Newman has gone so far as to claim that they marked "the beginning of the turn toward science at Cambridge."[29] Whereas in 1859 only three candidates had sustained the new tripos, fourteen did so in 1864.[30] By then, however, the Natural Sciences School at Oxford, which from the outset enjoyed constitutional equality with the schools in mathematics and in law and modern history, had also effectively attained it with respect to the school of *Literae Humaniores.* After 1863, Oxford honors candidates no longer had to pass two final schools, and *Literae Humaniores,* hitherto required of all B.A. candidates, became just one of the four final schools in which one could take an honors B.A.[31]

Had Foster now turned from the new examinations to other science-related legislation or to more tangible expressions of the place of science in university life, he would have found equally little basis for choice. By 1870, when a few Cambridge colleges had begun to offer scholarships and fellowships for achievement in the natural sciences, some Oxford colleges had already introduced scholarships and other prizes "for the encouragement of natural and physical sciences in schools."[32] If St. John's College at Cambridge had a scien-

[26] See Engel, "Clergyman to Don," pp. 233-236. Cf. Ward, *Victorian Oxford,* pp. 221-223.

[27] See Winstanley, *Early Victorian Cambridge,* esp. pp. 198-216.

[28] Winstanley, *Later Victorian Cambridge,* p. 190.

[29] Charles Newman, *The Evolution of Medical Education in the Nineteenth Century* (London, 1957), p. 284.

[30] Winstanley, *Later Victorian Cambridge,* p. 190.

[31] Engel, "Clergyman to Don," pp. 233-236.

[32] See Winstanley, *Later Victorian Cambridge,* pp. 191, 198.

tific laboratory of sorts, so too did Christ Church and Magdalen College at Oxford.[33] If Cambridge had a new chair in zoology and comparative anatomy (1866), Oxford could claim three new professorships in the natural sciences: the Waynflete Chair in Chemistry (1854), the Linacre Chair in Anatomy and Physiology (1859), and the Hope Chair in Zoology (1861).[34] And by 1868, when James Clerk Maxwell persuaded the Cambridge Senate that the Mathematical Tripos should include questions relating to the exciting new developments in heat, electricity, and magnetism, Oxford had already begun to build its Clarendon Laboratory of Physics, designed to accommodate Robert Bentley Clifton's innovative teaching in laboratory physics.[35]

Turning next toward Downing Street in Cambridge, Foster would have seen a block of large new science buildings, built between 1863 and 1865 at a cost exceeding £20,000. But had he then looked toward the parks at Oxford, he would have found an even more imposing university museum for natural science, constructed between 1855 and 1860 at a cost nearly fifty percent higher than that of the more recent Cambridge science buildings. And the story of how these buildings had come to be built offered no firmer ground for distinguishing between the place of science in the two universities.

In the middle years of the nineteenth century, when both Oxford and Cambridge began seriously to consider the building of new science museums, these projects had encountered stiff opposition primarily on the grounds of their immense expense. At Oxford, the eventually successful campaign to provide a new home for the natural sciences had been launched in the late 1840's, its "life and soul" being young Henry Acland, then Lee's Reader in Anatomy at Christ Church. The museum that arose a decade later stood as a monument to his energy, ability, persistence, and tact. For all of his personal élan, however, Acland got nowhere so long as his campaign depended on the voluntary generosity of individuals or colleges at Oxford. Not until £60,000 in uncommitted Clarendon Press profits had been transferred to the University Chest, and not until the Royal Commissioners had strongly advised execution of his scheme, did Acland manage to secure the £30,000 needed for the new museum.[36] No college funds were in-

[33] Cf. ibid., p. 191; Engel "Clergyman to Don," pp. 147ff.; A. E. Gunther, *Robert T. Gunther: A Pioneer in the History of Science* (Oxford, 1967), pp. 22-25.

[34] Adamson, *English Education*, p. 421; Winstanley, *Later Victorian Cambridge*, p. 193.

[35] D. M. Turner, *History of Science Teaching in England* (London, 1927), p. 129.

[36] Cf. Atlay, *Acland*, esp. pp. 202-204; and Ward, *Victorian Oxford*, pp. 151-152.

volved, and Acland apparently never even considered asking the colleges, as such, to make any sacrifices.

But neither had the colleges at Cambridge displayed any great enthusiasm for the new symbols of encroaching science. There, the question of building new science buildings came before the University Senate as early as 1853, primarily in response to the recommendations of the Royal Commissioners. The plans then drawn up led to an estimated cost exceeding £23,000, of which the university itself could contribute only about £5,000. Despite an earnest official appeal, the colleges proved unable or unwilling to make up the difference, and the new science buildings remained nothing but an architect's dream for some time to come. By 1860, the resources of the University Chest had increased considerably, £13,500 now being available for the proposed buildings. Thanks chiefly to loans from the surplus funds of the University Press and Library, this amount had doubled by the end of 1861. Even then, the project encountered spirited opposition, and despite the stimulus of the newly completed museum at Oxford, it was not until 1863 that the Cambridge Senate approved a more modest set of plans at an estimated cost slightly above £22,000.[37] Thus at Cambridge, as at Oxford, the new science museum had gone up without a shilling of support from the colleges.

Had Foster now walked into the new science museums, he would have discovered only minor differences between the facilities for science at Oxford and Cambridge. At both universities, the arrangements in the new buildings corresponded as nearly as possible to recommendations that the science professors had been asked to provide before construction began.[38] And at both universities, the results reflected an emphasis on facilities for teaching at the expense of the needs of research. Both museums had been designed primarily to increase the number of lecture rooms and, as their name implies, for the storage and display of acquired scientific apparatus and other materials.[39] The importance of teaching laboratories for chemistry students received tangible recognition at Oxford,[40] but neither for

[37] Robert Willis, *The Architectural History of the University of Cambridge*, ed. J. W. Clark, 4 vols. (Cambridge, 1886), III, pp. 169-181.

[38] Cf. ibid., III, pp. 165, 170-173; Atlay, *Acland*, p. 200; and H. W. Acland and John Ruskin, *The Oxford Museum* (London, 1859), p. 24.

[39] For the architectural plans of the Oxford museum, see Acland and Ruskin, *Oxford Museum*, pp. 32-33, insert. For a verbal description of the plans finally adopted for the Cambridge museums, see Willis and Clark, *Architectural History*, III, pp. 171-175.

[40] See Acland and Ruskin, *Oxford Museum*, pp. 38-39. At Cambridge, the need for a laboratory for chemistry students was repeatedly urged in the deliberations over the new museums, but the laboratory was not built until June 1872. See Willis and Clark, *Architectural History*, III, pp. 162-165, 167, 175, 184.

faculty nor for students were any special facilities provided for the prosecution of original research. At both new museums, emphasis was placed on the old and relatively settled sciences of astronomy, mineralogy, geology, descriptive anatomy, and taxonomy, while the newer experimental sciences were poorly provided for.

Nor, finally, could Foster have detected any striking differences between the quality and orientation of his prospective biological colleagues at Oxford or Cambridge. At Oxford the election of George Rolleston to the new Linacre Chair in Anatomy and Physiology (1860) and of J. O. Westwood to the new Hope Chair in Zoology (1861) had confirmed the traditional emphasis on descriptive anatomy and nonexperimental natural history. Westwood, an entomologist who never could bring himself to believe in Darwinian evolution,[41] hardly represented an accommodation to the newer currents in the biological sciences. Rolleston was a more impressive figure, who pioneered in the teaching of zoology on the "type system" and who read with understanding the experimental physiology of his day. But an experimental biologist he most certainly was not. His published papers dealt almost exclusively with comparative anatomy, often with the classification of skulls, and in no area did he make a significant original contribution.[42] Like his anatomist colleague Henry Acland, who deprecated experiments made on animals "for the mere discovery of fresh knowledge," Rolleston was more than a little ambivalent toward animal experimentation.[43]

At Cambridge, as we shall see more fully below, the biological situation was scarcely different under "Beetles" Babington, professor of botany from 1861 to 1895, and Alfred Newton, an ornithologist who held the new professorship in zoology and comparative anatomy from 1866 to 1907. To be sure, Babington and Newton ultimately proved willing to accept and even to make way for the innovations Foster already had in mind, but one would have had difficulty predicting their response in advance. On the more strictly medical side, George Paget, who became Regius Professor of Physic in 1872, shared many of Foster's educational priorities but sometimes challenged his expansive ambitions on the grounds that they prevented proper accommodation for medicine per se. From Peter Wallwork Latham, who was

[41] *DNB*, *20*, 1292-1293.

[42] See G. L. Geison, "George Rolleston," *DSB*, XI (1975) 513-515. Cf. George Rolleston, *Scientific Papers and Addresses*, arranged and edited by William Turner, with a biographical sketch by Edward B. Tylor, 2 vols. (Oxford, 1884).

[43] See Rolleston, *Scientific Papers*, I, lix-lx; Atlay, *Acland*, p. 422; and Richard D. French, *Antivivisection and Medical Science in Victorian Society* (Princeton, 1975), pp. 154-155.

promoted to the Downing Professorship of Medicine in 1874, Foster met only opposition and personal hostility.[44] Of the biomedical faculty, only G. M. Humphry, professor of human anatomy, was to exert himself actively and effectively on Foster's behalf. But as essential as Humphry's support proved to be, he had himself done little to promote the new biology before Foster arrived.

Wherein the difference lay: tradition and Trinity College

On the surface, then, Foster would have found little to choose between Oxford and Cambridge. But there were important, even profound, differences. For all of their outward similarities, Victorian Oxford and Cambridge retained vestiges of their quite different pasts. Like the fiddler on the roof, each university maintained a precarious equilibrium by keeping in touch with its own tradition. But the silent weight of tradition—sometimes inspiring, sometimes deadening, and never absent—made its mark only haphazardly and incompletely on bylaws and buildings. It is precisely for that reason that Rothblatt's objections to "whig" or "class conflict" interpretations of Victorian Cambridge strike a responsive chord. The most articulate products of Oxford or Cambridge, who felt in their marrow the distinctions so elusive to outsiders, would have had difficulty expressing those distinctions in legal or structural terms.

Through all the shifting political and clerical fortunes of the two ancient universities, Oxford had generally enjoyed a more intimate alliance than Cambridge with the established church, the state, and the crown. That traditional alliance had perhaps never been stronger than it was during the early nineteenth century, but a comparison of Oxford and Cambridge statutes would have offered little insight into their informal political differences. The distinction would have emerged clearly only when the clerical and political careers of Oxonians were compared with their counterparts from Cambridge. And though it was known to almost everyone that Oxford emphasized classics to the relative exclusion of other fields, while Cambridge placed a similar premium on mathematics, that traditional distinction too found only minor expression in the statutes in force by 1870. Partly because of its mathematical emphasis, Cambridge had always attracted a few religious dissenters—indeed J. J. Sylvester, Second Wrangler in the Mathematical Tripos of 1837, was Jewish.[45] At Ox-

[44] See Appendix I below.

[45] Cf. Ward, *Victorian Oxford*, pp. 242-243. On Sylvester, see P. A. MacMahon, *Proc. Roy. Soc.*, *63* (1898), ix-xxv.

ford, meanwhile, dissenters had been almost unknown. But while Cambridge colleges might admit dissenters, the university statutes differed not at all from those of Oxford in restricting B.A.'s (until 1856) and college fellowships (until 1871) to those who professed the Anglican creed.

If, however, such traditional distinctions had little statutory basis, they were scarcely lost on alumni then contemplating and planning their sons' futures. A graduate of either university might make a distinguished career in any field, but the Victorian father who hoped to see his son become Archbishop of Canterbury or Prime Minister would tend to favor Oxford, while the father who hoped to see his son become president of the Royal Society or Astronomer Royal would tend to favor Cambridge. One exaggerates, but does not deceive, by insisting on a profound if informal distinction between the university of Francis Bacon, William Harvey, Isaac Newton, and Charles Darwin on the one hand, and the university of Cardinal Wolsey, Edward Gibbon, Robert Peel, and Bishop Wilberforce on the other. And for every Charles Lyell that Oxford could claim, Cambridge could claim a John Ray, a Stephen Hales, and a John Herschel.[46]

As a matter of fact, even while outside critics accused it of neglecting science, Cambridge had kept its rich scientific tradition at least partly alive through its general emphasis on mathematics and through its prestigious Mathematical Tripos in particular. In the middle years of the nineteenth century, the work of G. G. Stokes, John Couch Adams, William Thomson (later Lord Kelvin), and Clerk Maxwell gave signs that Cambridge had entered a new golden era in mathematical physics and astronomy.[47] To be sure, the more expensive laboratory sciences were then no better situated at Cambridge than at Oxford, but the experimental scientist of 1870 who had ambitions for his subject and who respected the power of tradition would have leaned toward Cambridge. Indeed, such a scientist might well have favored Cambridge over the two new colleges in London or Owens College in Manchester, despite the curricular emphasis of these institutions on science. For quite apart from the general attraction and

46 From a rapid informal analysis of *A Biographical Dictionary of Scientists*, ed. Trevor I. Williams (London, 1969), I find about three Cambridge graduates represented for each Oxford product.

47 See Arthur Schuster and Arthur E. Shipley, *Britain's Heritage of Science* (London, 1917), esp. pp. 117-128. From Schuster and Shipley's account, among others, it is clear that optics, mathematical and observational astonomy, and mathematical physics never had suffered as much as the laboratory sciences at Oxford and Cambridge, and works of merit in these more mathematical sciences were produced at both universities throughout the 18th and 19th centuries.

prestige of Cambridge, it offered the advantages of wealth and of a respect for intellect that, however curious its exercise might then have seemed to an experimental scientist, at least avoided the overweening utilitarianism of King's or University College. In the event, Clerk Maxwell and Michael Foster moved to Cambridge almost simultaneously, and the great schools of physics and physiology that they established ultimately overshadowed, without overwhelming, their counterparts elsewhere in England. By the time World War I broke in upon the serenity of its laboratories, Cambridge had developed a tradition in laboratory research so impressive as to fortify, if not to replace, its more remote and more general scientific reputation.

From the sixteenth century onward, the histories of the two old English universities had been virtually indistinguishable from those of their constituent colleges, so it was only natural that the general distinctions between Victorian Oxford and Cambridge had collegiate parallels. To be sure, each of the colleges in turn had its own distinct heritage, and the differences between the colleges within one university were often more marked than the differences between the two universities as a whole. Just as the fraternities in an American university are known to differ in tone or style, some tending to attract athletes while others attract poets, so each of the Oxford and Cambridge colleges tended to have a characteristic flavor. That distinctive tone, rarely so strong as to justify a stereotype, could vary over time, and the fortunes of individual colleges certainly waxed and waned as fashions changed, but the basic phenomenon was and is real enough.

At Oxford, c. 1870, otherwise fashionable Oriel College remained in the shadow of the famous Tractarian movement, which had its center there and had already scandalized England by leading the future Cardinals Newman and Manning to embrace the Roman Catholic church. Balliol College, under the influence of Benjamin Jowett, had already begun its intellectual "colonization" of other Oxford colleges. All Souls College, though richly endowed, took no part in undergraduate education and scarcely differed from a purely social club.[48] At Cambridge, meanwhile, Downing College, "which had no past and seemed to have no future," tried to meet its goal of training physicians and lawyers. Trinity Hall supposedly favored students "in delicate health," and Clare College remained a haven for canon law. King's College had only recently opened its doors to non-Etonians, while Magdalene and Queen's had become "strongholds of the evangelical party." Poor St. Catherine's attracted little but pity, and Peterhouse entertained at college expense anyone who carried off high

48 See Ward, *Victorian Oxford*, passim.

honors on the Mathematical Tripos. Emmanuel invited comparison with Eton as "a very idle, though a very gentlemanlike college," while Corpus was "famous for its ale and nothing else."[49]

But in the midst of such diversity and much intercollegiate rivalry, Victorian Oxford and Cambridge each had its dominant college, which by virtue of its wealth, renown, and sheer size tended to set the tone for the university as a whole. Of the wealthy foundations at Oxford, only Christ Church educated undergraduates in any quantity.[50] At Cambridge, Trinity College had become the largest as well as the richest foundation, surpassing its ancient rival St. John's; and the distance between them continued to grow throughout the nineteenth century.[51] To some extent, of course, the very size of Christ Church and Trinity encouraged diversity among their students, and a student at Trinity could be, in Rothblatt's words, "a rowing blue or an intellectual *sans-culotte* without fear of interference."[52] Nonetheless, the two foundations had distinctive and quite different traditions, reflecting in microcosm the differences between Oxford and Cambridge as a whole.

By 1870, Christ Church had long been the leading symbol and most striking example of the intimate connection between Oxford, the established church, and the state. Unique among the foundations at Oxford or Cambridge for being at once a college and an Anglican cathedral, it represented the ultimate in classical and aristocratic education. It could claim among its products several members of the royal families of England and the rest of Europe, as well as ten men who became Prime Minister during the nineteenth century alone.[53] Under the deanship of the remarkable Cyril Jackson (1783-1809), Christ Church had gone beyond mere political tone to become an active force in Tory circles, and it remained one of the most conservative foundations at Oxford even after the appointment of the liberal H. G. Liddell as dean in 1855. Until 1867, its government rested entirely in the hands of the see of Oxford, and it retained its ecclesiastical character in the midst of the reforms that swept over the mid-Victorian university.[54]

[49] See Winstanley, *Early Victorian Cambridge*, pp. 385-386; and Rothblatt, *Revolution*, pp. 221-223, 235-236.

[50] Ward, *Victorian Oxford*, p. 295.

[51] See the statistical chart compiled by J. A. Venn of Trinity College, Cambridge, *The Entries at Various Colleges in the University of Cambridge, 1544-1906* (W. Heffer & Sons: Cambridge, 1908). I owe this reference to Lawrence Stone.

[52] Rothblatt, *Revolution*, p. 236.

[53] O. J. R. Howarth, "Oxford," *Encyclopedia Britannica*, 11th ed. (1911), *20*, 405-417, on 407.

[54] See Ward, *Victorian Oxford*, passim, esp. pp. 9-12, 212.

Trinity College, Cambridge, though it too attracted many wealthy and aristocratic students, had gained a reputation as a center of reformist sentiment, particularly with regard to religious tests. Graduates of Trinity were prominent among those who repeatedly petitioned Parliament to abolish the tests, from the early 1830's until the goal was achieved in 1871.[55] Although the college took a far more reverent attitude toward its own statutes before the late 1860's, its mildly reformist tradition gained force from its fame as a home for mathematical and scientific studies. Its founding statutes (1546) made special provision for fellowships in medicine, including a year devoted to physical science.[56] Moreover, Trinity could claim no lesser lights than Francis Bacon and Isaac Newton, and its success in the Mathematical Tripos—though eclipsed by St. John's and Pembroke between 1840 and 1860[57]—remained legendary. Even the crusty old Master, William Whewell, had been a Second Wrangler. Widely known for his writings on natural science, his support for the new Natural Sciences Tripos (if grudging) had been crucial to its success.

In the 1850's, Trinity had shown its willingness to contribute materially toward the expansion of scientific studies in the university, pledging £4,000 for the construction of the proposed new science museums with the proviso (never met) "that a sum sufficient for carrying out the scheme can be raised by the contributions of the [other] colleges."[58] The college also harbored the well-known geologist Adam Sedgwick (1785-1873), a Fellow since 1810, a leader from the outset of the campaign against religious tests, and long a general force in university reform. Sedgwick was still around to cast an approving eye on the Tests Bill of 1871, though his reforming ardor had otherwise cooled.[59]

By then, however, several younger Fellows of Trinity stood ready to take up the cudgels. The pace of internal reform quickened noticeably after the death of the aloof and autocratic Whewell, Master from 1841 to 1866. His crown-appointed successor, classicist W. H. Thompson, though not himself a leader of the liberal party, was vastly more receptive to newer currents.[60] So long as Whewell occupied the

[55] See Winstanley, *Later Victorian Cambridge*, Chapter 3.

[56] Arthur Rook, "Medical Education at Cambridge 1600-1800," in *Cambridge and Its Contribution to Medicine*, ed. Rook (London, 1971), pp. 49-63, on 52.

[57] Winstanley, *Early Victorian Cambridge*, pp. 384-385.

[58] Willis and Clark, *Architectural History*, III, p. 169, n. 2.

[59] See J. W. Clark and T. McKenny Hughes, *The Life and Letters of the Reverend Adam Sedgwick*, 2 vols. (Cambridge, 1890), esp. vol. II, p. 451.

[60] Cf. R. St.J. Parry, *Henry Jackson: A Memoir* (Cambridge, 1926), pp. 26-27, 294-296; Winstanley, *Later Victorian Cambridge*, pp. 241ff.; Rothblatt, *Revolution*, pp. 212-213; and G. M. Trevelyan, *Trinity College: An Historical Sketch* (Cambridge, 1972), pp. 102ff.

Master's Lodge, one could scarcely have guessed that new statutes of 1860, imposed upon Trinity by terms of the Cambridge Act of 1856, greatly increased the power of the Fellows at the expense of the Master. Upon Whewell's death, however, the younger Fellows launched a sweeping program of reform under the leadership of Henry Sidgwick, Coutts Trotter, and Henry Jackson, all three of whom had been elected to fellowships at Trinity within the previous decade. In November 1867, Sidgwick sent the Master and Seniority a thirteen-point position paper, later telling his mother that "the extent to which I am reforming mankind at present is quite appalling; the oldest inhabitant has never known anything like it."[61]

By 1872, Trinity had thoroughly revised the statutes of 1860, directing its efforts mainly toward the creation of full-time academic careers by allowing virtually all Fellows to marry, and by limiting all long-term fellowships to Cambridge residents who performed duties in the college or university. In 1867, the Master and Fellows agreed to offer a scholarship in natural science, and in 1868 decided to award at least one fellowship every three years to a graduate of the Natural Sciences Tripos.[62] In 1874, in a letter to the Clerk of the Privy Council, Thompson emphasized that the revised statutes embodied three basic objectives: "(1) to diminish, as far as possible, the number of sinecure or non-resident Fellows: (2) to assure to those Fellows, who take part in the government and instruction of the college and its students, a permanent professional career: (3) to enable the college, to a greater extent than at present, to make its funds available for the encouragement of literary, theological or scientific research, and of the higher scientific, literary or theological teaching in its own precincts or in the schools of the University."[63]

In their central thrust, the newly revised statutes amounted to nothing less than a formal manifesto for the "revolution of the dons" in Cambridge at large. Fellows of Trinity—including Jackson, Trotter, and especially Sidgwick—played a central part in that wider effort as well. To be sure, Fellows from Christ's, King's, St. John's, and other colleges also participated, as did John Robert Seeley, professor of modern history from 1869 to 1895, whose role Rothblatt particularly stresses.[64] But Trinity College, which in 1868 could be "accurately described as 'the most important educational corporation in England below the rank of a University,'" very definitely led the way.[65]

61 *Henry Sidgwick: A Memoir* (London, 1906), p. 172.
62 Winstanley, *Later Victorian Cambridge*, pp. 191, 242, 261.
63 Ibid., p. 246.
64 See Rothblatt, *Revolution*, p. 211 and esp. Chapter 5.
65 Winstanley, *Later Victorian Cambridge*, pp. 61, 216.

Ultimately, of course, the government took charge. In fact, by the time the new Trinity statutes were submitted to the Privy Council for approval, the Royal Commission of 1872 had just been named and approval was withheld for that reason.[66] The Royal Commission led in turn to the Universities Act of 1877 and to new statutes for each of the colleges at Oxford and Cambridge. These new regulations, which became operative in 1882, required each of the colleges to contribute a portion of their income toward university purposes and in other ways went somewhat beyond the provisions set forth in the earlier revised Trinity statutes. But the basic objectives were almost precisely those toward which the Trinity reformers had aimed in their statutes of 1872. In the decade that passed before the new governmental regulations came into effect, Trinity used its own revised statutes as a basis for further revision and as a guide to action. And when the government statutes of 1882 were published, the Executive Commission singled out Trinity as an example for other colleges to emulate, particularly for its willingness to contribute funds above and beyond the amount required by law.[67] The Universities Act of 1877 applied equally to Oxford and Cambridge, but Trinity College was already pointing Cambridge toward a "revolution of the dons" quite different in outcome and distinctly more beneficial to science than the "revolution" taking place almost simultaneously at Oxford.

Foster, Trinity College, and the rise of laboratory biology in Late Victorian Cambridge

It is hardly an exaggeration to insist that Foster was at once the living symbol and chief beneficiary of the new Cambridge being shaped under the leadership of Trinity College. If not the first, he must have been very nearly the first avowed dissenter elected to a fellowship in any college at Oxford or Cambridge, for the Universities Tests Bill became law only four months before he became a Fellow of Trinity in October 1871.[68] Even within Trinity, he was the only college officer who enjoyed the privilege of holding a fellowship while married.[69] Foster also promoted the new movement toward intercollegiate lectures then gaining strength informally both at Oxford and

[66] Ibid., p. 260.

[67] *Statutes for the University of Cambridge and for the Colleges within It, Made, Published and Approved (1878-1882) under the Universities of Oxford and Cambridge Act, 1877* (Cambridge, 1883), esp. pp. 753-754.

[68] Foster was elected a Fellow of Trinity College on 12 October 1871. Trinity College Minute Books (May 1870-May 1872), p. 191.

[69] Winstanley, *Later Victorian Cambridge*, p. 256.

Cambridge, though it remains unclear whether Trinity College brought him to Cambridge with that end specifically in mind. The assumption that it did is long-lived and persistent, and D. H. M. Woollam has recently gone so far as to claim that Foster was appointed "*with the proviso* that he should teach the undergraduates of the university as a whole," leading him to the further judgment that Trinity thereby performed "one of the best and most disinterested acts carried out by a Cambridge college before the end of the nineteenth century."[70]

But the Trinity College records provide no clear evidence that praelectors in general or Foster in particular had a special mandate to stimulate transcollegiate teaching. In fact, Sir Henry Dale and Robert Robson, who have examined this question in detail, conclude that Foster "was wanted [chiefly] . . . for College and not University purposes."[71] At the very least, the terms of Foster's appointment required him to submit his plan of lectures to the college authorities for their approval and to obtain their consent in order to admit students from other colleges to his lectures. Foster obtained that consent very early and without apparent difficulty, for his lectures were open to all members of the university from the beginning.[72] But it may have been less Trinity College than Foster himself who had university-wide ambitions for his subject.

In any case, Foster must have seemed to many older and more conservative dons the omen of an impending holocaust. To those who mistrusted or disdained science, who feared the divisive intrusion of religious dissent, and who believed that the greatness of Cambridge lay precisely in a close-knit collegiate system that required students and their unmarried tutors to live and work together—to such dons, the first Trinity praelector must have seemed an anathema. An ambitious experimental scientist, he had been brought in from the outside and could know little of Cambridge traditions. Descended from a family of leading evangelical dissenters, he was an intimate friend of the notorious "agnostic" Huxley. And though not married when

[70] D. H. M. Woollam, "The Cambridge School of Physiology, 1850-1900," in *Cambridge and Its Contributions*, ed. Rook, pp. 139-154, on 143. My emphasis.

[71] See Robert Robson, "Michael Foster and Trinity College," *Trinity Review*, Easter 1965, pp. 9-11; and Henry Dale, "Sir Michael Foster," *Notes and Records of the Royal Society (London), 19* (1964), 10-32, on 16-20. Quote from Robson, p. 11.

[72] Trinity College Minute Books (May 1870-May 1872), pp. 53, 58-59; cf. *Cambridge University Reporter* [hereafter *CUR*], 16 Nov. 1870, p. 89. During the first several years of its existence (1870-1873), the *Cambridge University Reporter* carried no volume numbers and so must properly be cited by dates. The first numbered volume was assigned the number 4 in order to indicate that three unnumbered volumes had already appeared.

appointed, his first wife having died in 1869, he remarried in 1872 without being obliged to give up his college fellowship. Obviously without loyalty, and possibly hostile to the Church of England, his commitment to his college must perforce be compromised by his loyalty to his family and to his discipline, which also had its institutional base outside the college walls. If they perceived in Foster a hint of things to come, the senior and more conservative dons might have predicted what another Cambridge man was much later sadly to recall:

> For centuries the centre of academic life had been the College (and a very good centre, too, with its diversities of types); now it was to be the "faculty!" (a group of people of one interest). We owed this to the scientific departments; they were each of them centered in huge buildings, more like government offices or factories than the old-time Colleges; their staffs were from every College; buildings were kept up and staffs paid largely by the taxation of College revenues which had been given for no such purposes. The 'lab' was really more to the new type of man than the College. . . . A college came to be a place where science men from the labs had free dinners, men very often who had been brought in from outside, and could not be expected to understand College feeling.[73]

Although such sentiments were doubtless common enough at Cambridge, and must have had their echo even within Trinity College, Foster also knew that he would not have to press the case for physiology all alone. Other dons at Cambridge had already expressed sympathy toward laboratory science and toward physiology in particular, and some of them were already prepared to act on his behalf. With the important exception of G. M. Humphry, professor of human anatomy, all of these allies shared (or had shared) one overriding bond with Foster—fellowship in Trinity College. In the absence of such allies, Foster would have found it extremely difficult to achieve his aims. During his first thirteen years at Cambridge, he held no university post and could not even vote in the University Senate. This right was conferred only on those who had earned an M.A. or higher degree from Cambridge, and though Foster was awarded an honorary M.A. as early as 1871, honorary degrees did not carry with them a legal voice in university affairs.[74]

[73] T. R. Glover, *Cambridge Retrospect* (Cambridge, 1943), pp. 110-111.

[74] Foster's name appeared on the tentative electoral roll of the university for the academic year 1874-1875, but was soon withdrawn after objections. *CUR, 4* (1874), 27, 52. For a statement of the regulations governing the right to vote in

Foster's notable allies included that familiar and influential trio in college and university reform—Henry Jackson (1839-1921), Henry Sidgwick (1838-1900), and Coutts Trotter (1837-1887). Jackson, a classicist and Fellow of Trinity since 1864, had once "thought that I should give my time to physiology and began to learn the bones";[75] but his support for Foster's aims probably stemmed mainly from his more general reformist stance. In 1875 he joined Foster as a Trinity praelector, his subject being ancient philosophy.

Sidgwick, who had won his Trinity fellowship in 1859, resigned it in 1869 on the grounds of religious conscience, probably thereby hastening passage of the Tests Bill of 1871. He also seems to have been chiefly responsible for reviving the dormant college statute under the terms of which Foster became praelector.[76] Thirty-third Wrangler in the Mathematical Tripos and Senior Classic in the Classical Tripos, Sidgwick vigorously defended the educational claims of natural science without denigrating those of the classics.[77] In 1875 he joined Foster and Jackson as Trinity praelectors, his field being "moral and political philosophy." He sat on the important 1875 University Syndicate that sought to balance the needs of various departments, an experience that served him well when he joined the new General Board of Studies in 1883, and continued the delicate task of arranging university priorities according to departmental needs and promise.[78] When a department's needs struck him as particularly worthy, Sidgwick gave generously from his own funds. He distributed his largesse across the whole range of university studies, but his largest single gift (£1,500) went to Foster's school of physiology in 1889-1890.[79] Knightsbridge Professor of Moral Philosophy from 1883 until his death in 1900, Sidgwick had been since 1876 brother-in-law to F. M. Balfour, the young embryologist and Fellow of Trinity College who was one of Foster's favorite and most favored students.

As we know from Foster's own obituary notice of him, Coutts Trotter was an even more active and effective agent on his behalf. Unlike Jackson or Sidgwick, Trotter had himself been trained in science.

---

the University Senate as they stood in 1879, see ibid., *8* (1879), 550. After his election to the professorship in physiology, Foster was placed on the election roll for the ensuing year. Ibid., *13* (1883), 83. He remained on the roll until his retirement in 1903.

75 Parry, *Jackson*, p. 18.

76 Cf. ibid., p. 293; Sidgwick, pp. 157, 172-173.

77 See, e.g., "The Theory of Classical Education" [1867], in Henry Sidgwick, *Miscellaneous Essays and Addresses* (London, 1904), pp. 270-319, esp. 311ff.

78 *Sidgwick*, pp. 327-328, 371-374.

79 Ibid., p. 373, nn. 1 and 2. Cf. Chapter 10 below, p. 308.

Elected Fellow of Trinity in 1861, he went to Heidelberg in 1865 to study physics and physiology under Kirchhoff and Helmholtz. After returning to Cambridge, he led the administrative battle to bring science and scientific research to the university. In 1869 be became lecturer in natural science at Trinity, a position he held until elected vice-master in 1885. A member of the University Council from 1874 until his death, he took a leading role in virtually every committee formed to deal with scientific or medical matters. John Willis Clark, another Fellow of Trinity who served on many of those committees with Trotter, wrote that "no one has been so completely identified with what may be termed modern Cambridge." Indeed, the shaping of the new Cambridge was "sometimes called in jest 'the Trotterization of the University.' "[80] The physicist J. J. Thomson, who came up to Trinity College in 1876, also believed that Trotter took "a larger share than anyone else in the great development in the opportunities for study and research which took place between 1870 and his death."[81]

In his obituary notice for *Nature*, Foster emphasized the constant support Trotter had given to his efforts to introduce laboratory biology and physiology. Calling Trotter's death at fifty "a calamity," Foster recalled that he and Trotter had formed an instant friendship upon their first meeting "a year or so" before the question of Foster's appointment as Trinity praelector had been decided.[82] Whether or not Trotter actually recommended that the praelectorship go to physiology and to Foster, he did thereafter support and encourage Foster's aims in every possible way. At Trinity, he was largely responsible for introducing the principle that original research (rather than examinations) should be the leading consideration in elections to Trinity fellowships in the natural sciences. This principle, so dear to Foster's heart and unique among the Cambridge colleges for some time to come, found its way into the college regulations just in time to ensure the 1874 fellowship election of Foster's protégé, F. M. Balfour.[83] In his will, Trotter left about £7,000 to be used by Trinity College to establish "a studentship for the promotion of original research in Natural Science, more especially physiology and experimental physics." Among the early winners of this studentship was Henry Dale, who came up to Trinity toward the end of Foster's career and who

[80] J. W. Clark, "Coutts Trotter," in *Old Friends at Cambridge and Elsewhere* (London, 1900), pp. 314-318, quotes on 314, 316.

[81] J. J. Thomson, *Recollections and Reflections* (London, 1936), pp. 280-283, quote on 280.

[82] M. Foster, "Coutts Trotter," *Nature, 37* (1887), 153-154.

[83] See Trinity College Minute Books (June 1872-June 1874), pp. 5, 14, 32, 42, 65, 272. On Balfour, see Chapter 5 below, "Foster and Balfour."

went on to win the Nobel prize in physiology or medicine.[84] Foster summarized his indebtedness to Trotter by saying that, "All through the thirteen years during which, while working within the University, I was really outside the University, my every move was made by and through Trotter; and since I have been Professor my every movement has been made with him."[85]

Before and after Trotter's death, Foster could also rely on the support of yet another Trinity man, John Willis Clark (1833-1910), known familiarly as "J." The son of William "Bone" Clark (1788-1869), professor of anatomy at Cambridge from 1817 to 1865, "J" became a Fellow of Trinity in 1858, having won a first class in the Classical Tripos two years before. From 1866 to 1891, he served as superintendent of the new Museum of Zoology and Comparative Anatomy and as secretary to the Museums and Lecture-Rooms Syndicate, placing him in a superb position to defend and promote Foster's claims on university grounds and buildings. In fact, Clark saw eye to eye with Trotter and Foster on virtually every issue. And since he himself instituted "practical" or laboratory instruction in zoology as early as 1871, there can be no doubt that he shared Foster's belief in the value of laboratory training in biology.[86]

With allies as sympathetic and well-placed as these, it becomes less surprising that Foster was able to win so much support in the university at large. In fact, as we shall see more fully below, Foster managed to secure nearly £10,000 in university funds for the accommodation of his and Balfour's ever expanding classes even before he became professor of physiology in 1883. As we shall also see below, Foster achieved that goal not only through the constant support of his Trinity College allies but also through the aggressive advocacy of G. M. Humphry, the influential professor of human anatomy. Throughout the 1870's, in fact, Humphry repeatedly sought a way to create a university professorship for Foster. Thus, in the spring of 1875, when the Jacksonian chair in experimental philosophy fell vacant, Humphry made an ingenious attempt to interpret the ambiguous provisions of the founder's will in such a way that the chair might go to a teacher of physiology—in other words, no doubt, to Foster.[87] He was unsuccess-

[84] Dale, "Foster," p. 17. Besides Dale, at least two other Nobel laureates have held the Trotter studentship—Lords Rayleigh and Rutherford. Thomson, *Recollections*, p. 283.

[85] M. Foster, "Coutts Trotter," *Nature, 37* (1887), 153-154, on 154.

[86] A. E. Shipley, *"J." A Memoir of John Willis Clark* (London, 1913), esp. pp. 256-343. On Clark's introduction of "practical instruction, see ibid., p. 270-274; and *CUR*, 20 Nov. 1872, p. 73.

[87] See *CUR, 4* (1875), 267-268. The founder's provisions are given in ibid., pp. 253-256.

ful, the chair going instead to the distinguished pioneer in spectroscopy and cryogenics, Sir James Dewar, who held it until his death in 1923. In 1876, the Board of Medical Studies, on which Humphry also served, recommended the founding of a chair in physiology,[88] but no means could be found for financing it and nothing came of the recommendation.

By 1883, when Foster finally did attain his professorship, the casual observer might well have thought that Humphry's campaign had succeeded long ago. Foster's numerous students came from colleges throughout the university. His large new laboratory had been built on university land at university expense. His name was associated throughout the world with "the Cambridge School of Physiology." And yet his only official position at Cambridge was still that of Trinity praelector. He himself never forgot the nature and status of his position—nor, indeed, could he. In a letter to Sir Joseph Hooker, probably written in 1880, Foster emphasized that he had received "*not one farthing from the University* except gas, water and rooms. . . ." For some time he had been required to rely solely on the fees paid by his students for "all current expenses and all new apparatus," and only in the past year or two had the fees been enough to meet these expenses. Foster had paid the rest out of his own pocket.[89]

If Foster was disappointed that his position gave him no greater claim on university resources, he expressed only gratitude for the way he had been treated by Trinity College. Besides his fellowship (worth £275 to £300 per year, depending on the income from college investments), his annual stipend (about £250 in 1880) came entirely from Trinity, deriving partly from investment income and partly from undergraduate fees in the college tuition fund. The college also provided much of the funding for his staff and apparatus. When Foster arrived in 1870, Trinity made him an initial grant of £400 to equip his lecture hall-cum-laboratory with the most primitive necessities: "simple furniture, some tables . . . , some bottles and reagents, a few microscopes and so on." Also from the beginning, the college gave him £80 annually for the wages of laboratory assistants. In 1873 he was allowed to use the college coat of arms on the first volume of his *Studies from the Physiological Laboratory in the University of Cambridge*. The next year the Master and Seniority awarded him £60 for the purchase of a set of resistance coils and a "recording apparatus," probably a kymograph. From 1875 the college paid £50 toward J. N.

[88] Ibid., *5* (1876), 325-326.

[89] Foster to Hooker, 23 February [1880]. Letters to J. D. Hooker, Royal Botanical Gardens, Kew.

Langley's salary as demonstrator to Foster, the remainder being paid by one of Foster's former students, almost certainly A. G. Dew-Smith, also a Fellow of Trinity who was Foster's benefactor in other ways. About 1878 Trinity also began to pay £100 per year for a second demonstrator, probably A. S. Lea.[90]

Although these funds were far from magnificent—indeed, by German standards, they were pathetic—Foster was much impressed by Trinity's willingness to contribute to an enterprise that benefited not only its own students, but also the university at large. In the first volume of *Studies from the Physiological Laboratory in the University of Cambridge,* Foster was glad to have "this or any other opportunity of expressing my warmest thanks to the Master and Fellows of Trinity College, whose Praelector in Physiology I have the honour to be, for their singular liberality in providing me with apparatus and the cordiality with which they have supported me in every way." He asked them "to look on these pages as simply the first, cotyledonary, leaves which may assure them that the seed they have sown has germinated, but which can tell little what kind of plant it will grow to or what manner of fruit it will bear."[91] Within a decade those "cotyledonary leaves" had given rise to a research school of international significance. Yet that school had been absolutely dependent on Trinity College from the beginning. Even Foster's students, as we shall see, came overwhelmingly from Trinity College. Take away the Trinity praelector, his Trinity students, his Trinity funds, and his Trinity allies—and the "Cambridge School" would have vanished. During its crucial first decade, at least, it might almost have been called "The Trinity School of Physiology."

As a matter of fact, Trinity College nearly formalized that bond even as it sought to elevate Foster to a university professorship. Under the terms of the Universities Act of 1877, each college was given until the end of 1878 to frame its own statutes (subject to the approval of an Executive Commission), after which time the commission would impose regulations to take effect no later than 1882. Trinity used this period of grace to pledge the endowment of chairs in physiology and history. Initially, the pledge to endow a chair in physiology rested on the condition that Foster be appointed its first occupant. Although Foster himself opposed this condition, which was then withdrawn,

[90] Cf. ibid.; Trinity College Minute Books (May 1870-May 1872), pp. 14, 58-59, 137, 182; ibid. (June 1872-June 1874), pp. 115-116, 205; Dale, "Foster," pp. 19-20; and Robson, "Foster and Trinity." p. 11.

[91] [M. Foster], ed., *Studies from the Physiological Laboratory in the University of Cambridge,* Part I (1873), p. 6.

no one in Cambridge had any doubt that he would be elected if the chair materialized.[92] In November 1878, Trinity placed before the University Senate a proposal to establish a university chair "to be called the Trinity Professorship of Physiology," its occupant being entitled to a college fellowship and an additional annual stipend of £500 to be paid by the college.[93]

As D. A. Winstanley has pointed out, this offer was not quite as generous as it seemed. For though it pledged Trinity to endow a chair at £500 a year, it also assumed that Foster would be named first occupant and that the expenses of his assistants, demonstrators, and apparatus would then become the responsibility of the university rather than the college.[94] Under these circumstances, the new Trinity chair might even save the college money, at least in the short run. When the proposal came before the University Senate, no really serious opposition developed. George Paget, Regius Professor of Physic, lamented the exclusion of the Medical Board from any connection with the election of the proposed professor, and repeated his familiar theme that the claims of the natural sciences were being exalted over those of medicine per se. Professors Newton and Babington objected to the title proposed for the chair. They thought the word "College" should be inserted after "Trinity," Newton's grounds being that otherwise no German or French physiologist would take the title seriously, and Babington's motive being to prevent any offense to "people outside."[95]

The proposal passed the Senate in December 1878, but the scheme fell through when Trinity found its income reduced and overextended as a result of the agricultural depression then plaguing England and her two land-rich universities.[96] In the end, the professorship was taken out of the hands of the college. The statutes of the Executive Commission, requiring the colleges to contribute to university purposes, became effective in June 1882. In the same month, it was announced that some of those funds would go toward the establishment of university chairs in physiology, pathology, and mental philosophy —in that order.[97] The stipend for the chair in physiology was to be £800, subject to a deduction if the occupant held a college fellowship,

[92] Trinity College Minute Books (Nov. 1876-Oct. 1878), p. 152. Cf. Winstanley, *Later Victorian Cambridge*, p. 345.

[93] *CUR, 8* (1878), 141-143.

[94] Winstanley, *Later Victorian Cambridge*, pp. 344-345.

[95] *CUR, 8* (1878), 179-180.

[96] That the depression was confined mainly to agriculture, and therefore should not be called 'The Great Depression," is the burden of A. E. Musson, "The Great Depression in Britain, 1873-1896," *Journal of Economic History, 19* (1959), 199-228.

[97] *CUR, 11* (1882) 795 806

and the appointee was not to practice medicine or surgery. The chair was officially created in May 1883, when the electors were announced. Of the eight electors, at least four were Foster's close friends, allies, or former students—G. M. Humphry, John Newport Langley, Sydney Howard Vines, and T. H. Huxley himself. With this board of electors, Foster's already anticipated election was now a certainty. He was elected professor of physiology, the university's first, on 11 June 1883.[98]

Two months later Foster wrote a long and revealing letter of gratitude to the Master of Trinity College, W. H. Thompson:

My dear Master

The University having done me the honour to appoint me to the newly established chair of Physiology, my connection with the College as Praelector comes to an end though I rejoice that I am still counted among the Fellows of the Society.

I cannot let this opportunity pass without making some attempt to thank you, and, through you, the College for all you have done for me during the thirteen years of my Praelectorship. You called me, a comparatively unknown young man, to the College in 1870; you not only at once gave me leave to follow out my own views as to what I ought to do, but from that time onward have constantly supported me not simply with cordial approbation but also with most material assistance.

I have reason to believe that many persons, not conversant with the organization and working of the University are under the impression that the necessary expenses which my work here has entailed have been provided out of University Funds. But I am sure that the authorities of the University would be the last to wish anything done by the College should be considered as done by the University. . . . Not only my own remuneration has come from the College but all the, really large, expenditure involved in my teaching Physiology, save what has been met by the fees of the students, has been provided for, in one way or another by the College.

At the outset, the College gave me a large grant of money for apparatus, and some years afterwards, a second smaller grant. During the whole thirteen years I have received from the College an annual sum for the payment of my Laboratory servants; and, for

[98] Ibid., *12* (1883), 746, 828. On the same day, Alexander Macalister was elected to the chair in anatomy, replacing G. M. Humphry, who had resigned to take (without stipend) a new chair in surgery. If, by this device, Humphry hoped to slip a permanent chair of surgery into the university, the hope was not fulfilled. See Humphry Davy Rolleston, *The Cambridge Medical School: A Biographical History* (Cambridge, 1932), p. 221.

several years past, two demonstrators (one at a comparatively high salary) as well as during the past year, three assistant demonstrators, have been paid partly from the Tuition Fund of the College partly by funds which though furnished by private liberality, cannot be wholly dissociated from the College. I think I may fairly say that I have never asked anything of you in vain. I might add that what you have done for me, did not prevent you from also assisting our lamented Balfour, working in a closely allied branch of science, or, in carrying on the work which he left behind, through aid given to Mr. Adam Sedgwick.

Let me assure you that I fully appreciate all the College has done for me; but perhaps after all I feel still more keenly the sympathy and kindness with which, a stranger I was first received among you, and which has made the thirteen years of my Praelectorship, the brightest as well as the best years of my life.

Yours ever truly,
M. Foster[99]

This letter perhaps reveals what Langley called Foster's "charm of manner," a quality which, together with his clarity of aim, "gained him instant support."[100] In a similar vein, E. A. Sharpey-Schafer suggested that Foster owed his success as much to "the power he had of influencing senior members of the University" as to anything else.[101] Perhaps, after all, Foster did not pose such a vital threat to the dons who remembered or revered Cambridge before 1850. He was, to be sure, a dissenter, an experimental scientist, and a married outsider. But he was also a gentle, humorous, and thoroughly respectable man who had himself been a classical scholar of no mean distinction, and who displayed a quite unexpected but genuine affection for his college. And if his personal qualities were not enough to disarm older members of Trinity who might otherwise have resented his expansive ambitions for his subject, his contributions to the efficiency and reputation of the college might have carried the day alone. He obviously played an important, almost paternal role in the lives of many students at Trinity. Ultimately, the college benefited from Foster at least as

99 Foster to W. H. Thompson, 28 July 1883. Archives of Trinity College, Cambridge. My thanks to Rick French, who alerted me to the existence of this letter, and to Dr. Robert Robson, tutor in history at Trinity College, who kindly sent me a photocopy. Only later did I discover a published version. See *Nature, 28* (1883), 374.

100 J. N. Langley, "Sir Michael Foster. In Memoriam," *J. Physiol., 35* (1907), 233-246, on 238.

101 E. A. Sharpey-Schafer, *History of the Physiological Society during its First Fifty Years, 1876-1926* (London, 1927), p. 25.

much as he benefited from it, for through him Trinity became, as D. H. M. Woollam put it in 1971, "a world centre of physiology and particularly of physiological research, right up to and of course including the present day."[102] Had they known this outcome, even those Trinity dons who initially feared what Foster represented might have decided that his appointment had been a wise one, and that there was more than one way to promote the interests of the college they cherished in common.

Conclusion

Foster could never had achieved what he did at Cambridge in the absence of "the revolution of the dons." This "revolution" depended, in turn, on powerful outside influences—on governmental action or pressure, obviously, but no less importantly on the German example, at once inspiring and threatening. It can be no accident that so many participants in English university reform, including Foster himself and his allies Sidgwick, Trotter, Clark, and Humphry, had experienced or admired the noncollegiate, professorial, state-supported German university system.[103] Yet the "revolution of the dons" by no means destroyed the distinctive collegiate system that Cambridge shared with Oxford, and even the governmental Royal Commissions left considerable latitude for local initiative and for the nuanced operation of local traditions.

The local response at Cambridge differed strikingly from that at Oxford, despite the virtual identity of the outside forces operating on them, despite their social and religious homogeneity, and despite the outward similarity of their statutes and provisions for natural science around 1870. By about 1900 even a superficial observer of English academic life could have perceived a dramatic difference between the place of science at Oxford and at Cambridge. This difference has no obvious economic explanation. Arthur Engel argues that the hostility to science in late Victorian Oxford derived largely from the effects of the agricultural depression,[104] but there is no evidence that the Cambridge colleges fared better during that depression, and they may

[102] Woollam, "The Cambridge School," p. 144. Cf. John C. Eccles, "British Physiology: Some Highlights, 1870-1940," in *British Contributions to Medical Science*, ed. W. C. Gibson (London, 1971), pp. 173-193, on 173-174.

[103] With the possible exception of Humphry, all of these reformers visited Germany between 1860 and 1870. Of Humphry's admiration or envy for the German system, there can be no doubt; see Chapter 6 below, p. 158.

[104] Engel, "Clergyman to Don," Chapter 5.

very well have fared worse.[105] In a period of relative economic stringency, both universities and their colleges had to make difficult choices in keeping with their priorities. Science clearly ranked higher on the list of priorities at Cambridge than it did at Oxford. For though Oxford did spend a considerable amount on science during the late Victorian period, Cambridge devoted an even higher proportion of its resources to the same end. F. M. Cornford, the Cambridge classicist whose witty *Microcosmographia Academica* first appeared in 1908, observed there that scientists are "dangerous, because they know what they want . . . all the money there is going"; and because these "cave dwellers" lack the refinement of classical scholars, "they succeed in getting all the money there is going."[106] In the *Quarterly Review* of 1906, "A Plea for Cambridge" took the very similar position that though "science had [already] emptied the University Chest," it remained " 'hungry and aggressive.' "[107]

In choosing to devote more of its resources to science, Cambridge merely reinforced the traditional distinction between the two universities. Among the Cambridge colleges, Trinity had long been and still remained the leading center and advocate for scientific studies. It is in local traditions, and above all in the traditions of Trinity College, that the scientific dominance of late Victorian Cambridge over Oxford is chiefly to be sought. That point may gain in credence from a brief comparison between the rise of biology and the rise of physics in Cambridge during this period.

At first sight, any parallel between the two may seem remote. James Clerk Maxwell, who launched Cambridge physics on its glorious path, had a brilliant reputation as a scientist in his own right; Foster did not. Moreover, Maxwell, unlike Foster, was on familiar ground at Cambridge, having studied there in the 1850's and having served for several years as an examiner for the Mathematical Tripos. Finally, whereas Foster came as a "simple College lecturer," Maxwell returned as a university professor assured of a new laboratory. As a matter of fact, the committee that had originally recommended the creation of Maxwell's chair had also insisted that a new and separate laboratory for physics was indispensable—indeed, more indispensable than the professorship itself, since provision was made to terminate the chair with the first professor's tenure of office.[108]

[105] See J. B. D. Dunbabin, "Oxford and Cambridge College Finances, 1871-1913," *Economic History Review, 28* (1975), 631-647, esp. 639, 641. I owe this reference to Dr. Roy Porter.

[106] F. M. Cornford, *Microcosmographia Academica: Being a Guide for the Young Academic Politician*, 4th ed. (Cambridge, 1949), pp. 4-5.

[107] "A Plea for Cambridge," *Quarterly Review, 204* (1906), 499-525.

[108] See *CUR*, 16 Nov. 1870, pp. 94-95.

These recommendations, made in February 1869, had generated a prolonged debate. Perhaps partly because the Mathematical Tripos now included questions on heat, electricity, and magnetism, and partly because the Clarendon Laboratory of Physics was already going up at Oxford, few dons disputed any longer the need for a laboratory of experimental physics at Cambridge. But the colleges demonstrated no greater willingness to contribute their funds to this project than they had to the new science museums nearly two decades earlier, and the debate was still going on when Foster accepted the Trinity praelectorship in May 1870. Five months later, the committee's proposal received a new lease on life when the chancellor of the university, William Cavendish (the seventh Duke of Devonshire) expressed his willingness to do what the colleges would not.[109] Early in 1871, Maxwell was appointed first Cavendish Professor of Experimental Physics. By autumn, he and Coutts Trotter had completed a tour of the newest physics laboratories in England and abroad, and the famous Cavendish Laboratory was soon being constructed in accordance with their recommendations at the then extravagant cost of £8,500.[110] Although this figure exceeded the original estimate by more than £2,000, "the Duke paid up without a murmur."[111]

Unlike the Cambridge School of Physiology, the Cambridge School of Physics has attracted considerable historical attention. In his book on the Cavendish Laboratory, J. G. Crowther suggests that it arose in response to the industrial needs of late Victorian England.[112] More subtly, Arnold Thackray once asked incidentally "whether the Cavendish does not represent a phenomenon common among the new universities, if rare at Cambridge—the support of the new organs of scientific research by the new wealth of successful middle-class entrepreneurs and manufacturers."[113] Certainly William Cavendish was an industrialist with a concern for the "promotion of science in its practical implications,"[114] and the Cavendish Laboratory did take on something of an industrial tone under Maxwell's successor, Lord Rayleigh.[115] But it seems odd to link the new industrial and commercial

[109] See ibid., 1870-1872, passim; Willis and Clark, *Architectural History*, III, pp. 181-184; Winstanley, *Later Victorian Cambridge*, pp. 194-198; and Sviedrys, "Rise of Physical Science."

[110] *CUR*, 6 December 1971, pp. 95-96; 6 March 1872, p. 182.

[111] Egon Larsen [pseudonym], *The Cavendish Laboratory: Nursery of Genius* (London, 1962), p. 14.

[112] J. G. Crowther, *The Cavendish Laboratory, 1874-1974* (New York, 1974), esp. introduction and Chapter 1.

[113] A. Thackray, "Commentary [on Sviedrys' 'Rise of Physical Science']," *Historical Studies in the Physical Sciences, 2* (1970), 145-149, quote on 149.

[114] J. G. Crowther, *Statesmen of Science* (Bristol, 1966), pp. 213-233, quote on 219.

[115] See Sviedrys, "Rise of Physical Science," pp. 142-143.

wealth with a man of such impeccably aristocratic lineage as the seventh Duke of Devonshire.[116] Surely it is crucial to emphasize that Cavendish was not only an industrialist, but also chancellor of Cambridge University and, first if not foremost, a Trinity man. In making his gift, Cavendish may have been motivated as much by loyalty to his university and to his college as by any concern for British industry.

In any case Maxwell, like Cavendish, was a Trinity man, and so in fact were the next four Cavendish Professors—John William Strutt (Lord Rayleigh), who was (like Henry Sidgwick) brother-in-law to Foster's student, F. M. Balfour; Sir J. J. Thomson, who eventually became Master of Trinity; Ernest Rutherford, who became Cavendish Professor and a Fellow of Trinity in 1919; and Sir Lawrence Bragg, who resigned the Cavendish Professorship in 1953.[117] All four men won Nobel prizes. Sir William Bragg, father of Sir Lawrence, who shared the 1915 Nobel prize for physics with his son, was yet another Trinity man. If, when he made his gift to the university, Cavendish hoped to increase its glory and that of his college, he succeeded famously. In short, Cambridge physics has long shared with Cambridge physiology a brilliant tradition deeply rooted in the traditions of Trinity College itself.

Other intriguing parallels can perhaps be drawn between the rise of biology and the rise of physics in late Victorian Cambridge. Like the Cambridge School of Physiology, which descended from and crystallized around the "ancestor problem" of the heartbeat, the embryonic Cambridge School of Physics apparently also focused on one problem to the relative exclusion of others. According to Sviedrys, at any rate, "Maxwell's laboratory was in effect an incipient electrical standards testing laboratory," and Lord Rayleigh, "who wanted to identify the Cavendish with a concrete area of research, uniting the laboratory workers around a common research theme, . . . selected the redetermination of electrical standards for this purpose."[118] The most obvious parallel between the two schools is chronological. They were coeval; they developed together; and they achieved maturity and renown almost simultaneously. Only Sviedrys' inattention to Foster's early efforts can justify his claims that the Cavendish "was the first [Cambridge] laboratory to stress research," and that Foster developed his school "*in the wake of* the example set by Maxwell and the Cavendish

116 For Thackray, "Commentary," p. 148, Cavendish is "the exception that proves the rule"; for Crowther, *Statesmen*, p. 213, it is an example of an "Aristocratic Superstructure on an Industrial Base."

117 Cf. Trevalyan, *Trinity College*, p. 104.

118 Sviedrys, "Rise of Physical Science," p. 142.

Laboratory."[119] In the end, though, the deepest parallel between the two schools lay in their common link with Trinity College. And if, as Winstanley insists, "college sentiment" posed one of the major impediments to university reform,[120] that same sentiment—under appropriate stimuli—could become a powerful engine for reform and even for the rise of natural science in the university at large.

[119] Ibid., p. 144. My emphasis. Indeed, Sviedrys himself elsewhere recognizes the virtual simultaneity; ibid., p. 137, n. 20.

[120] Another being the deification of formal mathematics; Winstanley, *Early Victorian* Cambridge, p. 168 et passim.

5. The Transformation of Biology in Late Victorian Cambridge: Foster, Huxley, and the Introduction of Laboratory Biology in England

> Cambridge men, and all who hope for the restoration of the English Universities to their legitimate place in the academic sisterhood of Europe, must feel proud of Mr. Balfour and the steadily working school of biologists which has risen around the Trinity Praelector on the banks of the Cam. The Cambridge biologists are now a power in the scientific progress of the country, and it is from Cambridge that the new men come to fill positions as teachers of the biological sciences in the colleges of Manchester, Birmingham, Dublin, Eton, and elsewhere.
>
> E. Ray Lankester (1881)[1]

EVERYONE who has examined Foster's career has been struck by the range of his students' interests. Gaskell and Sharpey-Schafer, among others, have emphasized the pains he took to ascertain the bent of mind in each of his students and then to groom each one for that branch of biology to which he seemed best adapted.[2] In the inaugural issue of Foster's *Studies from the Physiological Laboratory in the University of Cambridge*, papers on histology, embryology, and physiological chemistry stand alongside more purely physiological investigations. This diversity persisted in the next two issues (1876, 1877), and by the end Foster thought he should perhaps have called the publication *Studies from the Biological Institute in the University of Cambridge.*[3] In fact, precisely because Foster's influence ranged wider, "covering all branches of general biology," Fielding Garrison found some grounds for exalting him above Ludwig as a teacher.[4]

The breadth of Foster's influence had its intellectual base in his conviction that all branches of biology were united by certain fundamental principles, most notably Darwin's theory of evolution. This

[1] E. Ray Lankester, "Balfour's Comparative Embryology," *Nature, 25* (1881), 25-27, quote on 27.

[2] See W. H. Gaskell, "Sir Michael Foster, 1836-1907," *Proc. Roy. Soc., B80* (1908), lxxi-lxxxi, on lxxiv; and Edward Sharpey-Schafer, *History of the Physiological Society during its First Fifty Years, 1876-1926* (London, 1927), pp. 24-25.

[3] [Michael Foster], ed., *Studies from the Physiological Laboratory in the University of Cambridge,* Part III (1877), unnumbered end sheet.

[4] See Fielding H. Garrison, "Sir Michael Foster and the Cambridge School of Physiologists," *Maryland Medical Journal, 58* (1915), 106-118, on 111.

broadly biological perspective, as we shall see, probably stemmed mainly from T. H. Huxley, though it also had deep roots in Foster's training under William Sharpey. But this unifying view had the impact it did only because of a very specific institutional base—Foster's pioneering course in elementary biology.

During his first two years at Cambridge, Foster followed closely the lecture plan he had learned from Sharpey at University College.[5] In the Michaelmas term the lectures dealt with "the elements of physiology," the physiology of digestion, and the general properties of blood. The lectures in the Lent term focused on the physiology of the nervous system. In the Easter term, Foster lectured on "the elements of embryology."[6] By autumn 1872, however, he was beginning to develop his own program of lectures. During that term he gave a course on the physiology of the sense organs, which was continued in the Lent term. Then, in the Easter term of 1873, he announced the creation of a new "practical course of elementary biology." With the announcement, Foster sent a brief description of its nature and purpose:

> This course is intended as an introduction to the study of both Anatomy and Physiology. A short lecture of about half-an-hour will be given at each meeting, followed by practical work for about 1½ or 2 hrs.
>
> The Subjects considered will be somewhat as follows. The structures and functions of Elementary organisms—Yeast, Protococcus, Amoeba, and the phenomena of protoplasm in animals and vegetables. The elementary tissues of animals and plants compared. The structure and functions of lower animals, such as hydra, &c. The anatomy, histology, and elementary physiology of the frog and rabbit. The Course will accordingly include instruction in the use of the microscope and the art of dissection, and it is hoped will be found useful both to those intending to take up Comparative Anatomy, and to those intending to study Physiology.[7]

Simple as it is, this announcement marks the beginning of a new epoch in the teaching of biology in the English universities. By bringing elementary laboratory biology to Cambridge in April of 1873,

[5] On Sharpey's lectures, see D. W. Taylor, "The Life and Teaching of William Sharpey (1802-1880): 'Father of Modern Physiology' in Britain," *Med. Hist., 15* (1971), 126-153, 241-259, on 143-153; and esp. Chapter 3 above, pp. 53-54.

[6] See *Cambridge University Reporter* [hereafter *CUR*], 16 Nov. 1870, p. 89; 1 Feb. 1871, p. 164; 26 April 1871, p. 281; 10 May 1871, p. 340; 18 Oct. 1871, p. 11; 7 Feb. 1872, p. 153, and 12 April 1872, p. 225. The subject of the lectures was not invariably announced in the *University Reporter*, but this general plan seems clear from the available announcements.

[7] *CUR*, 22 April 1873, p. 19.

Foster apparently became the first to teach such a course in a university setting. To be sure, the value of laboratory teaching had long been recognized, thanks above all to the demonstrated success of Liebig's famous teaching laboratory in chemistry at Giessen. In several branches of the biological sciences, too, laboratory teaching had become more or less standard, especially in the German universities, where students had been learning histology and physiology in laboratories for many years. Even in England, "practical" teaching was not unknown in some areas of biology. As Jeffrey Parker pointed out long ago, "botanists had always been in the habit of distributing flowers to their students, which they could dissect or not as they chose; animal histology was taught in many colleges under the name of practical physiology; and at Oxford an excellent system of zoological work had been established by . . . Professor Rolleston."[8]

But none of these antecedents quite matched Foster's course in scope, for he included both plants and animals in his purview, and he insisted on making his course a truly elementary, introductory course in general biology, while the others (and especially the Germans) tended to specialize in one branch of biology or another. As we shall see more fully below, Foster's new course did have very definite roots in one earlier course, from which it scarcely differed at all in scope or aims—namely, T. H. Huxley's celebrated course in elementary biology at South Kensington. But Huxley's course, in the creation and success of which Foster himself played a crucial role, had not been transferred to a university setting until Foster brought it to Cambridge in the spring of 1873. And perhaps no one, not even Huxley himself, did so much to make elementary biology a systematic and integral part of his institution's biological curriculum. Foster made sure that his course in elementary biology served its announced function as the gateway to other biological courses at Cambridge, and above all to his own classes in histology and physiology.

In the Michaelmas and Lent terms of the academic year 1873-1874, Foster gave a practical course of physiology and histology "intended for those who have attended a course of Elementary Biology similar to that given last term." Then, in the Easter term of 1874, he repeated the course in elementary biology.[9] For the next decade, Foster repeated this sequence every year. He established the rule that elementary biology should be taken by first-year men, his introductory course in practical physiology by second-year men, and the advanced course in

[8] See Leonard Huxley, *Life and Letters of Thomas Henry Huxley*, 2 vols. (London, 1900), I, p. 377.

[9] *CUR*, 7 October 1873, pp. 11-12; 14 October 1874, p. 26.

physiology (under the triumvirate of W. H. Gaskell, J. N. Langley, and A. S. Lea) by third-year men.[10] In 1883, when he was appointed to the chair in physiology, Foster turned the course in elementary biology over to the morphologist Adam Sedgwick (grandnephew of the geologist by the same name) and botanist Sydney Vines. It then became a two-term course, as Foster had long insisted it should be. At the same time, he expanded his introductory practical physiology into a full-year course, thus again reaching a goal long desired.[11]

In other words, Foster devoted his own teaching efforts mainly to elementary and introductory courses, while delegating the advanced classes to his former students. This practice, so uncommon today, gradually diluted his influence over the research carried out in his laboratory, but it also allowed him to interact with students who became botanists, morphologists, pathologists, or even psychologists and anthropologists, as well as those who became physiologists. Among those who studied at the university between 1873 and 1883, when Foster was teaching his elementary course, were anthropologist A. C. Haddon; psychologist James Ward; pathologist J. G. Adami; botanists S. H. Vines, F. O. Bower, and H. Marshall Ward; and morphologists A. Milnes Marshall, Adam Sedgwick, D'Arcy Wentworth Thompson, and A. E. Shipley. Of these, Vines, Haddon, Ward, Sedgwick, and Shipley became leading teachers at Cambridge; and through them, Foster remained a living influence on Cambridge biology long after his direct role had come to an end. With their help and the vital contributions of embryologist Frank Balfour, Foster engineered a revolution at Cambridge as sweeping as that engineered by Huxley in English biology as a whole. The nature, direction, and origins of this revolution can be clarified only through an examination of Cambridge biology as it was when Foster appeared on the scene, followed by a brief discussion of Balfour's dramatic career, and then a more extensive analysis of Foster's debt to and role in Huxley's famous course in elementary biology at South Kensington.

Biology at Cambridge when Foster arrived: Babington and Newton

When Foster arrived at Cambridge in 1870, the biological sciences were represented chiefly by men of the old school, field naturalists and systematists not disposed by training or personal interest to appreciate the value of laboratory biology. The professor of botany was Charles Cardale Babington (1808-1895), a gentle, pious don known as "Beetles

[10] See, e.g., ibid., *6* (1878), 622.
[11] See ibid., *4* (1875), 479; *13* (1883), 77.

Babington" because of his interest in entomology.[12] Like Charles Darwin, Babington had studied botany at Cambridge under Professor J. S. Henslow. When Henslow died in the spring of 1861, Babington announced his candidacy for the vacant chair and was elected without opposition. He held the chair until his death in 1895 at age eighty-seven, and from beginning to end remained loyal to Henslow's style and methods of teaching. That botany had in the meantime undergone a revolution, that especially in Germany it had been changed from a dead to a living science through microscopy, experiment, and a physiological mode of thought, Babington seemed entirely unaware.

Since he lectured only during the Easter term, each course consisting of about twenty lectures, Babington found time to develop an impressive collection of insects, as well as to indulge a variety of interests outside science, including archeology and missionary work. F. O. Bower tried Babington's lectures in the 1870's, but found them "wanting both in spirit and in substance." Confined to descriptive and systematic botany in the old style, they were unaccompanied by laboratory demonstrations. Babington apparently neglected even field work. The university calendar announced "herborizing excursions, should the circumstances permit," but so far as Bower could determine, "the circumstances never did permit." Under Babington, "the official teaching of botany in Cambridge University was moribund in the summer and actually dead during the winter."[13]

With the founding of a chair in zoology and comparative anatomy in 1866, Cambridge had a chance to widen its conception of biology, but the opportunity was largely lost by the appointment of Alfred Newton (1829-1907), a Fellow of Magdalene College.[14] Professorships were at this time decided by vote of the Cambridge Senate, and in a contested election, as this was, the candidates were required to canvass for votes in the manner of a politician. Newton's most impressive weapons were testimonials from Richard Owen and George Rolleston; his only opponent, W. H. D. Drosier, M.D., a fellow of Caius College and an exponent of natural theology, countered with the only slightly exaggerated charge that Newton was a "mere ornithologist." Since G. M. Humphry and J. W. Clark favored Newton, it seems likely that he was the better candidate, but such issues were really of little consequence. Most members of the Senate were unable to decide the ques-

[12] See *Memorials, Journal and Botanical Correspondence of Charles Babington* (Cambridge, 1897).

[13] F. O. Bower, *Sixty Years of Botany in Britain (1875-1935): Impressions of an Eyewitness* (London, 1938), p. 13.

[14] See A. F. R. Wollaston, *Life of Alfred Newton* (London, 1921).

tion on the basis of merit and simply voted for the candidate whom they preferred as a friend. Newton had 110 friends; Drosier, 82.[15]

An expert on the extinct dodo, Newton was himself among the last representatives of a certain species of Cambridge don. In his later years especially, the sight of his hulking figure, bent by lameness and attired in an old-fashioned black tailcoat and stovepipe hat, evoked an earlier age. Opinionated and obstinate, he embellished Cambridge conversation by virtue of his eccentric conservatism. He resisted all attempts to bring change into the daily schedule of life. When music was introduced in the Magdalene College chapel, Newton showed his displeasure by snapping the hymnal shut before the music ended and sitting "ostentatiously down with an air of relief." When the matter of admitting women to the chapel came up for debate, he vowed never to set foot in a chapel occupied by members of a sex that had no business there.[16]

Although capable in rare instances of changing his mind, Newton was almost as set in his views as in his ways. When the Special Board of Biology and Geology was created at Cambridge in 1881, he was selected as its first chairman. Although he presided admirably, he "never approved of the existence of the body he presided over, and nothing would induce him to vote either for or against so new-fangled an idea as a Doctor of Science."[17] Newton published extensively, almost always on ornithological subjects, but was so leisurely and cautious in his writing that the publisher who had engaged him to edit the fourth edition of Yarrell's *History of British Birds* replaced him with another when, after ten years, Newton had only half finished the task.[18] With a small group in his book-cluttered rooms or on an outdoor excursion, Newton could inspire interest and even enthusiasm, but in the classroom he read verbatim the same lectures year after year. Not surprisingly, students found them incredibly dull and dry. A. E. Shipley, who came up to Cambridge in 1880, was at times the sole auditor:

> Not that that made the least difference to the Professor. He steadily and relentlessly read on—"the majority of you now present know," "most of my audience are well aware," and similar phrases left me in considerable doubt as to what parts of me were "the majority" and which the "most."[19]

[15] Ibid., pp. 133-134. Cf. Cambridge University Rec. 39.28, Anderson MS Room, Cambridge University Library, for a list of testimonials and a breakdown by college of individual votes.

[16] Wollaston, *Alfred Newton*, pp. 345-346.

[17] Ibid., p. 107. [18] Ibid., pp. 213-214. [19] Ibid., pp. 104-105.

Obviously neither Babington nor Newton was likely to remold Cambridge biology or to awaken in their students an appreciation for laboratory research. Essentially field naturalists, they observed life in its natural setting and scarcely thought of wrenching it from its environment so as to interfere with it in a laboratory. Then, as now, a holistic view of organisms in their natural habitat had great value and appeal; but the experimental approach was ripe with exciting possibilities in 1870, and Cambridge would remain isolated from the European mainstream unless it introduced laboratories and experiments into its limited conception of biology. Babington and Newton were themselves incapable of leading such a change.

But neither were they blind obstructionists. On one burning issue, Newton demonstrated a surprising flexibility of mind; an early convert to Darwin's theory of evolution, he eventually even dared to teach it to his classes. More importantly, Babington and especially Newton showed a willingness to support laboratory biology once it had come to Cambridge. It may not have engaged their intellectual sympathy, but it did win their approval after its advocates demonstrated that they were men of serious purpose whose new approach might contribute to the general development of biology at Cambridge.

In 1895, on Babington's death, the Museums and Lecture-Rooms Syndicate paid tribute to his services to the university. Referring to his steadfast advocacy in the cause of the new science museums (1863-1865), they also saluted his later contributions, saying that "he did all in his power . . . to promote the study of physiological as well as systematic Botany."[20] His power may have been limited but his support was welcome.

Less reluctantly than Babington, Newton gave support and encouragement to courses he would never have thought of teaching himself. As early as 1872, when it was first proposed to expand accommodation for laboratory work in physiology and comparative anatomy, Newton was among those who spoke out in favor of the proposal. In particular, he offered "grateful testimony to the results of the practical teaching given by Mr. Clark,"[21] that is to say, Foster's friend and ally J. W. Clark, who had been appointed superintendent of the Museum of Zoology and Comparative Anatomy in 1866. Early in 1871, with the unofficial assistance of his young friend T. W. Bridge, Clark had introduced a course of practical work in comparative anatomy. When a demonstratorship in comparative anatomy was established in March

[20] See Babington, *Memorials*, p. L.

[21] *CUR*, 20 November 1872, p. 73.

1873, Newton nominated Bridge, thus ensuring that Clark's innovation would be continued.[22]

In 1875 Newton gave up his private room in the new museums so that Frank Balfour could establish a class in laboratory morphology, a subject of which Newton was ignorant but which he thereafter consistently encouraged.[23] Newton's basic attitude toward his younger colleagues is beautifully captured in this excerpt from a letter written in 1888:

> . . . the self-abnegation of our biologists—many of them, be it borne in mind, young men of ambition only equalled by their capacity—in regard to the interests of the University, hampered as they now are by financial difficulties, is beyond any praise that I can bestow.[24]

No doubt Newton was motivated more by loyalty to colleagues and to the university than by any profound intellectual conversion. But instead of erecting obstacles, as he might have done, he smoothed the way for those who were to transform the character of Cambridge biology. To some extent, this transformation may have taken place whether or not Foster had come to Cambridge when he did. If so, it doubtless would have been led by G. M. Humphry, who belonged more to the medical than to the biological faculty, and by J. W. Clark, who held no official faculty position whatever. Already before Foster arrived, Humphry was giving a course in histology.[25] And though Clark did not inaugurate his course in practical comparative anatomy until after Foster had arrived—in fact, in the same term that Foster first opened his physiological laboratory[26]—there is no evidence that Foster had anything to do with Clark's innovation.

In the event, however, Clark and Humphry did not so much lead the movement toward laboratory biology as support the man who did. As early as 1867, in his article on scientific education, Foster had insisted

22 See A. E. Shipley, *"J." A Memoir of John Willis Clark* (London, 1913), pp. 270-274.

23 Cf. Wollaston, *Alfred Newton*, pp. 103-104, 252; *CUR, 5* (1876), 275; Robert Willis, *The Architectural History of the University of Cambridge*, ed. with large additions and brought up to the present time by J. W. Clark, 4 vols. (Cambridge, 1886), III, pp. 184-185; and esp. Alfred Newton, "The Late Professor Balfour," *Nature, 26* (1882), 342, where Newton emphasizes that the lectures on animal morphology were Balfour's own idea, and that he (Newton) "merely" gave Balfour his private room in the new science museum.

24 Wollaston, *Alfred Newton*, p. 258.

25 See Michael Foster, "Henry Newell Martin," *Proc. Roy. Soc., 60* (1897), xx-xxiii, on xx.

26 Cf. Shipley, *"J."*, pp. 270-274; and *CUR*, 1 Feb. 1871, p. 164; 20 Nov. 1872, p. 73.

that the student of science must observe for himself in the laboratory.[27] He came to Cambridge after several years of experience with this method of teaching at University College. And though it was in physiology that he most directly established this approach, his influence was felt in every branch of the biological sciences at Cambridge.

Foster and Balfour: the rise of embryology and animal morphology

Apparently no record exists of the number of students who attended Foster's lectures during his first year at Cambridge, but among them was one who attracted his special attention—Francis Maitland Balfour, known to his remarkable family as "Frank."[28] At the time he came to Cambridge, the Balfour family fortune was estimated at £4,000,000.[29] "Balfour belonged," as D. H. M. Woollam puts it, "to what is perhaps the rarest social class to be found among university teachers, the aristocratic."[30] He was born on 10 November 1851, the sixth child and third son of James Maitland Balfour (1820-1856) and his wife Lady Blanche Gascoyne-Cecil (1825-1872), daughter of the second Marquess of Salisbury. His eldest brother, Arthur James Balfour, was to become Prime Minister, succeeding his uncle, Robert Gascoyne-Cecil, the third Marquess of Salisbury. After preparing at Harrow School, Frank followed his father and his brother to Trinity College, where he matriculated in the Michaelmas term of 1870. His arrival at Trinity therefore coincided exactly with that of Foster, the middle-class, dissenting outsider with whom he seemed to have little else in common.

Balfour had done little to distinguish himself in his studies at Harrow School, and no great expectations awaited him when he walked through the Great Gate of Trinity College. He did, however, have a long-standing interest in natural history, and he had written at Harrow a prize-winning essay on the geology and natural history of East Lothian, which attracted the attention and praise of T. H. Huxley.

27 [Michael Foster], "Science in the Schools," *Quarterly Review, 123* (1867), 244-258. Cf. Chapter 3, pp. 67-68.

28 The most extensive available account of Balfour's life and work is T. E. Alexander, "Francis Maitland Balfour's Contributions to Embryology," unpublished Ph.D. dissertation (University of California, Los Angeles, 1969). See also J. W. Clark, "Francis Maitland Balfour," in *Old Friends at Cambridge and Elsewhere* (London, 1900), pp. 282-291; and Michael Foster, "Francis Maitland Balfour," *Proc. Roy. Soc., 35* (1883), xx-xxvii.

29 Alexander, "Balfour's Contributions," p. 49.

30 D. H. M. Woollam, "The Cambridge School of Physiology, 1850-1900," in *Cambridge and its Contribution to Medicine*, ed. Arthur Rook (London, 1971), pp. 139-154, quote on 150.

Moreover, though it did not form part of his normal course work, Balfour had already acquired a reasonably extensive training in microscopy, dissection, and comparative anatomy from G. Griffith, who came to Harrow in 1867.[31] Perhaps it is not so surprising that he should have carried off one of the new Trinity College natural science scholarships by competitive examination in March 1871.

It was during that examination, as Foster later recalled, that he first took special notice of Balfour, who had already been attending his lectures: "From that time onward we became more and more intimate, and I took an increasing share in the direction of his studies."[32] According to Cambridge legend, Balfour decided on a career in embryological research through the following encounter with Foster:

> Sitting in the little room of the philosophical library at Cambridge, Balfour . . . asked Foster to advise him as to his future career. Gnawing on his moustache for a moment, Foster's eye fell upon an egg lying on a bench, which he cracked, showing the embryo inside, with the suggestion. "What do you think of working at that?"[33]

Presumably apocryphal, this legend nonetheless accurately records that it was Foster who awakened Balfour's interest in embryology and who inspired him to undertake a career in scientific research.

Left-handed, somewhat awkward, frail, and shy, Balfour was also intelligent, studious, and remarkably attentive to detail. He came to Cambridge with all the natural endowments of a research scientist, and his family's wealth allowed him to pursue the expensive and financially insecure field of embryology without concern for his future.[34] Almost immediately, Foster asked Balfour to help him prepare a published version of his lectures on embryology, in which Foster traced the day-to-day development of the chick embryo. Using eggs incubated by Foster's own hens,[35] Balfour undertook a series of studies to clarify some of the more obscure points of this development, and in 1873, while still an undergraduate, published three brief papers describing

[31] See G. Griffith, "F. M. Balfour," *Nature, 26* (1882), 365; and Alexander, "Balfour's Contributions," pp. 32-36.

[32] Michael Foster, "Francis Maitland Balfour," *Nature, 26* (1882), 313-314, on 313.

[33] Garrison, "Foster and the Cambridge School," p. 112. Cf. W. H. Gaskell, "Foster," p. lxxix; and J. George Adami, "A Great Teacher (Sir Michael Foster) and His Influence," *Publication no 7, Medical Faculty, Queen's University (Kingston, Ontario, Canada)*, June 1913, pp. 1-17, on. 10.

[34] Cf. Foster, "Francis Maitland Balfour," *Nature, 26* (1882), 313; and Alexander, "Balfour's Contributions," pp. 49-50.

[35] W. T. Thiselton-Dyer, "Michael Foster—A Recollection," *Cambridge Review, 28* (1907), 439-440, on 439.

his results.[36] Foster's lectures were published in 1874 under the title, *Elements of Embryology*, with Balfour listed as co-author. This deep and early involvement in research may have prevented Balfour from doing as well as expected in his degree work. In the Natural Sciences Tripos for December, 1873, he ranked second behind another Foster protégé, Henry Newell Martin. Immediately after taking his B.A. in January 1874, Balfour went to the new zoological station at Naples, where he plunged into an investigation of the elasmobranch fishes.

This opportunity, too, Balfour owed largely to Foster. At Foster's urging, Cambridge had just agreed to an arrangement proposed by the new station's director, Anton Dohrn, whereby the university was given access to two research tables at Naples for an annual payment of £75. Foster was placed in charge of the applications, and it was probably more than coincidence that the first two successful nominees, Balfour and A. G. Dew-Smith, were both from Trinity College and both students of Foster. Later, whenever the arrangement came up for renewal, Foster was prominent among those who urged its continuance. Between 1873 and 1900, nearly all those responsible for teaching biology at Cambridge spent at least some time at the Naples station, including Foster himself, who went there in the winter of 1876-1877.[37] In this way, Cambridge biologists kept abreast of the latest developments in marine zoology.

In the autumn of 1874, on the basis of the work he had done at Naples on the elasmobranchs, Balfour was elected a Fellow of Trinity College. His candidacy would never have succeeded but for the introduction of new regulations making original research an important factor in Trinity fellowship elections.[38] Among those who recommended Balfour for the fellowship was T. H. Huxley, who had been invited by Trinity to judge the candidates and who was notably impressed by Balfour's work.[39] In the academic year 1874-1875, Balfour

[36] F. M. Balfour, "The Development and Growth of the Layers of the Blastoderm," *Quarterly Journal of the Microscopical Sciences, 13* (1873), 266-276; Balfour, "On the Disappearance of the Primitive Groove in the Embryo Chick," ibid., pp. 276-280; and Balfour "The Development of the Bloodvessels of the Chick," ibid., pp. 280-290.

[37] On Cambridge and the Naples zoological station, see *CUR*, 22 April 1873, p. 16; 6 May 1873, pp. 31, 38; 18 Nov. 1873, p. 87; *4* (1874), 75; *10* (1881), 331-332; *13* (1883), 238; *15* (1886), 471-472; *20* (1891), 397; and *25* (1896), 495. For the announcement that Foster had been nominated to study at the Naples station during December 1876 and January 1877, see ibid., *6* (1876), 117. On Balfour and the Naples station, see also Alexander, "Balfour's Contributions," pp. 63-81.

[38] Foster, "Francis Maitland Balfour," *Nature, 26* (1882), 313; Alexander, "Balfour's Contributions," pp. 43-45; and Chapter 4 above, p. 104.

[39] Cf. Cyril Bibby, *T. H. Huxley: Scientist, Humanist, and Educator* (London, 1959), p. 185; Foster to Huxley, 21 Oct. [1874], Huxley Papers, IV. 188. From the

returned to Naples to continue his research, being accompanied this time by A. Milnes Marshall, a young student from St. John's College. After returning to Cambridge in the summer of 1875, Balfour taught "a short course on embryology" for Foster, and then introduced laboratory courses in embryology and animal morphology, at first in association with Milnes Marshall.[40] Trinity College appointed him lecturer on animal morphology in 1876.

In 1878, after completing his important monograph on the elasmobranchs, Balfour gained fellowship in the Royal Society. His two-volume *Treatise on Comparative Embryology* (1880-1881), translated immediately into French and German, won international recognition. The German anatomist Wilhelm Waldeyer considered it the first successful attempt at a complete text of comparative embryology.[41] Like all of his work, Balfour's *Treatise* was written from a thoroughly Darwinian perspective and reflected his commitment to von Baer's germ layer theory.[42] It appeared too early to take full account of the cytological developments that were just beginning to transform conceptions of heredity, and it antedated the rise of the neurone theory by perhaps a decade. Parts of it therefore became obsolete fairly quickly, and (as Theodore Alexander emphasizes) its account of the spinal nerves was soon discredited, but the *Treatise* nonetheless remained a standard reference work for several decades.[43] Upon its completion, Balfour was elected to the Council of the Royal Society and awarded its prestigious Royal Medal.

Meanwhile, at Cambridge, Balfour had become a quiet but effective force in university reform,[44] and the extraordinary success of his courses had helped to promote a major expansion of the university's provision for the biological sciences. From an initial class of seven students in Easter 1876, Balfour's course in embryology had grown to thirty-one by Easter 1881 and to forty-three by the Easter term of 1882, when he did not teach because of illness.[45] The "practical" course in animal morphology enjoyed an even more spectacular suc-

---

Trinity College Minute Books (June 1872-June 1874), pp. 242, 272, it would appear that the candidates were judged by only three men: Foster himself, Huxley, and "Prof. Liveing."

[40] Foster, "Francis Maitland Balfour," *Nature, 26* (1882), 313-314; and Shipley, "*J.*", p. 120.

[41] See F. B. Churchill, "Francis Maitland Balfour," *DSB*, 1 (1970), 420-422.

[42] Cf. Alexander, "Balfour's Contributions," p. 87 et passim.

[43] Ibid., pp. 145ff; and Churchill, "Balfour."

[44] See Clark, *Old Friends*, p. 289; and *Balfour Memorial: Undergraduate Meeting at the Union* (Cambridge, 30 October 1882), Cambridge University Rec. 100, Anderson MS Room, Cambridge University Library, pp. 5-6.

[45] See *CUR, 6* (1877), 426; *11* (1882), 573; and *12* (1883), 699-700.

cess. It, too, began modestly, attracting ten students in 1875 and twelve in 1876, but by 1879 it had become large enough to justify a division into elementary and advanced sections. In the Michaelmas term of 1881, fifty-five students took the elementary section and twenty students the advanced course in laboratory morphology. By 1882, the two sections together were attracting ninety to ninety-five students.[46]

Obviously, Balfour's renown as investigator and teacher could no longer go unrecognized. In January 1882 Foster wrote to the vice-chancellor of the university, informing him that Balfour had only recently declined an invitation to become a candidate for the Linacre Chair in Anatomy and Physiology at Oxford and was now being pressed to accept the chair in natural history at the University of Edinburgh. Insisting that Balfour's loss would cause Cambridge immense and irreparable harm, Foster implored that he be appointed to "some honourable University post." Without reference to his own identical situation, Foster pointed out that Balfour's only official position was still that of lecturer at Trinity College. On the basis of Foster's letter, the Council of the University Senate recommended the establishment of a professorship in animal morphology, to be held by Balfour and to terminate with his tenure of the chair. This method of creating a chair—for a predetermined occupant—was a notable departure from university policy, but the proposal met with no apparent opposition.[47] Balfour was appointed to the professorship in June 1882, one year before Foster himself was elected to a professorship at Cambridge. At the time of their respective appointments, Balfour was thirty years old, Foster, forty-seven.

A month after his appointment, Balfour was dead, the victim of a fall in the Swiss Alps. Years later, strangely enough, his erstwhile associate A. Milnes Marshall also fell to his death while mountain climbing.[48] Balfour's death brought forth a tidal wave of shock, grief, and tribute. For three days T. H. Huxley was "utterly prostrated . . . scarcely able either to eat or sleep." He called Balfour's death "the greatest loss to science—not only in England, but in the world—in our time."[49]

On Saturday, 21 October 1882, Huxley went to Cambridge to participate in a widely advertised and well-attended meeting in Bal-

[46] Ibid., *11* (1882), 509-510, 572-573; and *Nature, 26* (1882), 631, where Balfour's successor (Adam Sedgwick) reports an expansion in the course in animal morphology from ten to ninety students "in seven years."

[47] See *CUR, 11* (1882), 427-428, 532.

[48] "Arthur Milnes Marshall," *Proc. Roy. Soc.*, 57 (1895), iii-iv.

[49] Huxley, *Life and Letters*, II, 37, 397.

four's honor. Amidst a profusion of praise for Balfour as teacher, investigator, and personality, it was decided to establish a studentship in his name. As the man who had discovered Balfour, Foster had merely to rise to be greeted with prolonged applause. He insisted above all that the regulations for the studentship should be framed in accordance with "the primary object of the Fund, namely the furtherance of original research." On October 30th, he emphasized this point again before a large and enthusiastic group of Cambridge undergraduates who had gathered at the Union Society to pay their own tribute to Balfour. In a university keenly aware of tradition, it was noted that the rooms of the Union Society had never before been placed at the disposal of the whole body of junior members of the university.[50]

The tragic and dramatic character of Balfour's death led both Huxley and Foster to recall Milton's lines:

> For Lycidas is dead, dead ere his prime,
> Young Lycidas and hath not left his peer.[51]

Balfour's friends, wrote Foster in his obituary notice for *Nature*, "mourn for Lycidas and cannot be comforted."[52] Perhaps the very pathos of Balfour's untimely end somewhat magnified his real achievements. Theodore Alexander implies as much by denying Balfour's claims as an original thinker or discoverer, by insisting that his work "illustrates the inherent danger of dogmatism," and even by disputing his legacy as a teacher. Conceding that Balfour "was a good teacher, and perhaps even a great one," Alexander nonetheless argues that he failed to establish "a particular school in morphology," and that, except for Henry Fairfield Osborn and D'Arcy Wentworth Thompson, "none of his students produced work of significance."[53] If so, it remains impossible to deny that Balfour, while still alive, earned international acclaim and major honors for his research, and played a crucial role in the rise of laboratory biology at Cambridge and thus indirectly at other English and American universities as well. That his influence would remain alive at Cambridge was ensured when his courses were given over to his own student Adam Sedgwick (1854-1913), a Fellow and lecturer at Trinity College who had been acting as Balfour's demonstrator since 1878 and who now became successively

[50] *Balfour Memorial*, p. 3.

[51] See ibid., p. 17, where Huxley is said to have quoted Milton's lines; and Woollam, "Cambridge School," p. 151, where Foster is credited.

[52] Foster, "Francis Maitland Balfour," *Nature, 26* (1882), 314.

[53] Alexander, "Balfour's Contributions," pp. 166-169.

university lecturer (1883-1890) and reader (1890-1907) in animal morphology. Ultimately, he replaced Alfred Newton in the chair in zoology and comparative anatomy.[54] Had Foster done nothing else but launch Balfour and Sedgwick on their careers, he would have gone a long way toward repaying the "poor paper IOU" he had sent Huxley on his own appointment to Trinity College in 1870.

Foster, Huxley, and the South Kensington course in elementary biology

In fact, of course, Foster did much, much more to promote the general development of biological studies at Cambridge. From its roots in animal morphology and physiology, his influence eventually reached the departments of anthropology, botany, pathology, biochemistry, and experimental psychology as well. As suggested above, Foster's course in elementary biology played a fundamental part in this process of diffusion. Here, again, Foster owed much of his success to the example set by Huxley, though in this case Huxley would have been obliged to return part of the debt.

By 1870, Huxley had been teaching natural history at the Royal School of Mines in Jermyn Street, London, for more than fifteen years. He had accepted his appointment there in 1854 with some reservation, perhaps because he hoped to find a position in physiology instead.[55] In any case, he gave a broad interpretation to his course in "natural history." He had long believed, he said in 1875, that "the study of living bodies is really one discipline, which is divided into Zoology and Botany simply as a matter of convenience." He was convinced that "the scientific Zoologist should be no more ignorant of the fundamental phenomena of vegetable life, than the scientific Botanist of those of animal existence."[56] This conviction derived in part from Huxley's long-standing commitment to the leading doctrine of the cell theory, namely, that plants and animals are analogous by virtue of their common structural basis in the cell. By 1870, his belief in the kinship of all living organisms had been strengthened on the one hand

[54] John Arthur Thompson, "Adam Sedgwick," *DNB (1912-1921)*, pp. 487-488.

[55] In his autobiography, Huxley says that when he was offered the two positions vacated by Edward Forbes in 1854—paleontologist and lecturer on natural history —"I refused the former point blank, and accepted the latter only provisionally, telling Sir Henry de la Beche that I did not care for fossils, and that I should give up Natural History as soon as I could get a physiological post." Huxley, *Life and Letters*, I, p. 132.

[56] T. H. Huxley and H. N. Martin, *A Course of Practical Instruction in Elementary Biology* (London, 1875), p. v.

by his acceptance and advocacy of the main features of Darwin's theory of evolution, and on the other by his recent conversion to the theory that cellular protoplasm represented a unitary substance of life—or, as Huxley put it in a famous lecture of 1868, "the physical basis of life."[57]

But in addition to believing that the study of all living bodies belonged to the same discipline, Huxley also had very definite ideas about the way in which this one discipline should be taught. He insisted above all that "sound and thorough knowledge was only to be obtained by practical work in the laboratory." Long before 1870, he had thought of teaching a course based upon these fundamental principles. As he explained in 1875:

> The thing to be done, therefore, was to organize a course of practical instruction in Elementary Biology, as a first step towards the special work of the Zoologist and Botanist. But this was forbidden, so far as I was concerned by the limitations of space in the building in Jermyn Street, which possessed no room applicable to the purpose of a laboratory; and I was obliged to content myself, for many years, with what seemed the next best thing, namely, as full an exposition as I could give of the characters of certain plants and animals, selected as types of vegetable and animal organization . . .[58]

During these years, from 1854 to 1871, Huxley's students received no direct experience in laboratory techniques. In 1870 he complained before a Royal Commission about the lack of facilities for practical teaching at the School of Mines, emphasizing that he had to teach natural history "without a biological laboratory and without the means of shewing a single dissection."[59]

The passage of the Elementary Education Bill of 1870 gave Huxley an opportunity to organize a pilot version of the practical course he had so long wished to teach. Since this Bill provided that elementary education should be made available to everyone, it required a vast expansion of the educational system. At the same time, it raised the question of the place of science in the school curriculum, and focused

[57] T. H. Huxley, "The Physical Basis of Life," *Fortnightly Review, 5* (1869), 129-145. For the background to this lecture and its place in the vitalist-mechanist debate, see G. L. Geison, "The Protoplasmic Theory of Life and the Vitalist-Mechanist Debate," *Isis, 60* (1969), 273-292.

[58] Huxley and Martin, *Elementary Biology*, p. vi.

[59] P. Chalmers Mitchell, *Thomas H. Huxley: A Sketch of His Life and Work* (London, 1901), pp. 177-181, quote from Huxley on p. 180. Cf. however, Huxley, *Life and Letters*, I, p. 377, where his son suggests that Huxley did supplement his lectures at the School of Mines with demonstrations, apparently including simple experiments as well as dissections.

attention on the fact that competent teachers of science were not to be found. Huxley was entitled to an influential hearing on the problem. Quite apart from his general scientific eminence, he had long been an examiner in zoology and physiology for the government's Science and Art Department, and he had just been elected to the newly created London School Board. He urged upon the government the need for science teachers to have laboratory training and, to this end, proposed a short summer course in laboratory biology for selected teachers in the Science and Art Department. The government agreed, authorizing £750 with which Huxley was to pay himself and his assistants and to buy material and apparatus.[60]

From the beginning, Huxley planned to engage Foster as the first of his assistants and so kept him informed of the arrangements he was making with the government. "Be it known to you," he wrote Foster on 12 December 1870, "that your namesake with the 'r' in his name [i.e., W. E. Forster, the Liberal M.P. who carried the Education Bill] has sanctioned the Biology teaching in the summer. I shall soon have the details."[61] On New Year's Eve, Foster wrote Huxley that two able young men were eager to join as demonstrators in the summer course —William Rutherford (1839-1899), who had graduated M.D. from Edinburgh in 1863 and was now a professor at King's College London; and E. Ray Lankester (1847-1929), a recent product of Christ Church, Oxford, who had published contributions to marine zoology while still a teenager and who had just been awarded the Radcliffe Travelling Fellowship, which he used to study marine biology under Anton Dohrn in Naples.[62] Huxley recorded his satisfaction with this news and then, on 5 January 1871, wrote Foster that he had "just received from 'My Lord' the official sanction for the six weeks course in the summer. . . . There will be 45 schoolmasters to be taught." Huxley and Foster together decided that Rutherford, Lankester, and Foster himself should share equally the duty of demonstrating, and that each should receive a salary of £120.[63] By the first week in July, the course was underway. In a letter to Anton Dohrn, director of the zoological station at Naples, Huxley described it as "a course of instruction in Biology which I am giving to Schoolmasters—with the view of converting them into scientific missionaries to convert the Christian Heathen of these islands to the true faith."[64]

60 See Bibby, *Huxley*, pp. 139-140, 145.

61 Huxley to Foster, 12 Dec. 1870, Huxley Papers, IV. 25.

62 Foster to Huxley [31 Dec. 1870], ibid., IV, 184. On Rutherford, see *Nature 59* (1899), 590-591; on Lankester, see E. S. Goodrich, "Edwin Ray Lankester," *Proc. Roy. Soc., 106B* (1930), x-xv.

63 Huxley to Foster, 2 and 5 Jan. 1871, Huxley Papers, IV. 27, 29.

64 Huxley to Dohrn, 7 July 1871, ibid., XIII. 202.

The daily work for the course consisted of an hour's lecture by Huxley in the morning, followed by four hours' laboratory work in the afternoon. The demonstrators, who were responsible for the actual laboratory teaching, "did their best," says Foster, "to make each member of the class see for himself or herself, so far as was possible, the actual thing of which the master had spoken."[65] Along with its emphasis on laboratory training, the course promoted the so-called "type system," according to which only a few illustrative types of organisms were chosen for the students to dissect. Despite a long history of criticism, which had already begun before Huxley's death in 1895, this system persists today in many introductory courses in biology. The critics have always considered it unwise to concentrate so much attention on just a few organisms. Such an approach, they have argued, buries the student under a mass of trivial details and prevents him from grasping fundamental biological principles. "For some years," writes Ainsworth Davis, "the extraction of the minute ovaries of the earthworm was the goal of the student of elementary Biology."[66] A. E. Shipley, who graduated from Cambridge in 1884, complained that a student could finish such a course "believing that a whale was a fish and having but little or no knowledge of adaptation, symbiosis, natural selection, marine life. . . ."[67]

Clearly there were and are very real dangers in following Huxley's method of teaching biology. This is especially true if the illustrative types are treated as isolated organisms and if it is forgotten that they belonged in life to a dynamic natural system. Neither Davis nor Shipley accused Huxley himself of making this mistake. In their view, his once valuable approach ossified or became perverted in the hands of unintelligent teachers. Jeffrey Parker, who served for a time as Huxley's demonstrator, made the same point in these words:

> . . . in [Huxley's] lectures, these types were not treated as the isolated things they necessarily appear in a laboratory manual or in an examination syllabus; each, on the contrary, took its proper place as an example of a particular grade of structure, and no student of ordinary intelligence could fail to see that the types were valuable, not for themselves, but simply as marking, so to speak, the chapters of a connected narrative. Moreover, in addition to the types, a good deal of work of a more general character was done. Thus, while we owe to Huxley more than to anyone else the mod-

[65] Michael Foster, "Thomas Henry Huxley," *Proc. Roy. Soc., 58* (1896), xlvi-lxvi, on lx.

[66] J. R. Ainsworth Davis, *Thomas H. Huxley* (London, 1907), pp. 115-116.

[67] See Alan E. Munby, *Laboratories: Their Plannings and Fittings*, with an introduction by Arthur E. Shipley (London, 1921), p. xiv.

> ern system of teaching biology, he is by no means responsible for the somewhat arid and mechanical aspect it has assumed in certain quarters.[68]

Foster, too, insisted that Huxley's course was "a model of instruction in the general principles of biology." From the admirable primer Huxley later wrote on the crayfish—one of his illustrative organisms—it is clear that he, at least, did not forget that his "types" had once been part of a living, dynamic system.[69]

Whatever the defects of the type method of teaching biology, it was based on this unimpeachably logical principle: if the beginning student is to acquire any real experience in dissecting and observing for himself, then some drastic means of selection must be employed. As Huxley himself observed, in defense and explanation of his procedure, there are more than a quarter of a million different species of plants and animals.[70] Only a very few can be chosen for laboratory study and training. This carried with it a corollary: to reduce the emphasis on selected "types" is to reduce the emphasis on laboratory training. Foster clearly agreed with Huxley, adopting the type system in his course at Cambridge, but other biologists insisted that field trips or museum work offered greater benefits to the beginning student.[71]

Like laboratory training itself, the type system had a place in biological teaching even before Huxley built his South Kensington course around it. In England, at least, the teaching of practical zoology on the type method is usually traced to George Rolleston, Linacre Professor of Anatomy and Physiology at Oxford from 1860 until his death in 1881.[72] E. Ray Lankester, who studied and worked at Oxford during much of that period, gave the following account of Rolleston's course:

> Rolleston was the first to systematically conduct the study of Zoology and Comparative Anatomy in this country by making use of a carefully selected series of animals. His "types" were the Rat, the Common Pigeon, the Frog, the Perch, the Crayfish, Blackbeetle, Anodon, Snail, Earthworm, Leech, Tapeworm. He had a series of dissections of these mounted, also loose dissections and elaborate MS. descriptions. The student went through this series, dissecting fresh speci-

[68] See Huxley, *Life and Letters*, I, pp. 379-380.

[69] T. H. Huxley, *The Crayfish: An Introduction to the Study of Zoology* (London, 1880).

[70] See T. H. Huxley, "On the Study of Biology [1876]," in *Science and Education: Essays* (New York, 1899), pp. 262-293, esp. 271, 276.

[71] See, e.g., Shipley's introd. to Munby, *Laboratories*, p. xiv.

[72] G. L. Geison, "George Rolleston," *DSB*, XI (1975), 513-515.

> mens for himself. After some ten years' experience Rolleston printed his MS. directions and notes as a book, called *Forms of Animal Life* [1870].[73]

And though Rolleston almost certainly developed his course under the inspiration and guidance of Huxley, who had helped him to secure his chair in the first place,[74] his position at Oxford gave him "the earlier opportunity of putting the method into practice."[75]

What, then, were the distinguishing features of the course Huxley established in 1871 at South Kensington? Why did Parker, writing in the 1890's, consider Huxley's course the basis for "the biological laboratory, as it is now understood?"[76] What caused Foster to say that it "brought about a revolution in the teaching of biology in this country?"[77]

Part of the answer has to do with Huxley himself. For almost twenty years he had lectured along the lines later adopted at South Kensington, and no one was better qualified to put the system into practice in the teaching laboratory, especially since he was a gifted teacher and stimulating personality. Perhaps more important, he brought an unusual breadth of vision to the course. By this criterion, his version surpassed Rolleston's in at least two ways. As Huxley himself emphasized in 1875, Rolleston's course (and the resulting manual) was based exclusively on the animal world and on the collections housed in the museum at Oxford. For these reasons, Huxley saw the need for a course and a manual "of wider scope, for the use of learners less happily situated."[78] Besides placing more emphasis than Rolleston on

[73] See Huxley, *Life and Letters*, I, pp. 377-378, unnumbered footnote.

[74] Cyril Bibby, "Thomas Henry Huxley and University Development," *Victorian Studies, 2* (1958), 97-116, on 100. Moreover, Rolleston himself emphasized that his course and resulting manual owed more "to the biological teachings of Professor Huxley . . . than even my numerous references to his works would indicate." George Rolleston, *Forms of Animal Life: Being Outlines of Zoological Classification Based upon Anatomical Investigation and Illustrated by Descriptions of Specimens and Figures* (Oxford, 1870), pp. ix-x.

[75] See Huxley, *Life and Letters*, I, pp. 377-378, unnumbered footnote.

[76] Ibid., I, pp. 377-378.

[77] Foster, "Huxley," p. lx. For similar testimonials to the tremendous impact of Huxley's course at South Kensington, see Huxley, *Life and Letters*, I, pp. 376ff; Mitchell, *Huxley*, pp. 177-181; Shipley in Munby, *Laboratories*, pp. xiii-xiv; Bibby, *Huxley*, pp. 110-111; Davis, *Huxley*, pp. 115-116; F. O. Bower, *Sixty Years of Botany*, pp. v-vi, 30-31; Thiselton-Dyer, "Foster," p. 439; "The Centenary of Huxley," *Nature, 115* (1925), 697-752, passim, esp. pp. 709-715; and J. Reynolds Green, *A History of Botany in the United Kingdom from the Earliest Times to the End of the Nineteenth Century* (London, 1914), pp. 525-539.

[78] Huxley and Martin, *Elementary Biology*, p. vi. The more limited scope of Rolleston's manual is apparent from its full title (n. 74) and from its goal of

microscopy, Huxley chose his types so as to include plants. Thus, like the course Foster would soon establish at Cambridge, the South Kensington course was "on *biology*, on plants as well as animals, to illustrate all the fundamental features of living things."[79]

The talent and enthusiasm of Huxley's demonstrators was even more crucial to the success of his course, both immediately, and especially in terms of its wider and more permanent influence. In the summer of 1872, when Huxley insisted on repeating the course despite a worrisome illness, Foster, Rutherford, and Lankester returned to South Kensington to act as demonstrators once again. This time Foster brought with him additional assistance from Cambridge in the form of his own demonstrator, H. Newell Martin. One year later, Huxley seemed ready to drop the course. "I have made up my mind," he wrote Foster in April 1873:

> not to give any schoolmasters course this year. They can very well do without it, and it would be more prudent on my part to be rid of the work. I am much better but still horribly hypochondriacal and I mean to take a long holiday in the summer and autumn.[80]

In the event, Huxley took his holiday but the course continued under the general supervision of William T. Thiselton-Dyer (1843-1928), a product of Christ Church, Oxford, who had just become professor of botany at the Royal Horticultural Society at South Kensington. Foster apparently did not participate directly in the South Kensington course during this or succeeding summers, but Cambridge continued to be well represented by H. Newell Martin and, during the summers of 1875 and 1876, by Sydney Howard Vines of Christ's College, who then developed laboratory courses in botany at Cambridge.[81]

Through the influence of these and other assistants and demonstrators, Huxley's methods of teaching biology quickly spread throughout England. Huxley himself established a course in elementary laboratory biology as a regular part of the curriculum at the School of Mines in the autumn of 1872, after the biological department of the school had been permanently transferred from Jermyn Street to new buildings in South Kensington. In these new buildings provision had been made for a biological laboratory, so Huxley was at last able to teach his practical course to full-time students.

---

meeting "certain requirements which . . . are felt by students of Comparative Anatomy" (p. v).

[79] See Huxley, *Life and Letters*, I, pp. 377-378, unnumbered footnote.

[80] Huxley to Foster, 25 April 1873, Huxley Papers, IV. 51.

[81] See the sources cited in n. 77 above, esp. Green; and G. L. Geison, "William Turner Thiselton-Dyer," *DSB*, XIII (1976), 341-344.

Curiously enough, the impact of this more extensive course was relatively slight compared to that of the summer courses. The reasons, however, are not far to seek. For one thing, during the regular academic year, Huxley had to do without the services of Foster and the other demonstrators who did so much to ensure the success of the summer courses. Their absence was felt with special severity since Huxley was by this time so busy with other activities that he had little time to devote to actual laboratory instruction. More importantly, the School of Mines was an institution without a clear sense of purpose, and few of its students were destined for careers in research or teaching at the university level. According to Cyril Bibby, the School of Mines had long been "pulled this way and that by rival enthusiasts" as it sought to perform "a triple task as technical school of mining, geological museum, and centre for the scientific instruction of the general public."[82]

The purpose of the scientific departments of the School of Mines was only partly clarified in the wake of their transfer to South Kensington. In 1881, at Huxley's suggestion, the institution became the Normal School of Science and Royal School of Mines. By borrowing the word "Normal" from the École Normale in Paris, Huxley expressed his conception of the School of Science as an institution for training schoolteachers.[83] A teacher's college differs in goals from a university, and Henry Fairfield Osborn, among others, has suggested that "Huxley was not always fortunate in the intellectual caliber of the men to whom he lectured."[84] Huxley is so famous as a teacher of science that it comes as something of a surprise to find that extremely few eminent biologists were directly trained by him.[85] "Had Huxley been an University professor," writes Ainsworth Davis, "the results of his teaching would probably have been very different."[86]

If Huxley felt that his talents were going to waste at the School of Mines, he kept his disappointment private. In fact, he stayed there throughout his career, despite at least one offer from a major university (Oxford in 1881),[87] and he seems to have been quite satisfied with his position—partly, no doubt, because it kept him in London, which he once described as "*the* place, the centre of the world."[88] Huxley

[82] Bibby, *Huxley*, p. 109.

[83] Ibid., p. 140.

[84] H. F. Osborn, *Impressions of Great Naturalists* (New York, 1924), p. 79. Cf. Davis, *Huxley*, pp. 116, 248.

[85] Osborn, *Impressions*, p. 79, lists Saville Kent, C. Lloyd Morgan, George B. Howes, T. Jeffrey Parker, and W. Newton Parker as "representative biologists who were trained directly by Huxley."

[86] Davis, *Huxley*, p. 248.

[87] See Huxley, *Life and Letters*, II, p. 30.

[88] Ibid., I, p. 119.

may also have believed that he was playing a more vital role at the School of Mines than he could at a university. Quite apart from his general convictions about the immense importance of good elementary and secondary education, he probably believed that the educational reform he so strongly desired could be more easily won at this level. In a letter of 1864 he asked Sir Joseph Hooker to reconsider assisting in the science examinations of the Science and Art Department. "I have always taken a very great interest in the science examinations," wrote Huxley, "looking upon them, as I do, as the most important engine for forcing science into ordinary education. The English nation will not take science from above, so it must get it from below."[89]

Like Huxley, Thiselton-Dyer contributed only indirectly to the extension of the South Kensington methods into the universities. In 1875 he resigned his professorship at the Royal Horticultural Society in order to become assistant director of the Royal Botanic Gardens at Kew. Beginning in 1872, in fact, he had already been serving as private secretary and editorial assistant to the director, Sir Joseph Hooker. He married Sir Joseph's eldest daughter in 1877 and succeeded him as director in 1885. In the new Jodrell Laboratory at Kew Gardens, completed in 1876, Thiselton-Dyer provided for and oversaw the research of several of the rising stars of the new "physiological school" in British botany, including F. O. Bower, W. Gardiner, D. H. Scott, and Marshall Ward. But his heavy editorial and administrative duties at Kew increasingly diverted him from teaching and research. Before long, Thiselton-Dyer's efforts to develop economic botany and agriculture throughout the far-flung British empire eclipsed his achievements as educator; and his active participation in the South Kensington course was effectively confined to the summers of 1873 through 1876, though he did offer a final series of botanical lectures when the course had its swan's song in the summer of 1880.[90]

Thus, despite their manifest importance in the rise of laboratory biology in England, neither Huxley nor Thiselton-Dyer took a direct part in its transmission to the universities. That task fell instead to the other summer demonstrators, and Foster led the way. Of the other early demonstrators at South Kensington, William Rutherford remained only briefly at King's College London before returning to

[89] Ibid., I, p. 254. Before going to Cambridge, Foster apparently shared Huxley's view. In his *Quarterly Review* article of 1867, "Science in the Schools," Foster wrote (p. 258) that there was "no need to throw into confusion the time-honoured arrangements of our ancient Universities. It may be left for them to decide whether they will follow science or no; that they should ever lead it can hardly be expected." The lead, Foster suggested, would have to come instead from the schools.

[90] See the sources in n. 77 above, esp. Green; and Geison, "Thiselton-Dyer."

Edinburgh in 1874 as successor to his own mentor, John Hughes Bennett. Although Rutherford introduced courses in practical physiology at King's College and taught large classes at Edinburgh, he apparently never taught a systematic course in elementary biology. In the end, he left a rather modest intellectual legacy at both institutions.[91]

E. Ray Lankester carried the fruits of his South Kensington experience back to Oxford, where he became a Fellow and Tutor at Exeter College in 1872. Soon thereafter he established a laboratory course in zoology that was probably more microscopical and Huxleian in its thrust than the "practical" course Rolleston taught in the same university. Also at Oxford, at about the same time, M. A. Lawson (1840-1896), the Sherardian Professor of Botany, began to develop a similar course on the botanical side after assisting Thiselton-Dyer at South Kensington in the summer of 1873. But these zoological and botanical courses were apparently not combined into a general biology course until 1875, by which time Lankester had been elected professor of zoology at University College, and the combined Oxford course never matched the success of its counterpart on the banks of the Cam.[92]

At Cambridge, Foster's course in elementary biology met an enthusiastic response from its inception in the Easter term of 1873. The second time it was offered, in the Easter term of 1874, it attracted 40-50 students.[93] Almost immediately, it became an important agent in the transformation of Cambridge biology. F. O. Bower, whose youthful enthusiasm for botany was nearly destroyed by Professor Babington, came to Foster's course just in time:

> It was different with the class in elementary biology instituted by Dr. Michael Foster. . . . It was conducted after the method introduced in 1872 at South Kensington by Professor Huxley, and his book was used as the text, while assistance was given in the laboratory by a full corps of willing demonstrators. The daily introductory lecture, and the detailed personal help given in the lab-

[91] On Rutherford's career at King's College, see F. J. C. Hearnshaw, *The Centenary History of King's College London, 1828-1928* (London, 1929), pp. 295-296. In 1899, when Rutherford ended his career at Edinburgh, W. H. Gaskell wrote to his successor (E. A. Schäfer) that "It will be quite a novelty to see some scientific work coming from Edinburgh." See R. D. French, "Some Problems and Sources in the Foundations of Modern Physiology in Great Britain," *History of Science, 10* (1971), 28-55, on 42, n. 30. Perhaps Rutherford's efforts at Edinburgh, like those of other scientists there after mid-century, were hampered by the conditions (notably poverty) to which probing attention is drawn in J. B. Morrell, "The Patronage of Mid-Victorian Science in the University of Edinburgh," *Science Studies, 3* (1973), 353-388.

[92] See Green, *History of Botany*, p. 535. For more on Lawson, see ibid., p. 506.

[93] *CUR, 4* (1875), 480.

> oratory-class that followed it, formed a coherent whole; they were a revelation to me. For the first time in that summer term of 1875 I learned what it meant to be taught science in a rational way. . . . This course counterbalanced the other disappointments: it launched my cockle-shell boat on the tide of biological enquiry.[94]

Whether the criterion be individual student enthusiasm or course enrollments, Foster's innovation enjoyed remarkable success. By 1880, nearly 60 students were taking elementary biology; in 1882, just before Foster turned the course over to Adam Sedgwick and Sydney Vines, more than 80 students enrolled; and in October 1884, Vines and Sedgwick expected more than 120 students in each of the next two terms.[95] Thus, in its first decade, the course experienced a threefold increase in enrollments.

Foster, Martin, Huxley, and the development of elementary laboratory biology

In the creation and success of Foster's course in elementary biology. Henry Newell Martin (1848-1893) must have played a crucial role.[96] Despite his youth, Martin had already been associated with Foster for several years by the time they brought elementary biology to Cambridge. Martin, born in Ireland in 1848, entered University College at the age of sixteen with the intention of preparing himself for a career in medicine. About two years later, Foster gave his first course in histology there. Martin sought him out and addressed him in words to this effect:

> I am very sorry, sir; I should like to take your course if I could, but you see my parents are not very well off and I get my board and lodging with a doctor close by. . . . I have in return for my board, to dispense all the doctor's medicines, and that dispensing always takes me from 2 to 5; now your lectures begin at 4. I cannot come for the first hour. You go on to 6. May I come in for the second hour? I will work hard and will try to make up the lost time.[97]

94 Bower, *Sixty Years of Botany*, p. 13.

95 *CUR, 4* (1875), 480; *10* (1881), 552; *11* (1882), 572; and *14* (1884), 197.

96 On Martin's life and work, see Foster, "Martin"; Henry Sewall, "Henry Newell Martin," *Johns Hopkins Hospital Bulletin, 22* (1911), 327-333; C. S. Breathnach, "Henry Newell Martin (1848-1893): A Pioneer Physiologist," *Med. Hist., 13* (1969), 271-279; and C. E. Rosenberg, "Henry Newell Martin," *DSB*, IX (1974). 142-143.

97 Michael Foster, "Reminiscences of a Physiologist," *Colorado Medical Journal, 6* (1900), 419-429, on 421-422.

Foster was persuaded to admit Martin to his course; and, despite the other demands on his time, Martin did so well that Foster awarded him the course prize and invited him to become his assistant.

When Foster left University College for Cambridge in 1870, Martin went with him. Foster later gave this account of the circumstances:

> After we had been at University College together I think two or three years, Martin carrying on his studies and at the same time helping me, he came one day to me in great trouble because he could not make up his mind. He had obtained . . . a scholarship at Christ College at Cambridge and he could not make up his mind to accept it and go there. He said he didn't want to leave me. But I was able to tell him what nobody else knew at that time, that in the October in which his scholarship would take him to Cambridge, I was going to Cambridge too, having been invited to lecture there.[98]

Actually, Martin arrived at Cambridge slightly before Foster did, for he took up residence there in the summer of 1870 in order to teach a class in histology for G. M. Humphry, professor of human anatomy. When Foster did arrive in October, Martin renewed their arrangement, acting as Foster's demonstrator and "right hand" throughout his time at Cambridge.

Martin's stay was, however, fairly brief. In 1876, at the age of twenty-eight, he migrated to the United States to fill the chair in biology at the newly created Johns Hopkins University in Baltimore. Foster was reluctant to see him go, but Daniel Gilman, first President of the Hopkins, "appeared upon the scene"; and, says Foster, "his influence was so strong that I felt that my own interests were not to be considered, and that I ought to send that favourite across the waters to occupy the first chair of Biology in this new university."[99] Actually, it was really Huxley more than Foster who sent Martin "across the waters." From the day he met Huxley and toured his laboratory at South Kensington, Gilman wanted the new course at the Hopkins to be modeled on Huxley's. This, Gilman believed, was just the sort of introductory course that could serve both medical students and those who intended to make a career in biology. When Gilman asked him who might be willing to come to Baltimore to organize such a course, Huxley recommended Martin.[100]

98 Ibid., p. 422.

99 Michael Foster, "University Education," *Nature, 29* (1898), 283-285, on 283.

100 See the Gilman Papers, Thies MS Room, Milton S. Eisenhower Library, The Johns Hopkins University. The pertinent letters are: Gilman to Huxley, 20 December 1875; Huxley to Gilman, 20 Feb. 1876; and Gilman to Martin, 14 March, 25 April, and 30 May 1876.

Martin carried to America the methods of teaching he had learned from Huxley and Foster. The Hopkins became in turn the leading training ground for the next generation of American biologists and physiologists. Geneticist T. H. Morgan, embryologist Ross Harrison, cytologist E. B. Wilson, and physiologists W. T. Sedgwick, Henry Sewall, and W. H. Howell were among those trained at the Hopkins between 1876 and 1893, when Martin resigned his chair because of failing health and turned it over to his own student, Howell.

But if Baltimore became the chief center of Martin's influence, Cambridge was also affected by his brief presence there, though the tangible traces are few. During his Cambridge career, Martin published only one research paper, a brief histological note on the olfactory membrane (1873).[101] This literary reticence may be attributed in part to his rather enthusiastic pursuit of degrees (between 1870 and 1877, B.Sc., M.B. and D.Sc. at London and B.A. and M.A. at Cambridge). But even more of Martin's time was taken up with teaching, to which he was forced by necessity but to which he also brought enormous dedication. In the detailed elaboration of the laboratory system of teaching elementary biology, he may actually have taken almost as large a share as Foster or even Huxley himself. Foster certainly believed so; in fact, he urged Huxley to let Martin's name stand on the title page of the South Kensington course textbook when it appeared in 1875.

In one letter to Huxley on this question, Foster attempted to assess the relative contributions of Huxley, Rutherford, Martin, and himself to the course and the textbook in elementary biology:

> When we first started, you drew up, if my memory serves me right, notes of the practical work to be done, for the guidance of Rutherford, Lankester and myself. Out of these notes Rutherford and I, I think I may say, elaborated a system of teaching. This was taught to Martin, and by him developed into the form in which it stands in the book.
>
> Looking at the book as a volume of instruction, he has, of course, no share in it—it is all yours. But looking at it as a book of *practical dodges*, there is I think a good deal that is his in it—so much as might be called a share. You told Rutherford and myself what we were to shew; we had very largely to find out by experience the best way of showing what was to be seen—receiving of course much help from yourself. Martin has, all the way through the book, improved

[101] H. Newell Martin, "Note on the Structure of the Olfactory Mucous Membrane," *J. Anat. Physiol.*, *8* (1874), 39-44.

on the system which he learnt from Rutherford. Is not all that his?[102]

Huxley's reply may have been motivated as much by pique as by humor. "The logical outcome" of Foster's letter, he wrote, "would be such a title page as the subjoined:

A Course of Practical Instruction
in Elementary Biology
BY
Dr. Foster F.R.S. and Prof. Rutherford
ASSISTED BY
Dr. Martin
WITH HINTS AND SUGGESTIONS BY
T. H. Huxley
(in smallest [?] obtainable and printed in faint ink)"[103]

In the end, the title page listed Huxley as chief author, "assisted by H. N. Martin." Neither Foster nor Rutherford won a place, though both received credit in the preface, as did Lankester and Thiselton-Dyer. Huxley identified himself as responsible for "the general plan used."[104] What is certain from all this is that Martin knew how to teach elementary biology superbly well, and that Foster must have benefited greatly from his assistance at Cambridge.

Laboratory biology in England: where Foster and Cambridge belong in the web of influence

The rapid development of laboratory teaching transformed English biology in the 1870's and 1880's. That much is clear. But the web of influence uniting London, South Kensington, Kew Gardens, Oxford, and Cambridge makes it difficult to separate Foster's role in the process from such other leaders as Huxley, Thiselton-Dyer, or Lankester. The picture is further clouded by the fact that so many of those who introduced laboratory biology in England, including most of Foster's own students, also studied in Germany or other foreign centers. We may give up in despair when we appreciate how quickly Foster's former students became capable of exerting their own influence at Cambridge or elsewhere.

102 Foster to Huxley, 27 Jan. [1875], Huxley Papers, IV. 194. Cf. Foster to Huxley, 19 and 31 Jan. [1875], ibid., IV, 192, 200.

103 Huxley to Foster, 1 Feb. 1875, ibid., IV. 202.

104 Huxley and Martin, *Elementary Biology*, p. vii.

The careers of the botanists Sydney Vines (1849-1934), F. O. Bower (1855-1948), and Harry Marshall Ward (1854-1906) illustrate some of the difficulties. Vines entered the medical school at Guy's Hospital in 1869, but soon came into Foster's sphere of influence at University College and decided to read for a science degree, specializing in physiology. In 1872, he won an open scholarship at Christ's College, Cambridge, where he went (so he later wrote) "particularly to work at Physiology under Michael Foster."[105] But H. Newell Martin had entered Christ's College two years earlier, and Vines fell immediately under his influence as well. Joining Martin as a demonstrator in the summer course at South Kensington, Vines quickly established a bond with Thiselton-Dyer, whom he assisted in the botanical teaching in the summers of 1875 and 1876. After 1875, when he graduated B.A. from Cambridge, Vines devoted himself exclusively to botany. He was elected Fellow and lecturer at Christ's College in the autumn of 1876. With the support of Foster, who lent him one of the university rooms assigned to physiology,[106] Vines then introduced laboratory courses in botany at Cambridge. His classes achieved a lively following, despite the indifference or outright hostility of the aging Professor Babington, and Cambridge soon became the leading English center for the new morphological and physiological movement in botany.[107]

F. O. Bower entered Trinity College in 1874. His enthusiasm for botany having been rescued by Foster's course in elementary biology, he soon became an ardent disciple of Vines. He twice joined Vines in *Studienreise* to Germany, spending part of 1877 with Julius Sachs at Würzburg and part of 1879 with Anton de Bary at Strassburg. For the next several years, Bower assisted in the laboratory teaching at South Kensington and University College, devoting his free time to research under the watchful eye of Thiselton-Dyer at the Jodrell Laboratory in Kew Gardens. In 1885, at the insistence of Thiselton-Dyer and others, Bower somewhat reluctantly accepted the Regius Chair in Botany at Glasgow, which he held for the next forty years.[108]

H. Marshall Ward was nurtured under virtually identical influences. Attracted to botany at South Kensington, where he and Vines assisted Thiselton-Dyer in the summer of 1875, Ward entered Owens College, Manchester, the same year. In 1876, like Martin and Vines before him,

[105] S. H. Vines, "Some Account of My Relations with the late Sir W. Thiselton-Dyer," Thiselton-Dyer Papers, Royal Botanic Gardens, Kew; in bound volume of letters to Lady Thiselton-Dyer in connection with her projected memoir of her husband, pp. 235-239, quote on 235.

[106] Ibid. Cf. Green, *History of Botany*, p. 538.

[107] Green, *History of Botany*, pp. 535-539, 543-547.

[108] Bower, *Sixty Years of Botany*, passim.

he won an open scholarship at Christ's College, where he studied under Foster and Balfour as well as Vines. Between 1879 and 1885, Ward graduated B.A. from Cambridge, pursued research at the Jodrell Laboratory in Kew Gardens, joined Vines as a Fellow at Christ's College. and served as a demonstrator at South Kensington and at Owens College. In 1885, just as Bower was leaving for Glasgow, Ward accepted the chair in botany at the Royal Indian Engineering College, Cooper's Hill. He returned to Cambridge in 1895 as Babington's successor in the chair of botany, which he held until his early death in 1906. In the meantime, Vines had moved from Cambridge to Oxford, where he was Sherardian Professor of Botany from 1888 to 1919.[109]

Foster obviously exerted some influence on Vines, Ward, and Bower. He inspired Vines, in particular, to undertake a career in science, and later gave him encouragement and material support at a time when they were not to be found elsewhere in Cambridge. But how can Foster's influence be clearly distinguished from that of Thiselton-Dyer or, in the case of Bower and Ward, from that of Vines himself? It is equally difficult to separate Foster's influence from Frank Balfour's in the case of such Cambridge-trained morphologists as A. Milnes Marshall, Adam Sedgwick, D'Arcy Wentworth Thompson, or Arthur Shipley. Almost every student interested in biology arranged his schedule so that he could attend both Foster's and Balfour's courses.[110] Only when we know much more than we now do about the work and careers of these Cambridge biologists can we begin to make confident judgments about the nature of their intellectual debts. In the meantime, it would be well to heed the warning that allegations of intellectual influence must be based on something more substantial than superficial similarity of thought or mere coincidence in time and space.[111]

Carried to its extreme, however, such caution might itself lead to distortion. For though it is certain that Foster did not have a strong direct influence on every student and teacher of the biological sciences in late Victorian Cambridge, it is equally certain that he led a biological revolution there no less profound than the revolution Huxley led in English biology at large. And while both revolutions depended importantly on elusive personal qualities, whose real force can now only be imagined, both also rested on a few fundamental and ascertainable principles. One such principle was that biologists of whatever

[109] Green, *History of Botany*, pp. 548-554.

[110] *CUR, 11* (1882), 509.

[111] Cf. Quentin Skinner, "The Limits of Historical Explanation," *Philosophy, 41* (1966), 199-215; Dennis M. McCullough, "W. K. Brooks' Role in the History of American Biology," *J. Hist. Biol., 2* (1969), 411-438, on 413; and John R. R. Christie, "Essay Review: Influencing People," *Annals of Science, 33* (1976), 311-318.

specialty must recognize the kinship of all living organisms, in their common structural basis in the cell, their common functional basis in protoplasm, and their common history in the evolutionary sense of that term. A second fundamental principle was that the teacher had no more important function than that of inspiring others to undertake a career in scientific research. Finally, Huxley and Foster also agreed that biology must be taught "practically," in the laboratory, so that each student might see and judge for himself whether the accepted views deserved their status.

But if they shared these basic principles, Foster and Huxley were placed in strikingly different circumstances. At the School of Mines, Huxley taught intending schoolteachers. At Cambridge, Foster taught a select group of students in a more effective setting. It is fascinating, if ultimately inconclusive, to speculate whether Foster could have engineered his biological revolution in any other place. Cambridge offered very special advantages to one with Foster's aspirations, and it is at least suggestive that Cambridge probably remained the leading English center of animal morphology even after Balfour's death in 1882, and of the new "physiological" botany even after the departure of Bower, Ward, and Vines in the 1880's.

In both fields, the torch passed to able successors—in animal morphology, as we have seen, to Balfour's own student, Adam Sedgwick; in botany to Francis Darwin, who established successful classes in physiological botany when he returned to Cambridge after his father's death in 1882. Darwin became university lecturer in 1884 and university reader in 1888, when Vines left for Oxford. With the help of Walter Gardiner, Darwin kept the botanical flame alive at Cambridge until Ward returned in 1895 as Babington's successor in the university chair. In 1903, at a cost approximating £20,000, a large new set of botanical buildings went up at Cambridge.[112] Meanwhile, at Oxford, Vines had attracted few students and had achieved no expansion in his meager facilities for laboratory teaching and research. Vines' disappointment with his situation could only have been deepened by the continued success of the school of botany that he himself had done so much to build at Cambridge.[113]

Needless to say, the biological prominence of late Victorian Cambridge did not go unchallenged. From the newer universities springing up in the industrial provinces, and from the rapidly expanding metro-

112 Green, *History of Botany*, p. 554.

113 Ibid., pp. 556-558. For Vines' own admission that "I have a sense I have not been a success here at Oxford," see Vines to Thiselton-Dyer, 23 Feb. 1908, Thiselton-Dyer Papers, Royal Botanic Gardens, Kew.

politan London centers, Cambridge faced increasingly formidable assaults on its hegemony. It is even conceivable, though doubtful, that Cambridge lost its zoological leadership to University College at some point between 1875 and 1891, when Lankester held the chair in zoology there. Almost simultaneously, the development of laboratory botany at University College under F. O. Bower, D. H. Scott, and F. W. Oliver posed another threat to the dominance of Cambridge.[114] By the turn of the century, the competition had become both diffuse and severe, and we lack even the crudest charting of the comparative fortunes of biology in the various English universities.

Between about 1875 and 1900, however, the general biological preeminence of Cambridge seem clear enough. Whether one assigns the preeminence mainly to Foster, or chiefly to the Cambridge setting in which he worked, individual and institutional factors obviously reinforced each other, as they did in the simultaneous rise of the Cambridge School of Physics under the leadership of James Clerk Maxwell, Lord Rayleigh, and J. J. Thomson. During this period of general expansion in experimental science, Foster founded and developed the great school of physiology that remains his most direct and most impressive legacy. Compared to his more diffuse role in the general rise of Cambridge biology, Foster's role in the rise of the Cambridge School of Physiology can be specified with quite striking precision. And when that story is told in full, it should become clear that while he may not have produced so stunning an achievement elsewhere, it is equally unlikely that anyone else could have brought the Cambridge School of Physiology so far so fast.

114 See H. Hale Bellot, *University College London, 1826-1926* (London, 1929), pp. 392-394.

6. The Rise of Physiology in Late Victorian Cambridge: Ways and Means, 1870-1883

In England I am sorry to say physiological science is like other sciences in a bad way & we all turn our science into money. And everything is here so expensive that money must be got somehow. At last our universities (Cambridge at least) has made a move but the professor [*sic*] of Physiology (Foster) knows less Chemistry & Physics than Huxley or Owen so very little progress will be made by him at Cambridge.

Henry Bence Jones to Justus von Liebig (1870?)[1]

The time of Foster's arrival at Cambridge coincided with a remarkable development of scientific feeling in the University, and when the history of the rise of the school of British physiology is written the tale will be largely of what two workers, Michael Foster and John Burdon Sanderson, have done. . . . The debt that Cambridge owes to Michael Foster may be indicated by saying that his name will live not so much in the printed accounts of his own investigations as in the work of those whom he trained and inspired. The magnificent school that he watched and cared for from infancy to adult life is his Cambridge record, and all Foster's work in the organisation and administration of great movements initiated for the advancement of knowledge, for the study of disease, and for the internationalisation of all that is good and worthy in science had its origin in his Cambridge labours.

The Lancet (1907)[2]

APART from the striking difference between their institutional settings, Huxley and Foster differed in one other significant respect. Foster performed experiments on living organisms. According to Thiselton-Dyer, Foster tried to persuade Huxley to teach experimental physiology as part of the summer program at South Kensington, but without success.[3] Huxley's opposition probably derived partly from his conception of the function of the course. In his eyes, it had the same

[1] H. Bence Jones to J. von Liebig, undated, deposited under Liebigiana, Henry Bence Jones, letter no. 9, Bayerische Staatsbibliothek, Munich. Since Liebig died in 1873, Jones's letter must have been written about 1870, when Foster had just gone to Cambridge as Trinity praelector rather than as professor of physiology. I am grateful to Frederic L. Holmes for alerting me to the existence of this letter.

[2] "Michael Foster," *Lancet*, 1907 (1), 307.

[3] W. T. Thiselton-Dyer, "Michael Foster–A Recollection," *Cambridge Review*, *28* (1907), 439-440, on 439.

function as the School of Mines itself—to teach science to future schoolteachers. And since experimental biology, more particularly, experimental physiology, was considered too messy for schoolboys, their teachers presumably required no training in it.[4] To insist otherwise, and thus to advocate the use of demonstrations on living animals before young boys, was certain to arouse the antivivisection forces. "Do you want to be abolished altogether by the humanitarians?" Huxley asked Foster in April 1873, after learning that he had recommended an artificial respiration apparatus for use in experiments at South Kensington.[5]

But Huxley's opposition apparently had another source as well. As Thiselton-Dyer tells the story, animal experiments never became part of the South Kensington course because Huxley could not bring himself to perform or watch them. While he "never had the smallest doubts as to the legitimacy of vivisectional methods, he had a constitutional shrinking from them." And thus, "though the necessary equipment was installed, it was never used."[6] Foster had no such qualms. He both conducted vivisection experiments and taught experimental physiology. In fact, he exerted his most immediate and most powerful influence in that subject; his ambitions for biology at Cambridge paled beside his ambitions for his school of physiology.

Seizing the transient moment: Foster, Cambridge, and the transformation of English physiology

When Foster left University College for Cambridge, he tried to take his two favorite students with him. By happy coincidence, H. Newell Martin had just been offered a scholarship at Christ's College, and so joined Foster in his new enterprise. Edward A. Schäfer, on the other hand, declined the invitation, even after Foster had devised an unusual scheme to get him financial aid at Trinity College. Foster told Schäfer of his scheme in a letter probably written in the spring of 1871:

> I don't know whether you are of the same mind as in the autumn—and I have no desire to tempt you—but there are one *or more*

[4] Even Foster admitted that "the study of physiology in schools should be put forward with caution and made to occupy an entirely subordinate position." [Michael Foster], "Science in the Schools," *Quarterly Review, 123* (1867), 244-258, on 252.

[5] Huxley to Foster, 22 April 1873, Huxley Papers, IV. 50.

[6] Thiselton-Dyer, "Foster," p. 439. On Huxley's more general position vis-à-vis vivisection, see Leonard Huxley, *Life and Letters of Thomas Henry Huxley*, 2 vols. (London, 1900), I, pp. 427-441.

> Scholarships at Trinity (£80 per annum). This year they will be given for Physiology in particular, Biology in general. They are advertized I think as open to undergraduates of Cambridge or Oxford only, *but I could I think get a man entered at Trinity for the purposes of competing*—with return of the fees if he failed. *This is entirely private between ourselves.* There are (or will be) huge opportunities at this place in many ways for a man who can work.[7]

As this letter suggests, Foster was alert to every possible opportunity and advantage the Cambridge setting offered. He had no intention of waiting the "thirty or forty years" that an 1874 editorial in *Nature* hoped would suffice to bring Cambridge physiology (or physics!) up to the level of "a second-rate German University."[8]

The sheer speed with which Foster created the Cambridge School of Physiology is almost as remarkable as the result itself. Quite probably, they are related. Faced with local rivals, Foster and Cambridge had to move quickly to take the lead as English physiology emerged from its stagnancy toward a fully competitive and in some ways preeminent place in the physiological world. Between 1870 and 1900, in fact, English physiology underwent a transformation so profound as to arrest, if not to reverse, the once heavy flow of physiologists from the island to the Continent. The precise nature and causes of this transformation have received only the most preliminary analysis,[9] and they remain too obscure to be explored at length here. With little more at hand than scattered biographical notices and "in-house" institutional histories, we lack even the most primitive materials for building a solid comparative institutional history of physiology in late Victorian Britain. Nor do we have any full-bodied study of the conceptual dimensions of the transformation. Nonetheless, the available literature makes it possible to insist even now on a few fundamental generalizations.

In the first place, it seems clear that the rapid institutional expansion of British physiology can be traced in large part to a superficially minor change in the examination statutes of the Royal College of Surgeons. In 1870, the College introduced a requirement that candidates for its qualifying examination must have attended "a practical course of general anatomy and physiology . . . consisting of not less than thirty meetings of the class," during which "the learners them-

[7] Foster to Schäfer, undated letter marked PRIVATE, Sharpey-Schafer Collection, Wellcome Institute of the History of Medicine, London. Emphasis in original.

[8] "Physiology at Cambridge," *Nature, 9* (1874), 297-298.

[9] For a brief but perceptive summary, with extensive references to the pertinent sources, see Richard D. French, "Some Problems and Sources in the Foundations of Modern Physiology in Great Britain," *History of Science, 10* (1971), 28-55.

selves shall individually be engaged in the necessary experiments, manipulations, & c."[10] Although it has been curiously ignored and demands further study, this statutory change may have been the single most important factor in the transformation of late Victorian physiology. When adopted the following year by the degree-granting body known as the University of London, this measure required "practical physiology" of virtually all medical students, for it was through the College of Surgeons or the University of London that the overwhelming majority of would-be medical men sought the right to practice. The resulting creation and expansion of classes in practical physiology produced a rapid growth in the number of studentships, demonstratorships, lectureships, and other subprofessorial positions.[11] No one, it seems, has even begun to appreciate or examine the extent to which aspiring English physiologists quite suddenly found positions available to them after 1870. Here, at least, our usual preoccupation with university professorships obscures an immensely important process taking place at less visible levels.

The new requirement of the College of Surgeons had profound conceptual consequences as well. The traditional anatomical bias of English physiology, which contributed so tellingly to its mid-Victorian stagnancy, arose in large part from the peculiar features of English medical education, not least among them the dominant place of the autonomous London hospital schools in the system.[12] By 1870, however, King's and University Colleges, the leading rivals of the hospital schools, had turned their courses in "practical physiology" over to specialists with experimental training, and their teaching had come regularly to include experimental demonstrations on living animals—experimental physiology in the Continental sense.[13] Together with

10 Cf. *Lancet*, 1870 (2), 578; and F.W. Pavy, "The Teaching of Physiology," ibid., 1871 (1), 67, from the latter of which my quotation is taken.

11 At St. Bartholomew's Hospital in London, for example, no fewer than 25 men found at least temporary employment as demonstrators or lecturers in physiology between 1870 and 1900. See John L. Thornton, "The History of Physiology at St. Bartholomew's Hospital, London," *Annals of Science, 7* (1951), 238-247, esp. 245. When the Fourth International Physiological Congress met at Cambridge in August 1898, Great Britain could claim 96 registered participants. Of the 51 who listed their institutional positions, exactly one-third (17) were lecturers. Also represented were six demonstrators, two assistant lecturers, two assistants of unspecified rank, and one assistant professor. See *J. Physiol., 23* (1899), Supplement, pp. [3]-[5]. The growth in such positions is also evident from the steadily rising percentage of papers published in the *Journal of Physiology* by men and women below the professorial rank.

12 See Chapter 2 above, "Anatomical bias and English medical education."

13 At King's College, practical physiology was taught by William Rutherford, on whom see *Nature, 59* (1899), 590-591; and F. J. C. Hearnshaw, *The Centenary History of King's College London, 1828-1928* (London, 1929), pp. 295-296. Beginning in January 1871, Rutherford published a series of his course lectures in the *Lancet*,

the growing prestige of these institutions, the comparative examination success of students taught under these conditions, especially in the laboratory at University College, ensured their imitation at other English medical schools.[14] During the 1870's and 1880's, in the wake of the new requirement of the College of Surgeons and in the face of the particularly successful example set by University College, physiological laboratories sprang up everywhere in England. Hospital schools that had eschewed laboratory science found that the new conditions gave them a compellingly "pragmatic" reason for including it. As early as 1876, in testimony before the Royal Commission on Vivisection, William Sharpey agreed that "physiological laboratories have

---

complete with drawings of apparatus and descriptions of animal experiments graphic enough to arouse antivivisection protest. For the first ten lectures, see William Rutherford, "Lectures on Experiment Physiology," *Lancet*, 1871 (1), 1-3, 75-78, 183-185, 295-298, 437-439, 563-567, 705-707; ibid., 1871 (2), 665-669, 739-742, and 841-844. The antivivisection reaction can be gauged from ibid., 1871 (1), 656-657, 769, 843.

The course at University College, as we have seen (Chapter 3, "University College") had been organized by Foster himself in the late 1860's, when practical physiology had not yet become required of medical students. When Foster left for Cambridge, his place at University College went to John Scott Burdon Sanderson, on whom see Lady Ghetal Burdon Sanderson, *Sir John Burdon Sanderson: A Memoir*, ed. J. S. and E. S. Haldane (Oxford, 1911). In 1873, under Burdon Sanderson's editorship, the two-volume *Handbook for the Physiological Laboratory* appeared. Announced as a guide "for beginners in physiological work" and containing more than 100 plates vividly illustrating experiments on living animals, the *Handbook* quickly became a leading target of the antivivisection forces. See Richard D. French, *Antivivisection and Medical Science in Victorian Society* (Princeton, 1975), pp. 47-50. For a brief description of the University College course in October 1870, just as Burdon Sanderson took over from Foster and just after the new statutes of the Royal College of Surgeons had been promulgated, see the *Lancet* 1870 (2), 578.

[14] Indeed, scarcely had the College of Surgeons introduced its new requirement when it was predicted that the course at University College "will doubtless be more or less imitated at all our medical schools." *Lancet* 1870 (2), 578. For one striking example of the impact of the new examination requirements on the teaching and practice of physiology in the London hospital schools, see Zachary Cope, *The History of St. Mary's Hospital School: Or a Century of Medical Education* (London, 1954), pp. 45, 80-83. In the spring of 1884, under mounting pressure from St. Mary's students dissatisfied with their results on the licensing examinations, especially in physiology, the medical school authorities honored an earlier recommendation that the lectures in physiology "should be given by an expert in that Department . . . who devotes the whole of his time to the subject." On Foster's recommendation, the appointment went to Augustus Desiré Waller, whose father had gained fame in the 1850's for his studies of nerve degeneration. In May 1887, in the enlarged and newly equipped laboratory at St. Mary's, the younger Waller performed a famous experiment that laid the foundation for clinical electrocardiography. Among those who later came to the St. Mary's laboratory to witness Waller's work was Willem Einthoven of Leiden (1860-1927), whose imaginative use of the string galvanometer in extension of Waller's work won him the Nobel prize.

been established in a great measure . . . under the direction of the examining authorities and bodies of this country, such as the Royal College of Surgeons."[15]

With the creation of independent laboratories and teaching positions in physiology in medical schools throughout England, and with virtually every medical student now exposed to experimental physiology, the old anatomical bias lost much of its institutional support and raison d'être. Whatever the exact meaning and intent of the new statutory requirement of the Royal College of Surgeons,[16] it made "practical physiology" a requirement in the medical curriculum at the precise historical moment when the phrase no longer served as a disguise for courses in microscopical anatomy. To be sure, English physiology long retained a closer (and often highly productive) alliance with histology than its Continental rivals,[17] but the experimental approach had now also found a permanent and important place. In so pragmatic a setting as Victorian England, it should not surprise us that physiology could expand and ultimately secure its independence from anatomy only by exploiting its traditional role in medical education. In the end, just as the earlier stagnancy of English physiology had been at once an institutional and conceptual phenomenon, so too was its late Victorian revival.

For Foster and his program at Cambridge, the 1870 statutes of the Royal College of Surgeons had the advantage of immensely enhancing the importance of his subject almost simultaneously with his appointment as Trinity praelector. But it also meant that the school of physiology he hoped to build there would face competition from every institution of medical education in England, including the London hospital schools, numbering perhaps a dozen in 1870. The rise of physiology in these schools is one of the most remarkable and least explored aspects of the late Victorian transformation. Even now, our

[15] See French, *Antivivisection*, pp. 42-44, n. 15.

[16] Even those sympathetic to the measure found it hard to believe that the College of Surgeons really expected all medical students to conduct full-fledged physiological experiments themselves. See Pavy, "Teaching Physiology," p. 67. In fact, both at University College and at King's, the experimental portions of the courses in practical physiology consisted mainly of professorial demonstrations, only occasionally including students as assistants. Cf. *Lancet*, 1870 (1), 578; and ibid., 1871 (1), 707, where Rutherford of King's says he had decided that full-scale student participation in the experimental work was impracticable and undesirable. More generally, the new requirement created great confusion among medical students and their teachers, and the confusion was only partly allayed by the accumulating experience of those who took their lead from the Calendar of the University of London. See, e.g., *Lancet*, 1878 (2), 174, 205, 233, 318.

[17] See Chapter 11 below, "Toward a 'national style.'"

scanty knowledge of the hospital schools is sufficient to insist that they deserve an important place in the story. They produced or harbored physiologists as distinguished as Augustus D. Waller, Ernest Henry Starling, Charles Scott Sherrington, and Frederick Gowland Hopkins, to name but a few; and by 1897 the imposing new laboratory at Guy's Hospital (where Starling and Hopkins worked) offered the best-equipped facilities for physiological research in London.[18] Perhaps the role of the hospital schools has been so neglected partly through the tendency to forget that they were then autonomous institutions. Unless this basic fact is kept firmly in mind, it is easy to slip into the habit of ascribing their valuable contributions to their leading metropolitan rival, University College.[19]

For all of that, however, there can be little doubt that the late Victorian transformation of English physiology had its deepest and sturdiest roots outside of the London hospital schools. Nearly all of the physiologists in these schools readily took a university position if the opportunity arose. At Guy's Hospital, for example, the stunning new physiological laboratory had scarcely been completed before it lost its leading lights—F. G. Hopkins to Cambridge in 1898 and E. H. Starling to University College in 1899. In fact, few observers of late Victorian physiology, then or since, have had difficulty recognizing that its transformation took place mainly at University College, Cambridge, and (less so) Oxford, and chiefly through the triumvirate of Michael Foster, John Scott Burdon Sanderson (1828-1905), and Edward A. Schäfer (1850-1935). All three were protégés and admirers of William Sharpey at University College (Schäfer enough so to change his name to Sharpey-Schafer after World War I), and all three taught there for part of their careers. When Foster ended his brief but influen-

[18] Almost any obituary or biographical sketch of Starling will confirm that Guy's had the best-equipped London laboratory for physiology in 1897. But many of these same sources will also suggest that Starling and his distinguished brother-in-law, William Maddox Bayliss, conducted most of their important collaborative work in the laboratory of University College. In fact, their joint contributions to the *Journal of Physiology* are invariably described as coming "from the physiological laboratory of Guy's Hospital, London."

[19] Apart from Starling and Bayliss (see n. 18 immediately above), the following English physiologists are among those perhaps most likely to have their contributions identified exclusively with University College rather than partly with the hospital laboratories in which they sometimes worked: A. D. Waller, J. S. Edkins, Leonard Hill, and M. S. Pembrey. When E. A. Schäfer's highly regarded *Textbook of Physiology* appeared at the end of the century, its twenty contributors included five men who held positions at the London hospital schools (Starling, Hopkins, Edkins, Hill, and Pembrey), compared to a single representative from University College (Schäfer himself). See E. A. Schäfer, ed., *Textbook of Physiology*, 2 vols. (Edinburgh, 1898-1900), list of contributors opposite title page.

tial career at University College in 1870, his place was taken by Burdon Sanderson, who became the first Jodrell Professor of Human Physiology in 1874.[20] Schäfer served as his assistant professor, sometimes a bit impatiently, and replaced him in the Jodrell Chair when Burdon Sanderson became first Waynflete Professor of Physiology at Oxford in 1883. In 1899 Schäfer left University College to become professor of physiology at Edinburgh, and E. H. Starling was elected to succeed him in the Jodrell Chair.

Oxford under Burdon Sanderson, but more especially University College under Burdon Sanderson, Schäfer, and Starling, constituted worthy and productive local rivals for Foster and the Cambridge School of Physiology. Nonetheless, Foster and Cambridge led the way. Richard D. French, who has examined the comparative fortunes of physiology in Victorian institutions, shows no hesitation in choosing Cambridge as "the leading school of physiology in the country during the last part of the century."[21] Fielding Garrison, writing in 1915, put no geographical limits on his judgment that "since the early 1880's the achievement of the Cambridge laboratory has not been surpassed."[22] Burdon Sanderson himself failed to conceal his envy of Foster's success at Cambridge, and perhaps even because of it exchanged the Waynflete Chair in Physiology for the Regius Chair in Medicine at Oxford in 1895.[23]

This is not the place to repair our woeful ignorance of the careers of Burdon Sanderson and Schäfer, or to undertake a proper study of physiology in late nineteenth-century University College and Oxford. Until these and other major lacunae are filled, our understanding of Foster's achievement at Cambridge will remain incomplete through the absence of a firm comparative base. Especially because Foster ranked decidedly below Burdon Sanderson or Schäfer as a creative research physiologist, some will be tempted to trivialize his achievement by making it the inevitable outcome of his special institutional advantages. The preceding two chapters may well have contributed to that impression by specifying some of the benefits Foster derived from the Cambridge setting in general and from Trinity College in particular. In what follows, those advantages will grow even more apparent and will be joined by others, including Foster's powerful position in the Royal Society and as editor of the *Journal of Physiology*. One

20 See H. Hale Bellot, *University College London, 1826-1926* (London, 1929), pp. 315-319.

21 French, "Problems and Sources," p. 33.

22 F. H. Garrison, "Michael Foster and the Cambridge School of Physiologists," *Maryland Medical Journal, 58* (1915), 117.

23 Cf. French, "Problems and Sources," p. 43, n. 43.

hopes, however, that the emphasis will gradually begin to shift from a description of the advantages Foster enjoyed to an analysis of the particular ways in which he cultivated, managed, and frequently created them. The effort will suffer from the fact that it is easier (and perhaps more fascinating) to describe the exercise of power than to analyze the often subtle process by which it was achieved in the first place. In any case, had Foster failed to seize his transient opportunity, and had he employed different means, University College or even Oxford might well have swept past Cambridge on the race to equality with Continental physiology.

Foster's ambassador at large: George Murray Humphry, medical reform, and physiology

When Foster began his efforts to develop physiology at Cambridge, he faced what could have been one overwhelming institutional disadvantage—the long-standing moribundity of medical education there.[24] Indeed, unless the medical school underwent some dramatic reformation, it might cease entirely to exist, and all medical education gravitate instead to London. Clinical opportunities had always been greater in the large metropolitan hospitals, and medical graduates of the less expensive and less exclusive University of London now enjoyed all the legal rights and advantages of Oxford or Cambridge graduates. In view of the traditional link between physiology and medical education, made even more intimate by the 1870 statutes of the Royal College of Surgeons, Foster certainly recognized that his ambitions for physiology at Cambridge depended very heavily indeed on the fate of its dormant medical school. His decision to move to Cambridge from the security of University College, which had one of the three largest medical schools in England,[25] thus represented a bold gamble on the future of medical reform at the ancient university.

Foster's gamble paid off stunningly. By the end of his career, the Cambridge Medical School had probably become the largest in England.[26] Foster could scarcely have predicted this outcome, but his move,

[24] See Humphry Davy Rolleston, *The Cambridge Medical School: A Biographical History* (Cambridge, 1932); Walter Langdon-Brown, *Some Chapters in Cambridge Medical History* (Cambridge, 1946); and Arthur Rook, ed., *Cambridge and Its Contribution to Medicine* (London, 1971).

[25] See Michael Foster, *On Medical Education at Cambridge* (Cambridge, 1878, brochure of 36 pages), p. 23.

[26] In 1897, the Regius Professor of Physic, T. Clifford Allbutt, claimed that "the [Cambridge] School of Medicine is the largest in numbers in England." See H. D. Rolleston, *The Right Honourable Sir Thomas Clifford Allbutt K.C.B.: A Memoir* (London, 1929), p. 135. Without disputing Allbutt's claim, Bill Bynum and Arthur Rook have emphasized to me the difficulty of validating it with certainty and of

though bold, had not been reckless. For he knew when he went to Cambridge that the reforming impulse then pulsating through the university at large had reached at least one vital member of the medical faculty, George Murray Humphry (1820-1896), professor of human anatomy.[27]

By 1870 Humphry had been teaching in Cambridge for nearly thirty years, an altogether startling achievement for a man born into modest circumstances only fifty years before. The son of a barrister who distributed stamps for Suffolk, Humphry took the usual middle-class route to the medical profession by studying at one of the London hospital schools (St. Bartholomew's) and by qualifying as a general practitioner through the Royal College of Surgeons (1841) and the Worshipful Society of Apothecaries (1842). He won a gold medal in anatomy and physiology in the M.B. examination at the University of London in 1840, having studied that subject at St. Bartholomew's under the renowned surgeon James Paget. But the first big step along Humphry's unusual path to eminence came in 1842, when he was chosen to replace one of the three surgeons who resigned that year from the staff of Addenbrooke's Hospital in Cambridge. He thus became, at the age of twenty-two, the youngest hospital surgeon in England. Almost immediately, his clinical lectures and surgical instruction earned him a local reputation as an outstanding teacher, and in 1847 he was invited to perform some of the duties of William "Bone" Clark, professor of human and comparative anatomy in the university and father of Foster's Trinity College ally, J. W. Clark. After long service as Clark's deputy, Humphry was elected to the chair in human anatomy in 1866, the same year that he and William Turner founded the *Journal of Anatomy and Physiology*. In the meantime, he had belatedly secured the advantages of an advanced degree from Cambridge. He entered Downing College in 1847, graduating M.B. in 1852 and M.D. in 1859, just short of his fortieth birthday.

---

grasping exactly what it means. Among other things, most Cambridge medical students actually stayed there only during the preclinical phase of their studies, while students who began at one of the London hospital schools usually stayed there for all of their training. So while the Cambridge Medical School very probably had the largest number of total *enrollments*, it is less certain that it would have had the largest number of medical students actually resident and working there.

27 Humphry deserves more thorough study than he has thus far received. Perhaps still the best source is D'Arcy Power, "Sir George Murray Humphry," *DNB*, 22 (Supplement), 11-13. See also Humphry Davy Rolleston, "Sir George Murray Humphry," *Annals of Medical History*, 9 (1927), 1-11; Rolleston, *Cambridge Medical School*, pp. 66-74, 174-180; Charles Newman, *The Evolution of Medical Education in the Nineteenth Century* (London, 1957), pp. 288-290; and *Cambridge and Its Contribution*, ed. Rook, passim.

Humphry and Foster knew and admired each other even before the latter's appointment to the Trinity praelectorship.[28] If Humphry played a role in Foster's appointment, he could only have done so indirectly, for he did not belong to Trinity College. From the outset, however, he was conspicuously active on the newcomer's behalf, and Foster had doubtless seen premonitory signs of this support before accepting the Trinity offer. Although his own publications were anatomical and clinical in character, Humphry was a vigorous advocate of experimental research and laboratory teaching. In October 1870, the very month of Foster's arrival at Cambridge, the *University Reporter* published a brief article in which Humphry deplored the meager laboratory facilities in the university and chastised the colleges, in particular, for this sorry state of affairs. He established his point by describing the impression Cambridge had made upon two unnamed eminent professors, one German and one English, whom he had recently conducted on a tour of the university. According to Humphry, the German professor had said that if he were rich, "I would come to Cambridge and build a laboratory. You do want working laboratories very much. Yet there are plenty of riches here, why are not more laboratories built for your University?" Saying that he could think of no good answer to this question, Humphry implied that the generous provisions for science in the German principalities might have had something to do with the recent victories of the Prussian army in the Franco-Prussian War. He went on to insist that "scarcely another University in Europe" was so ill-provided with laboratory facilities as Cambridge and to warn that little could be done until, as his English visitor put it, "your Colleges are felt to be national instead of private institutions, and the public attention is directed more keenly to them." Humphry's ire had clearly been aroused by the failure of the colleges to contribute to the new Cavendish Laboratory of Physics.[29]

Naturally enough, Humphry's general support for experimental research and laboratory teaching took its strongest and most effective form on behalf of the laboratory subject nearest his own interests—physiology. It was chiefly through physiology, he told the British Medical Association in 1873, that medicine would achieve a rational foundation and thus "be rescued from the opprobrium of empiricism." Admitting that physiology had as yet contributed little of value to actual medical practice, he nonetheless argued that "the solution of the great problems of pathology and therapeutics" could be reached only

[28] By 1869, Foster certainly knew Humphry well enough to pay him a personal visit at Cambridge. See Michael Foster, "Coutts Trotter," *Nature*, *37* (1887), 153-154.

[29] *CUR*, 19 October 1870, p. 26.

through "repeated, thoughtfully planned, and carefully conducted experiments upon living animals."[30] Such sentiments obviously placed Humphry in sharp opposition to the antivivisection movement, and he took a leading and eloquent part in combatting it.[31] Here, then, was a man after Foster's own heart, with energy and strength of character at least his match. No less important, Humphry was a man with the institutional leverage to ensure that his convictions would often prevail, not only at Cambridge, where he was sometime chairman of the Natural Science Board, a member of the executive council of the University Senate and easily the most influential figure in the Medical School during Foster's career there, but also in the medical profession at large. A leading force in the British Medical Association, Humphry represented Cambridge on the General Medical Council from 1869 to 1889 and served on the executive council of the Royal College of Surgeons from 1864 to 1884.

In this latter capacity, Humphry was probably responsible for that immensely consequential decision of the College of Surgeons to require "practical physiology" of all candidates for membership. In the absence of his advocacy, the 1870 decision would become a startling anomaly in the generally conservative and pragmatic pattern of College requirements. Perhaps because few practicing surgeons would have sponsored the new requirement, H. Hale Bellot assumed that its leading architect must have been T. H. Huxley, who had long served as examiner in anatomy and physiology both at the College of Surgeons and at the University of London, where the new requirement was quickly adopted.[32] Without at all disputing Huxley's presumably important role in these efforts, it should be pointed out that Foster had replaced him as examiner for the University of London by the time practical physiology became required of medical students there.[33] Most probably, Huxley and Foster acted together and in concert with Humphry, a practicing surgeon with far greater stature in the medical profession. At the Royal College of Surgeons, in particular, Humphry as a member of the executive council would have carried more weight than Huxley as examiner.

In the early 1870's, when the new requirement took effect, its most immediate and obvious beneficiaries were the physiologists at the large metropolitan medical schools. In that sense, Humphry and Foster acted unselfishly when they campaigned on behalf of obligatory prac-

30 G. M. Humphry, "Address...," *Brit Med. J.*, 1873 (2), 160-163, quote on 161.

31 See, e.g., *Brit. Med. J.*, 1881 (2), 332-334.

32 H. Hale Bellot, *University College London, 1826-1926* (London, 1929), p. 315.

33 See University of London, *Minutes of the Senate*, VII [1867-1870] (London, 1870), p. 56. Cf. also the source in n. 34 immediately below.

tical physiology. Indeed, by pressing so vigorously for the extension and serious execution of the new requirement as examiner for the University of London,[34] Foster might have been accused of advancing the interests of experimental physiology in London at the expense of his own nascent school at Cambridge, where medical students scarcely existed. Such a judgment, however, would have ignored Foster's hopes for the future growth of medical education at Cambridge. Moreover, Humphry had already used his influence at the Royal College of Surgeons to secure at least one important advantage toward that end. For in the same momentous year, 1870, that practical physiology became required of candidates for membership by examination, Humphry persuaded the College that Cambridge students should automatically become members (and thus gain the right to practice surgery) simply by taking a set of specified courses at the university, including physiology.[35] This agreement assured Foster of at least a few students and augured well for the future development of medicine at Cambridge.

Both Foster and Humphry had specific plans as to how that future development might best be managed and secured.[36] Their respective visions derived from very similar convictions and often overlapped, but Humphry's plan seems rather more decisive and sharply focused. Perhaps misled by Humphry's earlier position on the issue, historians of the Cambridge Medical School have managed to describe his plan as precisely the opposite of what it was in fact. That Humphry ardently sought a "complete medical school" at Cambridge has been standard coin ever since H. D. Rolleston introduced it in 1932.[37] In truth, at least from the 1870's on, Humphry argued that Cambridge could *not* become a complete medical school because it did not offer sufficient facilities for adequate clinical teaching.[38] In a manner very

[34] On the examinations in anatomy and physiology at the University of London during the 1870's, see University of London, *Minutes of Committees, 1867-1880* (London, 1881), esp. pp. 184-187.

[35] See Rolleston, *Cambridge Medical School*, p. 25; and D. H. M. Woollam, "The Cambridge School of Physiology, 1850-1900," in *Cambridge and Its Contribution*, ed. Rook, pp. 139-154, on 141-142.

[36] Foster, *Medical Education*. Humphry's views emerge most clearly in the *Cambridge University Reporter*, on which the next two paragraphs are based. Cf. also Appendix I below.

[37] See Rolleston, *Cambridge Medical School*, pp. 27, 69. The claim is repeated in Langdon-Brown, *Some Chapters*, p. 89; Newman, *Medical Education*, p. 290; and in several of the essays collected in *Cambridge and Its Contribution*, ed. Rook.

[38] See *CUR, 4* (1875), 470-471; 7 (1878), 526-528; and *25* (1895), 514-517. Arthur Rook, who has been examining Humphry's career in some detail, believes that he initially favored a clinical school but gradually changed his mind.

similar to Henry Acland at Oxford,[39] Humphry sought to establish the university as a center for the cultivation of the preclinical sciences. Students so trained at Cambridge might then go to the large hospitals in London in order to acquire their clinical training.

Humphry stated his position clearly and succinctly during the important university-wide debates over departmental priorities in the spring of 1875. He insisted that additions to the staff were required "for the proper development of the Medical School."

> In this my aspirations are very moderate. I am strongly of the opinion that Cambridge should not attempt to form a complete Medical School. It should be content to aim at a high standard of teaching in the earlier branches of Medical Study—those which form the subjects of the first and second Examinations for M.B., viz., Chemistry, Physics, Botany, Anatomy and Physiology, and Pharmacology—and to carry on initiatory teaching in Medicine, Surgery and Pathology. Thus the students would all be obliged to resort, in the latter part of their time, to the greater practical opportunities afforded by Metropolitan Hospitals and Schools of Medicine.

On this basis, he specifically recommended the establishment of a professorship in surgery and of demonstratorships in pathology and practical histology.[40] But Humphry's main point throughout was that Cambridge should aim above all "to get men to appreciate scientific principles."[41] This firm belief lay at the bottom of his constant and enthusiastic support for Foster and his goals, though the intellectual bond shared pride of place with a personal admiration so powerful that Humphry once said of Foster, publicly and in his presence, that a "kinder, truer, better, nobler man . . . does not exist in the world, perhaps . . . not a man of my acquaintance to whom in the hour of trial and in the hour of need I would look for help with greater confidence . . . no man who binds others to him by the simple kindness and unselfishness of his nature more than he."[42] To speak of Foster as having "cultivated" this remarkable ally would perhaps do violence to Humphry's sentiments. Outside of Trinity College, certainly, he was Foster's main advocate at Cambridge. Together with Foster's Trinity allies, Humphry did everything in his considerable power to expand the accommodations and appreciation for physiology in the university.

[39] See J. B. Atlay, *Sir Henry Wentworth Acland: A Memoir* (London, 1903).

[40] *CUR, 4* (1875), 470-471.

[41] Ibid., *25* (1895), 515.

[42] See *Brit. Med. J.*, 1881 (2), 333.

Enrollments, bricks, and mortar: the growth of Foster's laboratory and courses, 1870-1883

When Huxley recommended to the Trinity Seniority that a praelectorship in physiology be established, and that Foster be appointed to it, he also emphasized that the new praelector would need a laboratory. The response from the Trinity Seniority was encouraging. They promised Huxley that if Foster were appointed, they would "set to work about establishing this Physiological Laboratory."[43] The "Physiological Laboratory" that Foster found upon his arrival was very modest indeed. Even the three-room arrangement he had left behind at University College would have looked large and well-equipped by comparison.

In the beginning, the university lent him the use of a single room in the new science museums—"or more accurately," wrote J. G. Adami, "half a room, a large chamber being divided by a wooden partition into two, the other half, if I mistake not, being occupied by an ancient and revered gentleman, the Plumian Professor of Astronomy."[44] With an initial grant of £400 from Trinity College, Foster equipped this space as best he could and undertook his expansionary campaign. Perhaps with tongue in cheek, Adami claims the campaign was launched as follows:

> The accommodation was all too small, but Foster did not complain: with a wise humor he suggested to one of his pupils a research into some of the rarer of the dissociation products of proteid metabolism; if I remember aright, the substance was uroerythrin. It was a research which, in order to obtain an adequate amount of material, demanded the boiling down over several days and weeks of many gallons of excrementitious fluid. The Plumian Professor did not merely vacate the premises: he spread about so dire a report of the disadvantages of close propinquity to the physiological laboratory that by a natural process Foster's laboratory accommodation was doubled. . . .[45]

[43] W. G. Clark to T. H. Huxley, 2 April [1870], Huxley Papers, IV. 172.

[44] J. George Adami, "A Great Teacher (Sir Michael Foster) and His Influence," *Publication no. 7, Medical Faculty, Queen's University (Kingston, Ontario, Canada)*, pp. 1-17, on 8. On 23 June 1870, the University Senate "gave leave to Mr. Michael Foster recently elected Praelector in Physiology in Trinity College to give lectures in physiology in one of the rooms in the New Science Building." Rolleston, *Cambridge Medical School*, p. 80.

[45] Adami, "Great Teacher," pp. 8-9.

By 1873, certainly, Foster had somehow secured the use of two rooms. In the first issue of *Studies from the Physiological Laboratory in the University of Cambridge*, published in that year, Foster wrote:

> I am also deeply indebted to the authorities of the University for having permitted me, a simple College Lecturer, to occupy, at some inconvenience I fear to others, the two University rooms in which my lectures are given, the practical teaching of my class conducted and the physiological work carried on. I have presumed on their kindness, and ventured to call these rooms the Physiological Laboratory in the University of Cambridge.[46]

By the time he wrote those words, Foster had good reason to expect that far greater provision would soon be made for physiology. In October 1872, the Museums and Lecture-Rooms Syndicate had reported to the University Senate their opinion that "the accommodation in the physiological department is insufficient and the present temporary arrangements extremely inconvenient . . . steps ought to be taken, with as little delay as possible, to provide this department with a separate and detached series of rooms such as those described by Dr. Michael Foster."[47] To their report, the syndicate attached a long letter from Foster to his Trinity ally, J. W. Clark. Because this letter conveys so well Foster's vision of how physiology ought to be taught, it is quoted below in full.

> My dear Mr. Clark,
>
> The branches of learning in which you and I are engaged, are in some respects unfortunate, inasmuch as they need for the proper teaching of them a large amount of space and extensive arrangements. Let me tell you in a few words what I want, and the reason of my wants.
>
> Physiology, in order to be taught well, must be taught *practically*. The student will gain but little good if he simply listens to lectures, however well illustrated with experiments, or merely reads books. In order to be able to form a sound judgement on physiological questions, he must have been trained to see the things which are to be seen; he must learn, by making his own observations and experiments, how physiological conclusions are arrived at. In no other way can he hope to acquire a just sense of what is right and what is wrong in Physiology.

46 [Michael Foster], ed., *Studies from the Physiological Laboratory in the University of Cambridge*, Part 1 (1873), pp. 6-7.

47 *CUR*, 30 October 1872, p. 36.

The facts of Physiology have to be learnt partly by observations with the microscope, partly by experiments requiring chemical apparatus and operations, and partly by experiments requiring physical apparatus, often of a delicate and complicated kind. Many of even the most fundamental observations are tedious, and require considerable time for properly carrying them out. Very often the arrangements needed for a single experiment take up considerable room, for instance, a space of some five or six square feet.

It is obvious that delicate microscopes, and delicate physical apparatus, cannot be used in the same room at the same time that chemical operations giving rise to moist or acid vapours are being carried out. It is equally obvious that several students, each requiring considerable space, cannot work together in a small room. It is also obvious that observations with the higher powers of the microscope, or exact measurements with apparatus, cannot be successful when attempted in a room in which there is a good deal of walking to and fro, and moving about is going on, and everything is vibrating. And the difficulties are increased, if the same room is used as a lecture-room, and is blocked with chairs or forms.

What I want therefore is, first, a lecture-room, which need not be used as a laboratory; secondly, a room well lighted and free from vibrations, which can be used as a microscope room; thirdly, a room where the chemical operations involved in physiological observations may be conducted; and fourthly, a room where the more delicate physical apparatus may be used with as little dread as possible of rust, or of interruptions by vibrations, or of want of space.

All this seems to me absolutely necessary for carrying on properly the ordinary teaching of Elementary Physiology.

In addition, I also vehemently desire some space in which a few advanced students might prosecute the higher branches of the science undisturbed by the elementary course. It is impossible for any one to work earnestly at a careful observation in a room occupied at the same time by elementary students. Without some additional accommodation of this kind for higher students, it is useless to hope that anything more than the mere a b c of Physiology can be taught.

As far as I can foresee, the plans which you have drawn up will serve my purpose admirably. I have already apparatus, I have students anxious to learn properly; but at present I am very much cramped and hindered by want of room. More room is all I want; and I anxiously trust that your appeal will be successful.

Ever truly yours,
M. Foster[48]

48 Ibid., pp. 36-38.

Foster's wants seem modest enough until one recalls that he was "a simple College Lecturer" who held no university position and who therefore had no apparent claim whatever on university resources. But when the syndicate identified him as "Lecturer in Physiology," rather than by his more limited official title of "Trinity College Praelector in Physiology," they better conveyed the sense and spirit of the function that Foster was already beginning to serve at Cambridge. Having gained in advance the required consent of the Trinity Seniority to admit students from other colleges to his lectures, Foster threw them open to all members of the university from the outset. Similarly, from the first day of the Lent term of 1871 (1 February), his "physiological laboratory" was open daily for practical instruction to any member of the university willing to pay a fee of two guineas for the term.[49]

That Foster deserved university support was simply assumed, without explicit elaboration, in the October report of the Museums and Lecture-Rooms Syndicate. On 16 November 1872 the report was opened to discussion by the university community. Scarcely had the discussion begun when Professor Humphry rose from his seat in the auditorium of the Arts School to speak enthusiastically in support of Foster's proposals. According to the account given in the *University Reporter*:

> Professor Humphry said that he need not enter on the importance of the study of Physiology, if not the highest, certainly one of the highest studies that could occupy the intellect of man. It was the only foundation for a rational system of hygiene and a rational system of medicine. The object now proposed was to cultivate it in a more practical and complete manner than heretofore. He believed there were some twenty students under Dr. Foster, of whom he was himself one. He remarked on the liberality shown by Trinity College in endowing a Praelectorship of Physiology, and throwing open to the whole University the advantages thus secured. The work now being done was of the best and most thorough description. If the sum of money asked for seemed large, it must be remembered that the University had allowed very heavy arrears to accumulate, for a few years ago there was scarcely one instrument in its possession for practical experimental work. And in relation to the importance of the work, the sum was really small. At Leipsic, Berlin . . . [at] almost every University in Europe, they were spending much more money on this object.[50]

49 Ibid., 16 November 1870, p. 89; and 1 February 1871, p. 164.
50 Ibid., 20 November 1872, pp. 73-74.

Humphry's oration ended the discussion, and when the report came before the University Senate later that month, it carried by a majority of fifty.[51] One year later, in November 1873, Foster found a public opportunity to acknowledge his indebtedness to Humphry. He dedicated the first issue of *Studies from the Physiological Laboratory* to his fifty-two-year-old "student" and champion. My dear Humphry," he wrote, "Will you accept this little trifle as a token of how much I feel the friendly way in which, ever since I came to Cambridge, you have guided and supported my efforts on behalf of the science which we both teach and whose advancement we have so much at heart?"[52]

By confirming the syndicate's report, the University Senate gave its tentative approval to the plans drawn up by J. W. Clark, the recipient of Foster's long letter. In these plans, provision had been made for greatly expanding the accommodation in comparative anatomy (Clark's subject) as well as in physiology. But since Clark foresaw a large new building "of three storeys," it is easy to understand why Foster found the plans suited to his own immediate needs. The land on which it was proposed to erect the new building belonged to Clark, who sought to expedite his scheme by offering to sell the land to the university "at the price I paid for it, viz., £500."[53] The university took advantage of this offer by purchasing part of the proposed site during the next year, but the major expenditures—for the construction and equipment of the building—still remained, and nothing further was done for several years.[54]

During these same years, however, expansion became increasingly imperative. Besides an increase in the number of students in all departments using the new museums, additions to the staff caused further crowding. In 1873 a demonstrator was appointed in comparative anatomy, and in 1874 the Strickland Curatorship of Birds was established. Also given temporary accommodation in the museums was James Dewar, who had been appointed to the Jacksonian Chair in Experimental Philosophy in the spring of 1875.[55] In the meantime Foster had created a one-term course in elementary practical biology,

[51] See Robert Willis, *The Architectural History of the University of Cambridge*, ed. with large additions and brought up to the present time by J. W. Clark, 4 vols. (Cambridge, 1886), III, p. 184, n. 3.

[52] [Foster] ed., *Studies*, p. 5. Foster's reference to Humphry as a fellow teacher of physiology should not be taken too seriously. Insofar as Humphry taught physiology, it was in the old sense of functional anatomy. Foster's introduction of laboratory physiology represents an entirely new departure.

[53] *CUR*, 30 October 1872, pp. 38-39.

[54] Cf. Willis and Clark, *Architectural History*, III, pp. 184-185.

[55] See ibid.; *CUR*, *3* (1873-1874), 208, 398; *5* (1876), 275; and A. E. Shipley, *"J." A Memoir of John Willis Clark* (London, 1913), pp. 272-274.

which in the Easter term of 1874 attracted 40-50 students. The number of students attending his two-term course in practical physiology had also increased to between 30 and 40.[56]

By April 1875, the need for increased accommodation in physiology was "painfully evident" to the Museums and Lecture-Rooms Syndicate. In their annual report published in that month, they said that Foster's classes already filled the small and poorly-lighted rooms assigned to him.[57] This point was made with greater urgency and in much greater detail when the "Additional Buildings Syndicate" published its second report in June 1875. This committee, created in December 1874, had been charged with the difficult task of ascertaining what accommodations were needed and wanted by all of the departments in the university, and of attempting to arrange these needs and wants in some order of priority. Among its members were Humphry and Coutts Trotter.[58]

In reporting on his own area of responsibility, Humphry announced that "the most valuable and successful new feature in connection with my department is the School of Physiology formed and maintained under the able auspices of Dr. Michael Foster; and the most immediate and important requirement is the provision of suitable rooms for the carrying out of this work."[59] In general, the syndicate reported that the most urgent requirement "in the departments of Natural Science" was the provision of "work-rooms for Comparative Anatomy and Physiology." To meet this requirement, it foresaw an expenditure of £6,000.[60]

To its report the Additional Buildings Syndicate attached all of the letters it had received in response to its inquiries.[61] Foster's reply, dated 6 March 1875, was by far the most expansive. He insisted at the outset that accommodation was required "for the prosecution of original researches" as well as "for ordinary class-teaching." The broad interpretation he gave to physiology is indicated by his division of the discipline into three departments: (1) microscopy, including histology and embryology, (2) physiological chemistry, and (3) experimental physiology.

Under the first heading, Foster emphasized that each student should have a table of his own so that his microscopic preparations could be left undisturbed when he was not at work on them. If the classes in elementary biology and practical physiology were to be taught as they ought to be taught, and at a conservative estimate of the number of students attending these classes, Foster saw a need for one hundred

56 *CUR, 4* (1875), 480.
57 Ibid., p. 341.
58 Ibid., pp. 447-452.
59 Ibid., pp. 469-471.
60 Ibid., p. 451.
61 Ibid., pp. 459-487.

tables. These tables should be arranged "in the form of a continuous bench, placed beneath a series of windows having north or east aspect, so that each student may face the [sun]light." Since the minimum width of each table was three feet, the room containing them must be at least three hundred feet long, though it could be as narrow as twelve feet—"in fact a mere corridor." For original research in microscopy, the work tables should be at least four feet wide and should be placed in a number of small rooms for the sake of quiet concentration.

Physiological chemistry would also require extensive accommodation. Foster suggested that a laboratory thirty to forty feet square, containing twenty-five benches, would suffice for the main classroom. As adjuncts to this room, two small rooms, one for weighing and the other for spectroscopic work, would be necessary. For original research, two or more rooms each fifteen feet square and another small room "for combustions" ought also to be provided and equipped.

Foster set forth very similar recommendations for accommodating experimental physiology. The main teaching room, like that for physiological chemistry, should be thirty to forty feet square. Three rooms would be required as adjuncts: a small galvonometer room, a room for special experimental demonstrations by the staff, and a small storeroom for apparatus. For purposes of original research, two or more small rooms (each 15′ by 20′) would be needed. It was "extremely important" that each experimental investigation be prosecuted in a separate room.

Foster had not quite yet completed his catalog of wants. The laboratory should also provide a small private room for its director, one or more rooms for demonstrators, a workshop, and a storeroom. "Lastly, and of least importance," he wrote, "there should be a lecture-room having tolerably easy access from the various class-rooms of the laboratory." In short, what Foster wanted was a physiological laboratory not unlike the magnificent institute built at Leipzig some five years earlier for Professor Ludwig.[62] Foster admitted that "the demand may appear a large one," but he insisted that this was no

[62] Compare Foster's scheme with H. P. Bowditch's description of the institute at Leipzig in *Nature, 3* (1870), 142-143. Note, in particular, that Foster's scheme (like Ludwig's new institute) involved about a dozen rooms, with separate rooms devoted to histology, physiological chemistry, and experimental physiology. Foster probably visited Ludwig's institute himself during his tour of German laboratories in the summer of 1870 (see Chapter 3 above, p. 78), but apparently on a day when Ludwig was not personally available. In a letter to E. A. Schäfer, presumably written in 1873, Foster said that he had never met Ludwig. See Foster to Schäfer, 21 July [1873], Sharpey-Schafer Collection, Wellcome Institute of the History of Medicine.

more than what "Physiology may fairly expect of the University." Nor did he shrink from advising the syndicate on matters not directly connected with his own subject. In particular, he recommended the establishment of "a small marine and a somewhat large fresh-water aquarium" and of "a laboratory for Vegetable Histology and Physiology" similar to (and in the same building as) his proposed laboratory for animal physiology.[63]

During the academic year 1875-1876 the science buildings became more crowded than ever, due partly to the introduction (in October) of a class in animal morphology by Frank Balfour, the young Fellow of Trinity College who was Foster's protégé. To accommodate this class, Alfred Newton, professor of zoology and comparative anatomy, gave up his private room in the museums.[64] But the administrative wheels had begun to turn at last. In the same month that Balfour began his new course, the Museums and Lecture-Rooms Syndicate appointed a sub-syndicate to study the feasibility of temporary additions to the museums. Upon finding that temporary classrooms would cost at least £1,000, they rejected the idea in favor of a modified version of the plans Clark had presented in 1872 for a permanent new building of three stories. In a special report published in March 1876, the syndicate described the revised plans, which called for a building of 300,000 cubic feet, and estimated that such a building, constructed "of plain brick, without ornament," would cost £8,750. If "for the purpose of meeting this urgent want" the university were willing to spend virtually all of the money deposited both in the General University Building Fund and in the Museums and Lecture-Rooms Building Fund (a total of about £9,200), then no other funds would be required. On this basis, the syndicate asked for authority to communicate with an architect.[65]

This report met with some opposition. A member of the Faculty of Arts doubted the priority of this want over others and claimed that his faculty was now the most neglected in the university. George Paget, Regius Professor of Physic, said he would be glad that such large sums were now being spent upon the natural sciences except that "he feared he would not live to see all the rooms built which were needed for purely medical purposes."[66] But Clark and Trotter defended the report, for which they were in large part responsible, and an amended report submitted a few days later differed very little indeed from the original.[67] After this amended report was confirmed

[63] *CUR, 4* (1875), 479-482.
[64] Ibid., *5* (1876), 275.
[65] Ibid., *5* (1876), pp. 275-277.
[66] Ibid., pp. 293-294.
[67] Ibid., pp. 365-368.

by majority vote of the University Senate, a bid of £8,500 was accepted from a contractor in June 1876, and construction began immediately.[68]

By October 1877, some of the rooms assigned to physiology were ready for occupancy.[69] In February 1878, as the building neared completion, one section collapsed, bringing construction to a halt while the cause of this accident was considered and measures taken to prevent a recurrence. (The floors, which had been concrete, were all removed and replaced by wood.)[70] Throughout the academic year 1877-1878 and during the autumn of 1878, Foster's students shuffled back and forth between the old and new buildings. The new building was finally ready for complete occupancy by the end of January 1879, and Foster moved his classes into the new rooms at the beginning of the Lent term.[71] The two rooms in which Foster had been teaching his courses were then converted into one, which became an examination room as well as the home of the central scientific library in the university.[72]

In the new building, the entire second story was given over to physiology and part of the third to microscopy,[73] so that during the last four years of Foster's praelectorship (1879-1883), physiology was pursued in surroundings that almost matched his expansive needs and wants. Three months after moving into the building, Foster seemed quite satisfied with his new situation. He wrote:

> We find the rooms—both those which are intended for the practical work of the regular classes receiving ordinary instruction, and those which have been arranged for the studies and investigations of myself, my assistants, and advanced students—to be in every way admirably convenient. I think I may say that the Physiological Laboratory of the University of Cambridge, as I have ventured to call it, though not so extensive as some of those existing in Germany, may be most favourably compared, as far as practical convenience and working arrangements are concerned, with any similar institution either in this country or abroad; and I cannot let pass this occasion of expressing my warmest thanks to the University

[68] Willis and Clark, *Architectural History*, III, p. 185.

[69] *CUR*, *7* (1878), 421.

[70] For a detailed account of the accident and its sequelae, see ibid., *8* (1878), 61-80. Cf. also ibid., *7* (1878), 411; and Willis and Clark, *Architectural History*, III, pp. 185-186.

[71] *CUR*, *8* (1879), 677.

[72] Cf. ibid., *7* (1878), 515-516; *10* (1881), 549, 563-569, 656, 677; and Shipley, "*J.*", p. 278.

[73] See, e.g., *CUR*, *5* (1876), 367. Cf. Foster's initial request, ibid., *4* (1875), 481.

for having thus so largely assisted in the development in Cambridge of physiological teaching and research.

He also reported that accounts of several investigations carried on in the rooms for advanced studies were already in press.[74]

Even now, though, Foster saw reason for concern. His class in elementary physiology had continued to grow. During the Michaelmas and Lent terms just ended, the number of students attending had been 83 and 66 respectively. During the same terms a year before (1877-1878), the corresponding numbers had been 48 and 36.[75] It was the accommodation for practical histology that caused the most immediate concern. Because room had been made for only 36 students at one time (as compared to the 100 tables Foster had requested), the classes in microscopy had been conducted only at considerable inconvenience and only by virtue of careful management. "Should the classes continue to increase," Foster wrote, "it will become impossible to carry on the practical work with any satisfaction in so limited a space."[76]

One year later, as the academic year 1879-1880 neared its end, Foster recorded again his general satisfaction with the new accommodations and said that he did not have "anything special to report upon."[77] Apparently he had still been able to accommodate his classes in the histology room. This held generally true also for the academic year 1880-1881, though it did become necessary in the Lent term to erect an additional makeshift bench in the back of the room. The additional students thereby accommodated worked under the handicap of inadequate light, but the new bench filled the need "on the whole very fairly."[78]

By the Michaelmas term of 1881-1882, however, the situation had reached a point of crisis. The back bench, which had made it possible to add somewhere between 10 and 16 additional spaces to the original 36, no longer sufficed. The overflow from the histology room had then spilled into the room intended for experimental demonstrations, to the great detriment of both. Meanwhile, Frank Balfour's class in animal morphology had also outgrown the room allotted to it in the new building. Provision had been made in this room for 30 students to work at one time, but 55 students were now attending the elementary class in practical morphology, and about 20 the advanced course. Those finding no accommodation in the designated room had to be

[74] Ibid., *8* (1879), 677-678.
[75] Ibid., *7* (1878), 421. [76] Ibid., *8* (1879), 678. [77] Ibid., *9* (1880), 559.
[78] Ibid., *10* (1881) 552.

placed in Professor Humphry's room and elsewhere. In the autumn of 1881, therefore, Foster and Balfour sent a letter to the Museums and Lecture-Rooms Syndicate asking that the accommodation for practical morphology and histology be doubled.[79]

In their letter, Foster and Balfour admitted that they would be able to solve the problem "by dividing the classes into sections, and repeating all the practical work with each section." But this would require a large increase in the number of demonstrators and would disturb their students' schedules, nearly all of which had been arranged so that the students could attend the classes of both men. More importantly, such a plan would greatly upset "the essential character of the teaching hitherto adopted, viz., that the student immediately after the lecture goes through the practical work belonging to the lecture." They were exceedingly reluctant to give up the benefits of this method of teaching.[80]

In May 1882, the syndicate published this letter from Foster and Balfour as part of a report in which it advocated the addition of a fourth story over the central portion of the new building. This fourth story, containing two private rooms and a classroom sixty feet long, was to be given over to Balfour, whose present classroom would then be used to meet Foster's needs. The syndicate also recommended the construction of a new lecture room for biological studies. The cost of all this was estimated at £1,500.[81] The University Senate accepted the plan a few days later, and work was begun on the new morphological laboratory in June. It was ready for use by the autumn term of the next academic year (1882-1883).[82] With the additional space afforded by Balfour's former classroom, Foster was able to accommodate 130 students in the physiological classes during each of the first two terms of that year.[83]

To no small extent, the figures of student enrollment tell their own eloquent story of Foster's success. In the short span of a decade, his classes in physiology had increased from 20 to 130 students. During roughly the same period, his class in elementary biology had grown from 40 to 82 students. He also deserves some share of credit for the equally successful courses in animal morphology and embryology established by Frank Balfour, his student and leading protégé.

Largely because Foster and Balfour enjoyed such demonstrable success in arousing student interest, the university had been persuaded during this period to spend well over £10,000 for new accommodations in the biological sciences. Much more was to be spent for the

[79] Ibid., *11* (1882), 509-510. Cf. ibid., pp. 572-573.
[80] Ibid., *11* (1882), 509.
[81] Ibid., pp. 510-511.
[82] Ibid., p. 552; *12* (1883), 699-700.
[83] Ibid., *12* (1883), 699.

same purpose during the next twenty years, as the Cambridge Schools of Physiology and Biology continued to grow and to prosper under the influence of Foster himself and of teachers who quite naturally followed the successful pattern he had laid out, especially since nearly all of them had once been his students. This pattern had its origin in the single room the university had allowed Foster to use upon his arrival; and "from this room," wrote Langley in 1907, "there arose by successive extensions nearly all the laboratories now devoted in Cambridge to Biological studies."[84]

Table 1

Number of students taking courses in physiology and in elementary biology at Cambridge, 1872-1883

	Michaelmas term	*Lent term*	*Easter term*	
Acadamic year	*Physiology*	*Physiology*	*Elementary Biology*	*Advanced Physiology*
1872-1873	20	*	*	*
1873-1874	*	*	40-50	*
1874-1875	30-40	30-40	*	*
1875-1876	*	*	*	*
1876-1877	*	*	*	*
1877-1878	48	36	*	*
1878-1879	83	66	*	*
1879-1880	*	*	58[a]	10
1880-1881	Elem. 75, Adv. 18 } 93	Elem. 86, Adv. 17 } 103	77	15
1881-1882	Elem. 86, Adv. 20 } 106[b]	Elem. 84, Adv. 16 } 100	82	16
1882-1883	Elem. 107, Adv. 26 } 133	Elem. 104, Adv. 26 } 130	*	*

* No data available

Data from *Cambridge University Reporter*, 20 Nov. 1872, p. 74; ibid., *4* (1875), 480; *7* (1878), 421; *8* (1879), 677-678; *10* (1881), 552; *11* (1882), 572; *12* (1883), 699.

[a] Beginning with this Easter term of 1880, Foster sometimes recorded separately the number of men and women students. I have added the separate figures together to arrive at the total number of students. Nearly all of the women students attended the elementary classes only and the average number attending any one of these classes was about 18.

[b] Cf., however, *CUR, 11* (1882), 509-510, where Foster indicates that the number of students attending the elementary class in the Michaelmas term of 1881 was only 70, and in the advanced class only 15. The discrepancy can be explained on the assumption that the latter figures exclude women students.

[84] J. N. Langley, "Sir Michael Foster—In Memoriam," *Journal of Physiology, 35* (1907), 233-246, quote on 237.

Students and fellowships: manpower and resources in the Cambridge School

Although much larger in scale than the general rise in university admissions, the rapid expansion of biomedical enrollments in late Victorian Cambridge could scarcely have occurred in its absence. That is especially true because the students admitted increasingly included members of the middle classes and religious dissenters,[85] traditional aspirants to a medical or scientific career. Paradoxically, one of the leading attractions of Cambridge for the ablest and most talented of these new students was that old bête noire of university reformers, the college scholarships and fellowships. The regulations under which these awards were now bestowed opened them up to general competition. The scholarships often enabled students to come to Cambridge who otherwise could not have afforded it, while the fellowships allowed some of them to remain there after graduation. For the building of a research school, in particular, the college fellowships were absolutely crucial, as W. H. Gaskell stressed in his obituary notice of Foster.[86] Those reformers who had once sought to abolish college fellowships, on the undeniable grounds that they were so often abused from the point of view of scholarship or learning, would have been chastened to discover how vitally the redesigned fellowships of the late nineteenth century contributed to the rise of scientific research at Cambridge.

Even under the old rules, which allowed fellowships to be held duty-free for life (so long as the Fellow did not marry or secure a ma-

[85] The precise extent to which the sons of the middle class and religious dissenters began to enter late Victorian Cambridge remains unknown, but the phenomenon is reasonably clear in qualitative terms. For data on the presumably similar social composition of late Victorian Oxford, see Lawrence Stone, "The Size and Composition of the Oxford Student Body, 1580-1909," in *The University in Society. Volume I. Oxford and Cambridge from the 14th to the Early 19th Century*, ed. Stone (Princeton, 1974), pp. 3-110, esp. 60-64, 73. For impressionistic contemporary evidence that Cambridge, too, was attracting more middle-class students, see G. M. Humphry, "President's Address," *Brit. Med. J.*, 1880 (2), 241-244, on 243. Neither for Oxford nor Cambridge do we have any quantitative data on dissenter admissions, but the Parliamentary acts by which dissenters became eligible for the B.A. (1854 at Oxford, 1856 at Cambridge) and for college fellowships and offices (1871 at both universities) obviously made the two ancient universities increasingly attractive to them.

[86] W. H. Gaskell, "Sir Michael Foster," *Proc. Roy. Soc., B80* (1908), lxxi-lxxxi, on lxxiv. College fellowships probably played an equally vital role in the rise of the Cambridge School of Physics. Cf. Romualdas Sviedrys, "The Rise of Physics Laboratories in Britain," *Historical Studies in the Physical Sciences, 7* (1976), 405-436, on 429, 434.

jor alternate source of income), there had always been a few men who used this financial independence to pursue productive research. Until the late 1860's, no fellowships went to students of the natural sciences per se,[87] but several of those who won fellowships for their performance on the rigorous Mathematical Tripos made important contributions to science—notably that distinguished mid-century group of mathematical physicists and astronomers, G. G. Stokes, J. C. Adams, William Thomson (Lord Kelvin), and James Clerk Maxwell. But the new rules adopted in the wake of the Royal Commissions of 1850 and 1872 transformed the college fellowships into veritable engines of research and learning. By reducing the tenure of prize-fellowships to six years in most cases,[88] the new rules increased their number and produced a continually fresh supply of young men who might hope to sustain an academic career until a permanent position could be found. And by linking regular fellowship in a college with the performance of collegiate or university duties, the new rules kept such Fellows in Cambridge and opened to them the prospect of a lifetime academic career there. At Oxford, too, collegiate scholarships and fellowships were awarded under these new rules, but a much smaller number and percentage went to students of the natural sciences; indeed, this difference is one of the more striking indices of the scientific dominance of late Victorian Cambridge over its ancient rival.[89] At the newer English universities, including University College in London, scholarships and fellowships were rare, and endowments for research were meager at best.[90]

Thus, under new regulations, the old and much-maligned system of college fellowships gave Foster one of his most important institutional advantages over his local competitors. In particular, the research-oriented fellowships of Trinity College brought him a small but vital nucleus of able research students and assistants. Among those who took a direct and leading part in the transformation of Cambridge biology and physiology, Frank Balfour, John Newport Langley, Adam Sedgwick, Walter Morley Fletcher, and of course Foster himself

[87] When Downing College in 1867 elected a Fellow on the strength of his performance on the Natural Sciences Tripos, it was an unprecedented event at Cambridge. See D. A. Winstanley, *Later Victorian Cambridge* (Cambridge, 1947), p. 191.

[88] See, e.g., ibid., pp. 347, 354-355.

[89] Cf., e.g., A. E. Gunther, *Robert T. Gunther: A Pioneer in the History of Science* (Oxford, 1967), pp. 45-53 et passim.

[90] For a bitter statement of this fact, together with an alleged historical explanation for it, see E. Ray Lankester, "Introductory Address. . . ," *Brit. Med. J.*, 1878 (2), 501-507.

held fellowships at Trinity. Walter Holbrook Gaskell and Arthur Sheridan Lea, who ultimately held fellowships at other Cambridge colleges, had earlier won scholarships at Trinity. It was also Trinity that (after 1887) offered the Coutts Trotter Studentship in physics or physiology.[91] In such ways, science in general and physiology in particular became the newest beneficiaries of the traditional wealth of Cambridge and Trinity College.

Nor should it be forgotten that Cambridge continued to attract men with ample resources of their own. We have already alluded to the immense family fortune that eased Frank Balfour's entry into the expensive but unremunerative field of embryology and allowed Foster to draw him immediately into a research career without any sense of guilt.[92] To speak only of those who joined Foster at least briefly in pursuit of the problem of the heartbeat (and whose contributions have yet to be described), we can also identify Albert George Dew-Smith, Francis Darwin, George John Romanes, and probably Walter Holbrook Gaskell as men of considerable wealth.

Yet another valuable and remarkably fertile award, the George Henry Lewes Studentship in Physiology, must often have seemed the special preserve of Cambridge men, though legally it was not. Established in 1879, this studentship provided an annual income of £250 "to advance the study of physiology by supplying students of either sex with the means of pursuing original investigations during the interval between their noviciate and their attaining the status of Professor."[93] The successful applicant was required to devote full time to research during his or her tenure of the award, which could extend from one to three years, or even longer in exceptional cases. Although open to candidates from anywhere, the studentship had its chief origins and associations at Cambridge. Its namesake, George Henry Lewes (1817-1878), man of letters and popular writer on physiological and psychological topics, has often been credited (jointly with George Eliot) with the original suggestion that brought Foster and physiology to Cambridge.[94] No clear evidence seems to exist for that claim, but Lewes did know Foster and spent at least a day or two in the Cambridge physiological laboratory in November 1875, when Foster and his students were focusing on the problem of the heartbeat in frogs

[91] For verification of these fellowships and scholarships, see the appropriate individual entries in *Alumni Cantabrigensis*. On the Trotter Studentship, see Chapter 4, p. 104.

[92] See Chapter 5, pp. 124-125.

[93] See Rolleston, *Cambridge Medical School*, p. 84.

[94] See Chapter 3, pp. 75-76.

and snails. Lewes worked with them on this research, had dinner at Foster's home with Martin, Balfour, and Langley, and left flushed with enthusiasm for the whole enterprise.[95]

By this time, Lewes had been living for twenty years with novelist George Eliot (born Mary Ann or Marian Evans) in a union "which she always regarded as a marriage though without the legal sanction."[96] Her alleged role in Foster's appointment also lacks documentary evidence and seems highly unlikely. In December 1875, as the Royal Commission on Vivisection was taking evidence, she wrote (to a friend shocked by some passages in Foster's primer on elementary physiology) that "Mr. Lewes knows Michael Foster, who is in many respects an admirable man—but men, like societies, have strange patches of barbarism in their 'civilization.' "[97] Perhaps her putative role in the founding of Foster's praelectorship has been confused with her definite role in the founding of the Lewes Studentship. In his recent biography of George Eliot, Gordon Haight offers the following account of the circumstances surrounding its creation:

> On 2 January 1879 Marian [i.e., George Eliot] received a letter from Dr. Michael Foster, volunteering to help her with any physiological points that might arise in her work on Lewes' [posthumous] manuscript. . . . Lewes had held Foster in high esteem. His letter set her ruminating on the establishment of some sort of foundation to give young students the training that Lewes had had to get by his own (often mistaken) efforts; in acknowledging Foster's letter, she expressed her interest. Henry Sidgwick [Foster's Trinity ally] came to London to discuss the plan with her on 9 March, and Dr. Foster a few days later. They drew up the terms of the George Henry Lewes Studentship in Physiology and selected as trustees Francis Balfour, W. T. T. Dyer, Huxley, Pye Smith and Sidgwick. Marian made over to them the sum of £5000.[98]

The trustees were to choose as director of the studentship "a physiologist of established reputation who is in charge of a physiological laboratory in Great Britain."[99] Especially in view of the composition of the original board of trustees, it comes as no surprise that they chose

[95] See Gordon S. Haight, *The George Eliot Letters*, 7 vols. (New Haven, Conn., 1954-1955), VI, p. 181, n. 4.

[96] Leslie Stephen, "Mary Ann or Marian Cross," *DNB*, *13*, 216-222, on 218.

[97] Haight, *Eliot Letters*, VI, p. 221.

[98] Gordon S. Haight, *George Eliot: A Biography* (Oxford, 1968), p. 522. Cf. Haight, *Eliot Letters*, VII, pp. 199, 215.

[99] Rolleston, *Cambridge Medical School*, p. 84.

Foster as first director, and in fact the directorship has remained in the hands of the professor of physiology at Cambridge ever since. Among those who held the Lewes Studentship before World War I were Charles Smart Roy, the first recipient and later professor of pathology at Cambridge, and the Cambridge-trained Nobel laureates C. S. Sherrington (1884), H. H. Dale (1901), and A. V. Hill (1911).[100]

Quite apart from the general attraction of the college fellowships at Cambridge, the lure of prizes and posts specifically for physiology (or closely related fields) eventually assured Foster of a regular supply of able manpower for his expanding research school.[101] At the outset, however, he had to depend mainly on the private resources of his students and the occasional research-oriented fellowships of Trinity College. Especially under these circumstances, judgments of ability had to be made quickly and decisively. Foster's talent for making such judgments was celebrated. Indeed, he had so much confidence in this talent as to presume that any chosen pupil who failed to meet his expectations had somehow changed in character, rather than that his original judgment had been mistaken.[102] Obviously, the most important decisions concerned those whom Foster invited to join him as teachers in the Cambridge biological laboratories. In choosing Frank Balfour to oversee the fortunes of animal morphology, Foster could scarcely have done better. At the time of his tragic death, Balfour had gained international recognition as a giant in his field, and D. H. M. Woollam could insist as late as 1971 that he remained "the greatest physiological embryologist the world has yet seen."[103] Nor is it easy to conceive of worthier choices than those Foster made along more strictly physiological lines. It is to those colleagues that we now turn.

100 Ibid. According to Woollam, "Cambridge School," p. 149, Cambridge retained its hold over the Lewes Studentship as recently as 1971, and I assume that is still the case.

101 By the time Foster retired in 1903, Cambridge offered seven university posts in physiology per se: the professorship itself, three lectureships, and three demonstratorships. See J. R. Tanner, ed., *The Historical Register of the University of Cambridge* (Cambridge, 1917), pp. 105, 119, 140-141. In pathology, there were a professorship, a demonstratorship, and the John Lucas Walker studentship (ibid., pp. 106, 142, 277-278). In animal morphology, there were a readership, two lectureships, a demonstratorship, and the Francis Maitland Balfour Studentship (ibid., pp. 114, 121, 141, 275-276). In other related fields, there were a readership in chemical physiology, a lectureship in organic chemistry, and a lectureship in physiological and experimental psychology (ibid., pp. 115, 123-124).

102 Cf. Gaskell, "Foster," pp. lxxviii-lxxix.

103 Woollam, "Cambridge School," p. 151. More generally on Balfour, see Chapter 5, "Foster and Balfour."

Dramatis personnae: teaching personnel in the early Cambridge School

As we have seen, H. Newell Martin moved from University College to Cambridge simultaneously with Foster in 1870, when he won an open scholarship at Christ's College. Although only twenty-two years old at the time of his transfer, Martin took a leading part in developing the course in elementary biology before leaving for the United States to become first professor of biology at the new Johns Hopkins University in 1876. During his last year in residence, he also taught separate courses on invertebrate comparative anatomy and on animal histology.[104]

If Martin conducted physiological research during his six years at Cambridge, there exists no published record of his activity. Nonetheless, it seems proper to include him among the strictly physiological personnel in the Cambridge laboratory. As Foster's "right hand" he doubtless acted as demonstrator in the physiological courses for the first several years. From the direction of his later research at Johns Hopkins, which focused on the isolated mammalian heart,[105] it is obvious that he had mastered physiological technique under Foster. In fact, by appointing Martin, President Gilman of Johns Hopkins may have got a little more of Foster's influence and a little less of Huxley's than he expected. For at Baltimore, Martin developed into a strong representative of the experimental and physiological bias he must have acquired mainly from Foster. His assistant at the Hopkins, William Keith Brooks, represented the morphological bias. Of the two, Martin exerted the more powerful influence.[106]

When Martin left for Baltimore, the gap was filled by two young physiologists who were to play a far more significant role in the development of the Cambridge School. These two, Walter Holbrook Gaskell (1847-1914) and John Newport Langley (1852-1925), were intimately associated with Cambridge physiology from the beginning to the end of their careers. Without them, or someone very like them, the Cambridge School would soon have lost its early momentum and would never have attained the international reputation that it enjoyed during Foster's last two decades there.

Gaskell was actually a year older than Martin, though his associa-

104 See *CUR, 4* (1874-1875), 36, 205, 322.

105 See C. S. Breathnach, "Henry Newell Martin (1848-1893): A Pioneer Physiologist," *Med. Hist., 13* (1969), 271-279.

106 Dennis M. McCullough, "W. K. Brooks' Role in the History of American Biology," *J. Hist. Biol., 2* (1969), 411-438.

tion with Foster was more recent, beginning soon after Foster's arrival at Cambridge. By then Gaskell had already taken his B.A. degree. Although descended from a prominent Unitarian family, he had come to Anglican Cambridge in October 1865 as a member of Trinity College, where he was elected to a scholarship in his third year. His father, a barrister, presumably inherited a substantial fortune, for he practiced his profession only briefly before retiring to private life. After graduating as Twenty-Sixth Wrangler in the Mathematical Tripos for 1869, Gaskell began to prepare for a medical career. His medical studies soon led him to Foster's course in physiology. Foster fired his enthusiasm and changed the course of his life. According to Langley, the conversion was a sudden one:

> In 1872 [Gaskell] went to University College Hospital, London, for clinical work. On his return to Cambridge, Foster, in the course of a conversation with him, suggested he should drop his medical career for the time and try his hand at research in physiology. Gaskell, I believe, adopted on the spot this suggestion, and instead of proceeding to the M.B. degree went to Leipzig to work under Ludwig.[107]

Gaskell spent a year at Leipzig investigating the vasomotor nerves and then, in 1875, returned to Cambridge, where he pursued the problem further. During the 1880's he turned to the heart and then to the involuntary nervous system, and in both areas published work of great significance. He became a Fellow of the Royal Society in 1882, its Royal medallist in 1889 and Baly medallist of the Royal College of Physicians in 1895. His last twenty-five years were spent in pursuit of the morphological and evolutionary problem of the origin of the vertebrates. His research, especially that on vasomotor action and the heartbeat, is discussed in detail in Chapters 8-10 below.

Gaskell probably began to teach and to assist Foster in the physiological laboratory as soon as he returned to Cambridge from Leipzig in 1875. His official status is, however, difficult to divine for several years. Not until 1880 was he recognized individually in the printed announcements of the physiology department.[108] From 1883 until his death in 1914, Gaskell was university lecturer in physiology, initially at the paltry annual salary of £50, which was raised to £150 in 1889.[109]

[107] J. N. Langley, "Walter Holbrook Gaskell," *Proc. Roy. Soc., 88B* (1915), xxvii-xxxvi, on xxvii.

[108] *CUR, 10* (1880), 28, where it is announced that "Dr. Gaskell will lecture on the Circulation on Fridays."

[109] Ibid., *18* (1889), 918.

In the latter year he was also elected a Fellow of Trinity Hall. At least until then, however, Gaskell must have depended for his income largely on private sources. Although he took an M.D. degree from Cambridge in 1878, he never practiced medicine.[110]

Langley, the son of a schoolmaster, prepared for Cambridge at Exeter Grammar School, where his uncle was headmaster. His financial resources must have been modest, for he held a sizarship when he matriculated at St. John's College in the Michaelmas term of 1871. He later won a scholarship at St. John's. During his first five terms he read mathematics and history in anticipation of a career in the civil service, at home or in India. But perhaps at the suggestion of A. Milnes Marshall, his close friend at St. John's and Frank Balfour's future associate in animal morphology, Langley took Foster's course in elementary biology during the first term it was given, in the spring of 1873. He immediately abandoned his former plans and began to read for the Natural Sciences Tripos, in which he took a first class in 1874. By the time he graduated B.A. in 1875, Langley was already deeply involved in original research. Quite probably, his election to a fellowship at Trinity College in 1877 depended (like Frank Balfour's before him) on the new regulations stressing research talent and promise. From 1875 to 1890 Langley focused on glandular secretion, successfully challenging the influential theories of Rudolf Heidenhain. Like Gaskell and A. S. Lea before him, he also studied in Germany, in the laboratory of Wilhelm Kühne at Heidelberg, where in 1877 he investigated salivary secretion in the cat.[111] From 1890 until his death in 1925, Langley worked mainly upon the involuntary nervous system, which he mapped out in exquisite detail. This work receives attention in Chapter 10 below.

Officially at least, Langley held a more visible position in the Cambridge School than Gaskell. After Martin's departure, he became Foster's chief demonstrator, part of his salary being paid by Trinity College.[112] In 1883, when Foster became professor of physiology, Langley was made university lecturer in physiology. At the same time, he became lecturer in natural science at Trinity College and Fellow of the Royal Society. He became the Society's Royal medallist in

110 Langley, "Gaskell," p. xxxv.

111 See J. N. Langley, "On the Physiology of Salivary Secretion," *J. Physiol., 1* (1878), 96-103, on 96n.: "The experiments leading to this paper were commenced and chiefly carried out in the Physiological Laboratory in Heidelberg. . . ." More generally, see W. M. Fletcher, "John Newport Langley–In Memoriam," *J. Physiol., 61* (1926), 1-27; and G. L. Geison, "John Newport Langley," *DSB*, VIII (1973), 14-19.

112 Trinity College Minute Books (June 1874-Nov. 1876), p. 119.

1892, thus joining Balfour and Gaskell in this honor. In 1903, having already served as Foster's deputy for three years, Langley was elected to succeed him as professor of physiology at Cambridge. Upon his death in 1925, the chair passed to Joseph Barcroft.[113]

Gaskell and Langley were, after Foster, the real backbone of the Cambridge School of Physiology during its crucial early years, but Albert George Dew-Smith (1848-1903) and Arthur Sheridan Lea (1853-1915) also deserve recognition for their contributions. Like Foster, Gaskell, and Langley, they were associated for part of their Cambridge careers with Trinity College.

Dew-Smith was admitted to Trinity as a pensioner in 1869. Actually he was then only "Dew," for he assumed the additional name of Smith on succeeding to some property in July 1870. He too was attracted very quickly to Foster and by 1873 had published a note on an insoluble ferment in *Penicillium* and a brief article proposing a new method for the electrical stimulation of nerves.[114] In the same year he graduated B.A. and went to the zoological station in Naples where, at Foster's suggestion, he conducted electrophysiological experiments on mollusk hearts. Upon Dew-Smith's return to Cambridge, he and Foster collaborated on two major papers dealing with the effects of electrical currents on the hearts of mollusks (1875) and of frogs (1876). These two papers are discussed in detail in the next chapter.

Dew-Smith, who apparently never taught physiology at Cambridge, abandoned his career in scientific research in 1876 and used his inheritance to launch at least two business ventures, the Cambridge Scientific Instrument Company and the Cambridge Engraving Company. Because of the need for well-made scientific instruments, Foster was more pleased than disappointed by this turn of events, especially since he had great admiration for Dew-Smith's business capacities.[115] When Foster founded the *Journal of Physiology* in 1878, Dew-Smith generously provided the necessary funds for this decidedly unprofitable enterprise.

Arthur Sheridan Lea came to Cambridge and to Trinity College in October 1872. Admitted as a pensioner, he won a scholarship in 1875, having taken a first class in the Natural Sciences Tripos in the same

113 K. J. Franklin, *Joseph Barcroft, 1872-1947* (Oxford, 1953); and F. L. Holmes, "Joseph Barcroft," *DSB*, I (1970), 452-455.

114 A. G. Dew-Smith, "Note on the Presence of an Insoluble Sugar-Forming Substance in Penicillium," *Studies from the Physiological Laboratory in the University of Cambridge*, Part I (Cambridge, 1873), pp. 33-35; Dew-Smith "On Double Nerve Stimulation," ibid., pp. 25-32.

115 See Foster to Huxley, 13 July [1876?], Huxley Papers, IV. 208. Cf. Gaskell, "Foster," pp. lxxvi-lxxvii.

year. Also about this time he went to Heidelberg to work in the laboratory of Wilhelm Kühne, thus preceding Langley's visit there by a year or two. In 1876, in a paper published jointly with Kühne, Lea argued, in opposition to the widely accepted view of Rudolf Heidenhain, that the pancreatic cells became less rather than more granular during the act of secretion.[116] Langley later extended this view to a wide range of glands, showing that granules were stored up during rest and released during secretion. Lea, after returning to Cambridge, concentrated on physiological chemistry. For Foster's *Textbook of Physiology* (first ed., 1877), he wrote a large chemical section, which was later expanded and issued as a separate volume under the title *The Chemical Basis of the Animal Body.* By 1890 Lea had published a dozen research papers, all but one of which were notes or brief articles. The longer paper, a comparative study of natural and artificial digestions, shows the continued strong influence of Kühne.[117]

Lea probably assisted in the physiology courses at Cambridge from the time he returned from Heidelberg. His salary, like Langley's, was probably paid for several years by Trinity College.[118] In 1883 he was elevated with Gaskell and Langley to the position of university lecturer in physiology. By 1895, however, Lea's health had broken down and he was obliged to resign his lectureship in 1896, at the early age of forty-three. He died in 1915 after a pathetically long illness that confined him to a wheelchair during the last two decades of his life.[119]

Compared to his choice of Martin, Gaskell, and Langley, Foster's decision to entrust physiological chemistry to Lea seems rather less inspired, though perhaps Lea's health prevented him from achieving his full potential. If Foster's judgment of Lea did somewhat miss the mark, he repaired the error in stunning fashion by replacing him with F. Gowland Hopkins, who was to win the Nobel prize for his work on vitamins and to become the leading intellectual and institutional force in British biochemistry.[120] Even earlier, under Foster, Gaskell, Langley, and Lea, Cambridge had become the leading center of phys-

[116] W. Kühne and A. S. Lea, "Ueber die Absonderung des Pankreas," *Verhandlungen des Naturhistorisch-Medicinischen Vereins zu Heidelberg,* n.s. 1 (1877), 445-450.

[117] A. S. Lea, "A Comparative Study of Natural and Artificial Digestions," *J. Physiol., 11* (1890), 226-263.

[118] In his letter of 28 July 1883 to W. H. Thompson (quoted in Chapter 4, pp. 109-110), Foster thanks Trinity College for its contributions "for several years past" to the salaries of two demonstrators. One was Langley. The other was almost certainly either Lea or Gaskell, and Gaskell's ample private resources lead to the presumption that it was Lea.

[119] J. N. Langley, "Arthur Sheridan Lea," *Proc. Roy. Soc., 89B* (1917), xxv-xxvii.

[120] See esp. Joseph Needham and E. Baldwin, eds., *Hopkins and Biochemistry* (Cambridge, 1949).

iological research in England, if not in the world. That result could scarcely have been achieved in the absence of the regular supply of high-quality students and personnel that the special advantages of Cambridge did so much to secure. But neither could it have been achieved in the absence of a powerful research ethos, and not even the Cambridge setting guaranteed any such outcome. It was in fact by creating, organizing, and continually nurturing this research ethos that Foster exercised his most important influence on the Cambridge School and most clearly distinguished his personal contributions from the institution in which they took place.

"Work, finish, publish": Foster, organs of publication, and the ideology of research

In retrospective accounts of Foster's career, several of his students sought to articulate the way in which he had inspired them to follow the path of scientific research. "His influence," wrote Langley, "was not exerted by way of exhortation." Rather, "the desire to undertake research was imbibed from him in lecture and in familiar talk. It was in the air we breathed."[121] Gaskell made the same point in these words:

> Not only would Foster point out the direction in which advance in any science was to be looked for, but by his earnestness, his lovable charm of persuasion, his entire freedom from any thought of monetary gain, or any kind of selfishness, the conviction was gradually borne in on his pupils that the particular line of research on which each was engaged was the one thing in life worth doing, and that the only place to do it was in Cambridge by Foster's side. As Foster used to say, the true man of science must feel with respect to his own research that "in this way only lies salvation." It was that feeling he had so pre-eminently the power of raising in a man.[122]

No institutional advantage can explain Foster's inspirational power over students, which he exercised in his brief time at University College no less than in his long career at Cambridge. Nor, deprived of his personal presence, do we stand much chance of recapturing that power in words. Only after emphasizing these points as strongly as possible can we turn to the more mundane expressions of his commitment to research and learning. And none but the hopelessly cynical will suppose that the sometimes highly pragmatic character of Foster's efforts on behalf of his students can go more than halfway toward accounting for the bond they shared.

121 Langley, "Foster," p. 238.

122 Gaskell, "Foster," pp. lxxiv-lxxv.

Perhaps Foster's students did not form a clan apart—to live, eat, and breathe research—as Liebig's students at Giessen had done,[123] but the embryonic Cambridge School did have a strong corporate sense and a well-developed spirit of camaraderie. Some of this camaraderie developed from a shared enthusiasm and talent for athletics among its members. Even the rather frail Frank Balfour was considered a fine tennis player and cyclist, while his skill as an alpinist gave support to other evidence that his experienced guide fell first in the accident that killed them both in the French Alps.[124] As an undergraduate, Gaskell rowed, swam, and played cricket and racquets. Langley's remarkable talent as a figure-skater overshadowed his considerable abilities in swimming, track, and crew. Among a somewhat later group of students, the diminutive Sherrington was a superb rugby player, while Walter Morley Fletcher was a distinguished performer in track and field whose athletic interests may even have led him into his important research on muscle metabolism.[125] Under Foster's leadership, these muscular bonds actually found a sort of institutional expression. At University College School, Foster had been captain of the cricket team, and he chose his own game to establish an annual competition between the laboratory staff and the students. This annual match was played on a field that Foster owned, and according to Gaskell was "a great success." Later, as the Cambridge School increased in size, the students no longer took part in the match, which then became a contest between the teaching staff and their research assistants. Always "elected" captain of one side, Foster "played in the match regularly up to 1895."[126]

But it was, of course, their common commitment to laboratory research that chiefly united the members of the Cambridge School. In a university that had never witnessed such a bond, Foster moved in special ways and with incredible speed to establish it. Perhaps no scientist ever made swifter judgments of research talent than Foster did in the 1870's, and perhaps none ever thrust his students more quickly into original research. Often enough, Foster made research scientists of mere undergraduates, sometimes of undergraduates who

123 See J. B. Morrell, "The Chemist Breeders: The Research Schools of Liebig and Thomas Thompson," *Ambix, 19* (1972), 1-46, on 36.

124 See *Balfour Memorial: Undergraduate Meeting at the Union* (Cambridge, 30 October 1882), Cambridge University Rec. 100, Anderson MS Room, Cambridge University Library, p. 9. For full accounts of Balfour's accident, see *The Times,* 31 July and 11 August 1882.

125 A. V. Hill, *Trails and Trials in Physiology* (London, 1965), p. 2. More generally on Fletcher, see G. L. Geison, "Walter Morley Fletcher," *DSB,* v (1972), 36-38.

126 Gaskell, "Foster," p. lxxi.

had scarcely left their schoolboy days behind them. Among those whom Foster drew into research before they had reached the B.A. were H. Newell Martin, Frank Balfour, J. N. Langley, A. G. Dew-Smith, morphologist A. Milnes Marshall, chemist Arthur Liversidge, and botanist Sydney Vines.

To ensure these neophytes a public outlet for their work, Foster in 1873 established and edited a journal called *Studies from the Physiological Laboratory in the University of Cambridge*. Apart from one paper by Foster himself, the inaugural issue contained eight papers by men who had begun their research as Cambridge undergraduates. Foster presented the results with great diffidence, saying that he did not think them terribly important contributions to knowledge. But it was the duty of a university to promote research, and he offered these pages "because I sincerely trust their publication may help towards the establishment of a habit of research among our students of physiology at Cambridge." He also complained, as he would throughout his career, that the emphasis on formal written examinations cramped his efforts to encourage research.[127]

In fact, Foster's *Studies* (of which the third and last issue appeared in 1877) was but the first and least significant of his efforts to assure his students access to channels of publication. From the mid-1870's on, Foster threw himself into a series of organizational efforts destined—the cynic might say calculated—to promote the interests of his own school. A century before J. B. Morrell wrote the following passage, Foster behaved as though he had already read it:

> If a [research] school wanted a more than local reputation it had to publish its work. The relatively easy access to publication opportunities, or best of all control of them, enabled a school to convert private work into public knowledge and fame. Publication was vital to the success of any ambitious research school. Otherwise its reputation remained restricted and its students lacked the spur of seeing their names in print. It was sufficient if the director had access, for instance, to journals published by learned societies of which he was an influential member or to proprietary journals published by colleagues or friends. However the most desirable situation clearly occurred when the director controlled and published his own journal. In that case he himself, his friends, and his better students published without obstacle in a journal which created and consolidated the

127 Foster, ed., *Studies*, Part I (1873), pp. 5-7. By the time the third issue of Foster's *Studies* appeared, an anonymous reviewer in *Nature* clearly believed that it had achieved his aim of establishing a research tradition at Cambridge. See *Nature*, *19* (1878), 145.

status of his school, and publicized its specific field and style of work. In short, an ambitious research school had to take the full measure of Faraday's maxim: "work, finish, publish."[128]

To some extent, Foster promoted the work of the Cambridge School through his famous *Textbook of Physiology*, of which the first edition appeared in 1877. Quickly recognized as the leading textbook in English, it went through six editions before 1900 and was translated into German, Italian, and Russian. Part of its reputation stemmed from the wide and international scope of the sources on which it drew and from its balanced discussion of controversial issues.[129] But Foster did not hesitate to make known his own best judgment on such issues, and the research of his Cambridge students clearly affected those judgments decisively. Even in the first edition, he cited unpublished research by Gaskell to settle a minor issue in vasomotor physiology,[130] and later editions became sufficiently identified with the work of the Cambridge School to attract protests from outside.[131]

But it was mainly by founding the *Journal of Physiology* in 1878 that Foster guaranteed his students direct and regular access to a publication of high repute. Foster created the *Journal*, with financial support from his Trinity benefactor A. G. Dew-Smith, even though his students already enjoyed relatively clear channels to the *Proceedings* and *Transactions* of the Royal Society through T. H. Huxley (biological secretary of the Society from 1871 to 1881, when Foster succeeded him), and to the *Journal of Anatomy and Physiology*, co-founded and co-edited since 1866 by another of Foster's leading friends and allies, Professor G. M. Humphry of Cambridge. For the latter journal, indeed, Foster himself helped to select papers, but he apparently felt the need for a periodical that would formally dissolve the traditional union between anatomy and physiology in England.[132] The *Journal of Physiology* filled that need. Published at Cambridge, edited by Foster himself until 1894 and by Langley through the next

128 Morrell, "Chemist Breeders," pp. 5-6.

129 For the immediate English reaction to Foster's textbook, see E. A. Schäfer, "Foster's 'Text Book of Physiology,'" *Nature, 16* (1877), 79-81; and *Brit. Med. J.*, 1878 (2), 364-365.

130 M. Foster, *Textbook of Physiology* (London, 1877), p. 143, n. 1. Other students of Foster who found a place in the footnotes of the first edition were H. Newell Martin (p. 1, n. 1; p. 299, n. 1). E. A. Schäfer (p. 27, n. 1), and George Romanes (p. 57, n. 1). Moreover, the 55-page appendix, "On the Chemical Basis of the Animal Body," was essentially the work of Foster's student, A. S. Lea.

131 See Chapter 10 below, pp. 301-302.

132 Cf. Gaskell, "Foster," p. lxxvi; Langley, "Foster," p. 239; and Foster to E. A. Schäfer, 25 March 1877, Sharpey-Schafer Collection, Wellcome Institute of the History of Medicine.

generation, the *Journal* was dominated by papers produced in the home laboratory, though never so obviously or so overwhelmingly as to become stigmatized as a mere "house organ" for the Cambridge School. Between 1878 and 1900, the *Journal* published 700 articles by 337 individuals representing more than 60 different institutions, including many foreign laboratories. Nonetheless, 20-25 percent of those articles can be reliably identified as products of the Cambridge laboratory, with Langley himself the most prolific contributor. The nearest institutional competitor, University College, could claim about 15 percent of the total, thanks mainly to the somewhat repetitive and theoretically undistinguished work of clinician-pharmacologist Sydney Ringer, of "Ringer's solution" fame.[133] From 1900 to 1937, incidentally, the Cambridge School continued to produce about 25 percent of the articles published in the *Journal of Physiology*.[134] Few research schools indeed can have enjoyed so consistent a domination for so long a period over a journal of comparable international repute.

Obviously, then, the institutional advantages of the Cambridge School included Foster's *Textbook* and more especially his *Journal of Physiology*. But to place too much emphasis on these and other advantages would be to distort our understanding of his achievement—and to do so precisely in the direction of what is at once the easiest, most common, and least satisfying sort of explanation for his success. Quite apart from the fact that Foster himself created several of his institutional advantages, any merely institutional or quasi-political explanation of his success ignores at its own peril vital issues of quality of mind and character. Above all, it threatens to perpetuate the myth that Foster created a great research school despite his own lack of research talent or experience.

In fact, though the mystery of Foster's achievement can never be entirely dispelled, the single most important step toward that end is to recognize that he actually did engage in original research during the early part of his career. From his great emphasis on the value of research, it would be rather surprising if he had not. After Balfour's tragic death, Foster urged above all else that his memorial take the form of a studentship for the encouragement of original research. To the undergraduates gathered at the Union Society, he emphasized that while Balfour had "loved his teaching and his scholars," he had loved research even more. "His real child" said Foster, "was the struggle

[133] See Appendix II below. On Ringer, see W. F. Bynum, "Sydney Ringer," *DSB*, XI (1975), 462-463.

[134] See Henry Barcroft, "The Cambridge School of Physiology, 1900-1937," in *Cambridge and Its Contribution*, ed. Rook, pp. 193-211, on 193.

with the unknown; the desire to extend the boundaries of knowledge. . . . And indeed without that love of research his teaching would have been bald and narrow."[135] Foster can scarcely have intended that this should apply only to Balfour. He was convinced that some experience in original work was essential to the development of a great teacher.

To emphasize that Foster engaged in research is not to insist that he was a prolific or major contributor to the physiological literature. He was not. Perhaps, as J. G. Adami suggests, Foster produced "no signal piece of research of the first order," but one may doubt Adami's further judgment that he lacked the requisite technical aptitude.[136] Certainly Foster did not accomplish in research all that he wished, but this may be at least partly ascribed to the demands made upon his time by lecturing, writing, and other public duties. More than once, he advised his students not to follow his example. To Edward Schäfer (later Sharpey-Schafer), Foster wrote in 1876:

> Take warning by me, who have been writing and teaching until all the juice has gone out of me and I am worth nothing more. If you have enough pocket money, don't spend a moment longer on teaching than you can help . . . a few years hence you will only have the ideas of your youth to fall back upon. Do *now* as much original work as you possibly can and let everything else go as it likes.[137]

"For Heaven's sake," Foster insisted in another letter to Schäfer, "don't do too much lecturing—it *destroys a man* as I know. I have been driven to lecturing from my youth upward. You are not *obliged* to; don't do it. Give all your energy to research."[138] And despite the great success of his own textbook, Foster warned Adami not to write one: "if it is a failure it is time thrown away and worse than wasted; if it is a success it is a millstone around your neck for the rest of your life."[139]

From his letters to Schäfer, it is clear that Foster's career in laboratory research had effectively ended by 1876. He published his last physiological paper in that year, and only a dozen or so in his career. Most of these were brief and unimportant, including several that were

[135] See *Balfour Memorial*, p. 11.

[136] Adami, "Great Teacher," p. 12. On the issue of Foster's technical competence, see esp. Chapter 9 below, "Gaskell's cardiological research and the Cambridge setting."

[137] Foster to Schäfer, 5 July [1876], Sharpey-Schafer Collection, Wellcome Institute of the History of Medicine.

[138] Foster to Schäfer, undated letter, Sharpey-Schafer Collection, Wellcome Institute of the History of Medicine.

[139] Adami, "Great Teacher," p. 13.

written before he came to Cambridge and which have already been described in Chapter 4 above. But in his last three papers (two of which were published jointly with Dew-Smith), Foster demonstrated his capacity to deal competently with a major issue in late nineteenth century physiology, an issue which had attracted his attention since his first excursion into research in 1859: the problem of the heartbeat.

This work was not only competent, but immensely influential in the early development of physiology at Cambridge. In fact, Foster's work on the heart imbues the embryonic Cambridge School with a unity of purpose not apparent on the surface, and provides a clear insight into just how Foster launched his early physiological students on their careers in research. He not only inspired and encouraged them to undertake research; he also exerted a powerful direct influence on their choice of topics, their methods, and their general approach. He did all this in a very brief period, and the whole process may have been rather unconscious. Even Walter Gaskell, whom Foster clearly influenced most strongly and directly, did not in the end realize the extent of his debt. In his obituary notice of Foster, Gaskell scarcely mentioned that Foster had done research, and said nothing at all about the connection between Foster's work on the heart and his own later research, all of which had its origin in the problem of the heartbeat. Langley, for his part, gave a valuable account of Foster's research, but without recognizing that his own work on glandular secretion had also grown out of the problem of the heartbeat. Like all of Foster's students, Gaskell and Langley remembered only that they were allowed to pursue their own research interests. This is true enough, but when they first came to Foster they had no clearly defined interests. He gave them their first problem.

As a research physiologist, Foster was no Ludwig. But during his brief career in research, he developed a competence in physiological technique and a capacity to judge critically the work of others. Moreover, he brought to the problem of the heartbeat his broadly biological and evolutionary approach toward physiology. All of this was passed on to his students, and particularly to Gaskell, whether they realized it or not. Had Foster never engaged at all in research, it is inconceivable that he could ever have been compared to Ludwig in any way, or that he could have overseen the transformation of biology and physiology in late Victorian Cambridge.

PART THREE

The Problem of the Heartbeat and the Rise of the Cambridge School

7. Foster as Research Physiologist: The Problem of the Heartbeat

Is there not a tendency in all living nature to go on for a while and then to stop, to go on again and then to stop once more? May not the heart in its simplicity unfold to our view what is elsewhere hidden, or but dimly seen in other intermittent vital actions: such as our inbreathing and outbreathing, our sleeping and waking, our working and resting? Is not our whole life one throb, the sequence of our fathers', the forerunners of our children's. . . .

The snail's heart is but a tiny bit of flesh, no bigger than a pea, but in the day when its riddle is read out full and clear, mankind will have a power over their own lives of which at present they have scarcely even dreamt.

Foster (1864)[1]

IN 1859, shortly after his graduation from medical school, Foster presented to the British Association the results of experiments he had performed on the snail's heart.[2] The central concern of these experiments was the cause of the rhythmic contractions of the heart. For centuries physiologists had debated whether the heartbeat had its origins in nervous influences or rather arose from an inherent rhythmicity in cardiac muscle. The issue was as old as Galenic physiology and as new as James Paget's Croonian Lecture before the Royal Society of London in 1857.[3]

The dominant view until the nineteenth century can be traced to Galen. He believed that the muscular heart possessed an innate faculty for rhythmic pulsation, which was in no way dependent on nervous action. In support of this view, he cited the capacity of hearts to continue beating after severe damage to the brain, after section of the spinal cord and vagus nerve, or even after excision of the heart itself from live animals.[4] That excised hearts could continue to beat was

[1] Michael Foster, "On a Snail's Heart," *Monthly Christian Spectator*, n.s. *5* (1864), 268-273, quote on 272-273. This article, signed only "M.F.," is attributed to Foster in ibid., p. 423, unnumbered footnote.

[2] Michael Foster, "On the Beat of the Snail's Heart," *Brit. Assoc. Rep.*, *29* (1859), Transactions, p. 160.

[3] James Paget, "On the Cause of the Rhythmic Motion of the Heart," *Proc. Roy. Soc.*, *8* (1857), 473-488.

[4] Galen, *De anatomicis administrationibus*, Bk. VII, in C. G. Kuhn, ed., *Claudii Galenii opera omnia*, 20 vols. (Leipzig, 1821-1833), II, p. 614; Galen, *De usu pulsuum*, in ibid., V, p. 186; Galen, *De Hippocrat. et Platonis decritis*, Bk. IX, in ibid.,

demonstrated again by Volcher Coiter in the sixteenth century, by Richard Lower in the seventeenth, and by Albrecht Haller in the eighteenth.[5] Their experiments clearly established that external influences were not essential for the production of the heart's rhythmic beat. If a heart completely separated from the rest of the body could continue its rhythmic pulsations, then the cause of the heartbeat must reside solely in the substance of the heart itself.

But this conclusion did not entirely resolve the question, for the possibility remained that the excised heart was not merely muscular. If nervous elements were found embedded in its muscular substance, might they not be responsible for the continued rhythmic beats? By the late 1830's, Johannes Müller had reached the conclusion that this possibility best explained the often contradictory evidence as to whether the heartbeat were muscular or nervous in origin. For though the excised heart might seem to beat independently of nervous action, it had long been obvious that the heart within the body was greatly affected by excitement, passion, and other influences of nervous origin. The rise of experimental neurophysiology during the first third of the nineteenth century had focused new attention on the role of the nervous system in heart action. Legallois and Pierre Flourens had emphasized how profoundly the heart was affected by destruction of the spinal cord. Alexander von Humboldt, Karl Burdach, and Müller himself had shown that a quiescent heart could be brought to pulsation by electrical stimulation of its cardiac nerves. An Italian histologist, Scarpa, had recently demonstrated that the cardiac nerves were distributed in great abundance to the muscular substance of the heart. In a less specific context, Müller also emphasized that all rhythmically contracting muscles seemed to require the intervention of the nervous system in order to do so. Although none of this evidence was really conclusive, Müller, in reviewing it, believed that it led as a whole to-

---

v, pp. 237-239, 261-267, 531. I am indebted to Jerome J. Bylebyl for these references, for the reference to Volcher Coiter immediately below, and for his suggestion that I examine Johannes Müller's summary of the debate up to the late 1830's.

[5] Volcher Coiter, "Observationum anatomicarum chirurgicarumque miscellanae varia," in *Externarum et internarum principalium humani corporis tabulae* (1572), ed. and trans. Niryens and Schierbeek in *Opuscula selecta Neerlandicorum de arte medica*, XVIII (Haarlem, 1955), p. 143; Richard Lower, *Tractatus de corde item de motu & colore sanguinis et chyli in eum transitu* (1669), facsimile edition with introduction and translation by K. J. Franklin (Oxford, 1932), esp. pp. 66, 74-75, 78-88, 96, and 122-127; and Albrecht von Haller, *First Lines of Physiology*, translated from the correct Latin edition (1767), printed under the inspection of William Cullen, M.D., and compared with the edition published by H. A. Wrisberg, M.D., ... two volumes in one, a reprint of the 1786 edition with a new introduction by Lester S. King (New York, 1966), pp. 58-60.

ward the conclusion that the excised heart must contain nervous elements embedded in its muscular tissue and that its continued beating must depend on "a specific influence" of these residual nervous elements. In the end, he argued, the rhythmic heartbeat depends on impulses in the sympathetic nervous system and is not independent of the nerves as Haller believed.[6]

During the next two decades, Müller's position received highly impressive support. About 1845, the Weber brothers, Eduard Friedrich and Ernst Heinrich, astonished the physiological world with their discovery that a motor nerve, the vagus or pneumogastric, possessed the remarkable capacity of inhibiting rather than provoking or accelerating contractions in the frog's heart. Hitherto it had been universally supposed that stimulation of a motor nerve must always provoke contraction in the muscle to which it ran. Quickly extended to other organisms, and to the smooth or involuntary muscles of the arterial and alimentary canals, the discovery of vagal inhibition opened a vast new field of research. Except for Magendie's earlier discovery of the distinction between motor and sensory nerves, it was the most important neurophysiological discovery of the nineteenth century. Its implications for the problem of the heartbeat were profound and, initially at least, straightforward. For vagal inhibition offered yet further and particularly dramatic evidence of the extent to which nervous influences governed the movements of the intact heart. Especially after Claude Bernard established the existence of inhibitory (or vasodilator) and vasoconstrictor fibers in the nerves of the arterial system as well, the entire cardiovascular system could now be seen as a sort of muscular puppet whose strings were controlled by agents in the sympathetic nervous system.[7]

Almost simultaneously with the Webers' discovery of vagal inhibition, Robert Remak and others excitedly announced the discovery of nerve cells or ganglia *within the heart itself*.[8] It was immediately al-

[6] See Johannes Müller, *Handbuch der Physiologie des Menschen*, 2 vols. (3rd ed. 1837-1840), I, pp. 187-198; II, pp. 63-80.

[7] For a summary and critical discussion of these issues as they stood in 1859, together with extensive references to the pertinent literature, see Henri Milne-Edwards, *Leçons sur la physiologie et l'anatomie comparée de l'homme et des animaux*, 14 vols. (Paris, 1857-1881), IV, pp. 120-168, 199-205.

[8] As early as 1839, in fact, Remak claimed that he had found ganglia within the human and other mammalian hearts, and specifically attributed the continued beating of excised vertebrate hearts to these nerve cells. Gabriel Valentin, however, denied Remak's claim. See Johannes Müller, *Elements of Physiology*, translated from the German by William Baly, arranged from the second London edition by John Bell (Philadelphia, 1843), p. 201, unnumbered footnote. By the late 1840's Remak's discovery had been generally accepted, thanks partly to the later and more careful descriptions of ganglia in the calf's heart by Remak himself and by the

leged that these ganglia were the nervous elements responsible for the continued beating of excised hearts, a notion that gained in plausibility from the presence of ganglia along the path of the cardiac nerves and more especially because the automatic rhythmicity of respiratory movements had already been ascribed to the action of similar ganglia in the medulla oblongata.[9] Furthermore, investigations of the frog's heart by several German histologists and physiologists revealed that its ganglia were concentrated in those regions of the cardiac tissue that were physiologically most significant. Ganglia were particularly prominent in the sinus venosus and in the tissue between the auricles and ventricles, precisely those regions where excision or ligature disturbed the normal rhythm of the heartbeat.[10] That a ligature between the sinus venosus and the right auricle could bring the heart to standstill had been demonstrated by the important experiments of Friedrich Stannius. He had also shown that rhythmic beating could be at least temporarily restored to such a quiescent heart by artificially stimulating the tissue (and thus presumably the ganglionic masses) between the auricles and ventricles.[11]

The remarkable coincidence between this histological and physiological evidence produced and sustained the mid-century view that the cardiac ganglia determined and regulated the rhythmic movements of the heart. Since it had never been demonstrated that ganglia existed in all rhythmic parts of the heart, or that they could give off periodic discharges, this view was only a hypothesis, but it came to dominate all other views as to the origin of the rhythmic heartbeat. By the 1850's, the prevailing assumption was that the rhythmic movements of the heart were determined primarily by nervous mechanisms.

In his lecture of 1857 Paget gave a special twist to the evidence

---

English worker Robert Lee. See Robert Remak, "Neurologische Erlaüterungen," *Müller's Archiv*, 1844, pp. 463-472; and Robert Lee, "On the Ganglia and Nerves of the Heart, with Postscript," *Phil. Trans.*, 1849, pp. 43-47.

[9] Cf. Paget, "Rhythmic Motion," pp. 478-479; and Carl J. Wiggers, "Some Significant Advances in Cardiac Physiology during the Nineteenth Century," *Bull. Hist. Med.*, *34* (1960), 1-15, on 4. In fact, no decisive evidence established a direct connection between the cardiac nerves and the intracardiac ganglia, but the connection was widely presumed to exist. E.g., William Rutherford, "Lectures on Experimental Physiology," *Lancet*, 1871 (2), 841-844.

[10] E.g., A. W. Volkmann, "Nachweisung der Nervencentra, von welchem die Bewegung der Lymph- und Blutgefassherzen ausgeht," *Müller's Archiv*, 1844, pp. 419-430; Carl Ludwig, "Ueber die Herznerven des Frosches," ibid., 1848, pp. 139-143; Gustavus Rosenberger, *De centris motuum cordis disquisitiones anatomico-pathologicae*, inaugural dissertation (Dorpat, 1850); and F. H. Bidder, "Ueber functionell verschiedene und raümlich getrennte Nervencentra im Froschherzen," *Müller's Archiv*, 1852, pp. 163-177.

[11] F. H. Stannius, "Zwei Reihen physiologischer Versuche," *Müller's Archiv*, 1852, pp. 85-100.

gathered by the Germans. He suggested that the presumed ability of cardiac ganglia to discharge impulses at regular intervals must depend, like all other time-regulated organic processes, on their rhythmic nutrition; that is to say, on strictly regular molecular processes involved in their activity, waste, and repair. But so far as the more immediate question of myogenicity vs. neurogenicity was concerned, Paget endorsed the view that the vertebrate heartbeat depended primarily on automatic nervous ganglia.[12]

Such, then, was the basic state of knowledge at the time Foster undertook his first experiments on the snail's heart. He began this investigation, he later wrote, by seeking in the snail's heart an inhibitory nerve comparable to the vagus nerve in the vertebrate heart.[13] This effort failed, but he did show that even very tiny pieces excised from the beating heart of a snail continued for some time to beat rhythmically. From this he concluded that the beat could not be "the result of any localized mechanism [such as ganglia], but is probably the peculiar property of the general cardiac tissue."[14] The implication was that in the snail at least the heartbeat is myogenic rather than neurogenic.

These experiments and this conclusion, abstracted in four sentences in the published report of the 1859 meeting of the British Association, created no discernible sensation. Nor is this surprising. In his lecture of 1857 Paget had already suggested a priori that since invertebrates possess at most a rudimentary nervous system, their heartbeat is probably myogenic rather than neurogenic as in the vertebrates. Foster had only demonstrated experimentally the validity of this suggestion. And even at that, very similar experiments had been described almost two centuries earlier by Richard Lower. He too had seen excised invertebrate hearts continue to beat even after they were cut into small pieces.[15]

But this should not be allowed to obscure the importance of Foster's work. Lower's experiments had very probably been forgotten by then, and Foster certainly directed his attention more pointedly than Lower had toward the question of myogenicity.[16] By far the most important

[12] Paget "Rhythmic Motion," pp. 479, 482-487.

[13] M. Foster, "Ueber einen besonderen Fall von Hemmungswirkung," *Archiv für die gesammte Physiologie des Menschen und der Thiere, 5* (1872), 191-195, on 191.

[14] Foster, "Beat," p. 160.

[15] See Lower, *Tractatus de corde*, p. 66. Even earlier, Volcher Coiter ("Observationum," p. 143) had shown that the beat persisted in excised hearts despite fairly drastic cuts in the cardiac substance.

[16] Lower used his experiments on the excised heart to argue that the heartbeat was independent of any processes taking place *in the blood*. Far from asserting the myogenic character of the heartbeat, his remarks on the role of nervous influences

result of Foster's investigation was that it led in time to much further research. It is in fact astonishing how much physiology ultimately owes to the investigation of this problem by Foster himself, and by students under his direction at Cambridge. Certainly no research problem had greater impact on the rise of the Cambridge School of Physiology.

Foster and the problem of the heartbeat, 1864-1869

Fifteen years separate Foster's first experiments on heart action from his focused attack on the problem at Cambridge between 1874 and 1876. During those intervening years he directed his attention from time to time to other questions in physiology, but the problem of the heartbeat remained even then his overriding concern.

In a popular article of 1864, Foster rehearsed the results of his 1859 experiments on the snail's heart, this time comparing them to the results of excision experiments on the frog's heart. While each excised piece of snail's heart continued to beat whatever the mode of excision, the frog's heart presented a more complicated situation. In the frog, the two auricles and the ventricle could be entirely separated from each other and yet continue to beat. But if the ventricle were isolated and divided longitudinally, the broad upper portion continued to beat while the lower tip immediately fell quiescent. And since the upper portion contained ganglia, while the narrow lower tip did not, "it has been supposed that these ganglia, these minute brains, are the cause of the heart's beat, working upon the heart's muscles, and making them contract, just as the great brain works upon the muscles of the body, and makes them contract." To invoke the same mechanism for the snail's heart, one needed only to assume that it contained numerous, widely scattered ganglia such that even tiny pieces excised from it would retain ganglia and hence the capacity to beat.

To Foster, however, the matter did not seem quite so clear as it did to others. Even assuming that these as yet undetected ganglia would one day be found in the snail's heart, or even forgetting entirely about

("spirits") in heart action seem directed *against* myogenicity. To give just one example, Lower suggested (ibid., p. 66) that eels' hearts could continue to beat after removal from the body because "their spirits are entrapped and entangled in the rather viscous matter and are unable to escape so quickly." Thus, even though Kenneth J. Franklin himself translated Lower's *Tractatus de corde*, he is open to challenge when he says that Lower gave "definite proof of the myogenic nature of the heartbeat," and that Foster in 1859 did "nothing more than Lower had done in the seventeenth century." See K. J. Franklin, *A Short History of Physiology*, 2nd ed. (London, 1949), pp. 63, 102.

invertebrate hearts, "the question of why the heart beats is not decided by referring it back to ganglia." For the problem remained how ganglia could produce so regular and rhythmic a motion. Why did they not cause contractions "irregularly at different times, and in obedience to varying circumstances, as the great brain moves the muscles of the body?" For Foster, that is to say, it was the *rhythmicity* of the heartbeat, even more than its seeming independence of the central nervous system, that chiefly required explanation. In his article of 1864 he offered two crude and hopelessly vague hypotheses toward that end. The first derived from the observation that an excised snail's heart, kept in pure water until it ceased to beat, could be roused to action again merely by being placed in a pool of snail's blood. Under these circumstances, the excised heart "feeds upon that blood just as it feeds upon the blood within the body," and resumes its rhythmic pulsations. By linking the rhythmicity of the heartbeat with the nutritive power of blood, this hypothesis recalls Paget's view that the heartbeat depends upon regular molecular processes involved in nutrition. Unlike Paget, Foster assumed that the cardiac muscle, rather than the ganglia, was the dominant site of these intermittent nutritive processes, but his position was no less hypothetical than Paget's own.[17]

Foster's second hypothesis rested on an analogy for which observational support was flimsy at best. The heart, he suggested, displayed an inherent and continuous tendency toward contraction, which however always met "a certain resistance tending to prevent that force from being exerted." And thus the force, though continuous, might manifest itself only periodically, just as when we blow through a pipe into a basin of water, our continuous effort may manifest itself only intermittently in the form of bubbles. "In the same way," Foster wrote, "we may easily comprehend how the force which brings about the beat, though arising continuously out of the vital labour of the heart's atoms, has to stay and gather itself together, from time to time, in order to overcome the resistance it meets with." But if we could "easily comprehend" such a process, we could not demonstrate it, for "these things are as yet hidden from us."[18] Had his concept of the heartbeat remained as crude, tentative, and unsubstantiated as these hypotheses suggest, Foster might well have been embarrassed to have

[17] Foster, "Snail's Heart," pp. 269-272. Cf. Paget, "Rhythmic Motion," pp. 482, 485, 487-488; and Milne-Edwards, *Leçons*, IV (1859), pp. 127-128, n. 1a, where he cites Paget's Croonian Lecture only to say that it is "useless to discuss [his] hypothesis" because "he gives no proof of the existence of intermittence in this nutritive work."

[18] Foster, "Snail's Heart," pp. 272-273.

his research resurrected. In fact, his campaign on behalf of myogenicity never did achieve complete consistency or cogency, but he did eventually produce highly suggestive evidence on its behalf.

In a brief paper not published until 1869, though he claimed that the basic observation had been made "a long time ago," Foster made his first serious attempt to show that the vertebrate heartbeat, like that of the snail, depended essentially on the properties of its cardiac musculature. He insisted that the quiescent lower portion of a frog's ventricle, deprived of ganglia, could nonetheless be induced to rhythmic pulsations by direct application of an interrupted electric current. These beats, "separated from each other by distinct intervals of complete rest," were totally unlike the tetanic contraction produced in an ordinary muscle when its nerve was stimulated by an interrupted current. Foster therefore concluded that "the cardiac muscular tissue itself differs for some reason or other from ordinary muscular tissue in a disposition toward rhythmic rather than continuous contraction." He supposed also that the ganglia in the frog's heart exerted a continuous rather than a rhythmic influence, "whatever the exact nature of that influence be." His final sentence asserted that these conclusions rested not only on the evidence set forth in this paper, but also on "other arguments drawn from various sources."[19]

The Royal Institution lectures of 1869: toward a more general context for Foster's views on the heartbeat

In February of the same year, 1869, Foster gave a series of three lectures on the "involuntary movements of animals" at the Royal Institution of Great Britain. Delivered just two months before his appointment there as Fullerian Professor of Physiology, these lectures very probably formed part of Foster's candidacy for the post. Although never published in English, the lectures have been preserved in a manuscript in Foster's own hand, deposited in the library of the Cambridge Physiological Laboratory.[20] This manuscript provides a reveal-

[19] Michael Foster, "Note on the Action of the Interrupted Current on the Ventricle of the Frog's Heart," *J. Anat. Physiol., 3* (1869), 400-401.

[20] Michael Foster, *Three Lectures on the "Involuntary Movements of Animals" delivered before the Royal Institution of Great Britain, February 1869,* Library of the Cambridge Physiological Laboratory. In this holographic MS, Foster uses Roman numerals to separate his three lectures, each of which he then paginates individually as follows: I, pp. 1-48; II, pp. 1-46; III, pp. 1-50. My footnotes to the MS follow Foster's own scheme. When quoting directly from the MS, I have sometimes altered Foster's punctuation and spelling, but never his words. Because of the frequency of direct quotation, only lengthy quotations are cited specifically.

Within weeks of discovering the above MS, I found a reference to a French trans-ing glimpse of the general context in which Foster developed his views on the heartbeat and allows us to perceive the central thrust, as well as several of the difficulties and ambiguities, of those views. Thanks no doubt to the diverse and chiefly lay quality of audiences at the Royal Institution, Foster here expressed his ideas with a simplicity and generality unmatched in his published research papers. And though his mature views differed in certain important respects from those conveyed in these lectures, chiefly in the direction of an increasingly consistent and aggressive campaign on behalf of myogenicity, we can nonetheless see him wrestling even here with the issues that would henceforth haunt him, and groping toward the resolution that would forever elude him. Indeed, because they do anticipate so many of the issues to be addressed in more technical terms below, these popular lectures seem worthy of extended discussion at this point.

In the first of his three lectures, Foster focused on two sorts of "involuntary movements," ciliary action and "amoeboid" movements. After defining or illustrating several technical terms—muscular contraction, nervous impulse, and artificial stimuli (whether electrical, mechanical, or chemical)—he pointed out that in ordinary muscles, "there can be no contraction without a stimulus." If a movement or contraction could be traced to conscious volition, it was commonly called "voluntary." If the exciting stimulus seemed independent of the will, the resulting movement earned the label "involuntary." But, as Foster soon demonstrated, some involuntary movements took place not only independently of conscious volition, but also independently of any apparent nervous apparatus whatever. Thus, if one excised a piece of mucous membrane from a frog's throat and attached it to a plate of paraffin, and if one then placed a piece of cork at one end of the membrane, the piece of cork moved slowly but steadily toward the other end. The "involuntary movement" thus illustrated had long

---

lation of Foster's lectures, published within a year of their delivery. See Michael Foster, "Mouvements involuntaires chez les animaux," *Revue scientifique, 6* (1869), 658-665, 677-684, 712-720. This reference, located in a scrapbook of Fosteriana owned by his grandson, Sir Robert Foster, has escaped the standard bibliographical sources, including the *Royal Society Catalogue of Scientific Papers*. The *Revue scientifique* frequently translated popular English lectures on scientific topics, and the quite literal translation of Foster's lectures (signed "Battier") must have been produced by someone working directly from his MS. Neither the MS nor the published French version contains any footnotes by Foster, but the published lectures are illustrated with a few diagrams. Although the French version will obviously be more accessible to scholars, I have decided to preserve the freshness of Foster's original words. Moreover, since my footnotes to his MS provide a fairly precise location of significant passages, no one should have any serious difficulty identifying their published French counterparts.

been familiar to physiologists, but no explanation had been forthcoming until Gabriel Valentin, Johann Purkyně, and Foster's own mentor, William Sharpey, independently solved the riddle in the 1830's.

These pioneers of microscopic anatomy had discovered that such movements across membranes were produced "by the action of certain minute organs, called *cilia*; tiny hairs 1/4000th of an inch long, numbering perhaps 12 to 20 or more in each cell." When alive and vigorous, these cilia move so swiftly that it is as impossible to detect an individual cilium as it is to see "the separate spokes of a swiftly revolving wheel." Each cilium beats (or bends and unbends) at the rate of at least 12 strokes a second, being "in fact a tiny oar incessantly engaged in lashing with exquisite rapidity the fluid into which it projects." Because the beating cilia bend more quickly than they unbend, the overall effect of their action is to push fluid continuously in the direction of their bending. And for this reason, the cork drifts from one end of the mucous membrane to the other.[21]

For Foster, though, the chief interest of the phenomenon lay in the question, "by what mechanism are these cilia kept thus at work?" The movement obviously took place independently of the frog's brain or central nervous system, which had been "long quieted in death," but could it not "fairly be imagined that there existed even in this simple membrane beneath the layer of ciliated cells some mechanism of great vital endurance, possibly of a nervous character, whose working was at once the cause and guide of the ciliary movement?" Foster's negative answer depended on the fact that the cilia on any ciliated cell continued to "beat" with only minor changes in speed or regularity even when separated from the membrane. Thus, he argued, "we must look in the cell itself for any mechanism to explain the movement." But, as Sharpey himself had shown, no special structural mechanisms could be detected in the ciliated cells, which were mere delicate masses of protoplasm. Admitting that no one had yet observed motion in an individual cilium separated from its cell, Foster nonetheless considered it likely that "each cilium bends or bows itself," and that any apparatus responsible for this action "is most probably stowed away in the transparent, homogeneous hair-like process itself." In fact, he suggested, "the process that gives rise to the swift dash of these tiny oars" must be considered as "a complex, vital and not a simple physical one." The dependence of ciliary motion on life could be inferred from its gradual slowing and ultimate cessation as the surrounding tissues died, from its sensitivity to temperature and other physical or chemical agents, and from its dependence on oxygen.[22]

21 Foster, *Three Lectures*, I, pp. 1-14.

22 Ibid., pp. 14-23.

After noting that cilia responded to most agents in a manner very similar to ordinary muscles, Foster begged leave "to speak of things being muscular, which nonetheless have not even the shadow of the structure of muscle." That permission granted, he could describe the cilium "as a long slender muscle, or perhaps rather as made up by a special arrangement of muscular molecules." What, then, is the stimulus which causes this "muscle" to contract? Such a stimulus ought to be rapidly intermittent, acting and ceasing to act twelve times in each second. But no such stimulus could be discovered outside the cilium or ciliated cell, and so "the stimulus, whatever it be, must come from within the cell or cilium itself, must be generated out of that tumult of the molecules of the tissue which we call its nutrition or its life."[23]

If the "efficient cause" of ciliary motion thus remained mysterious, Foster found it easy in most cases to identify "its final cause, its purpose." In many organisms, especially lower marine forms, cilia serve as the chief or sole means of respiration, of locomotion, or of securing food—or even of all three at once (e.g., in the tadpole).[24] In the mucous membrane of the frog's throat, ciliary action carries food toward its stomach, while the cilia lining our own bronchial tubes transport mucus constantly upward away from the lungs towards the throat. To accomplish such ends, the cilia obviously must work together in harmony; if not, if each ciliated cell acted independently, "the result would be a mass of eddies, a whirl of confused currents." And yet, despite its delicacy, this harmony was achieved in the absence of any nervous apparatus or other structural mechanism, "as if each cell just felt by its primeval protoplasmic sensibility the throb of its neighbour cells, and as if that throb were the key note by which all its own molecular processes were pitched." In all cases, whatever the particular function of ciliary action in whatever organism,

> we should see the same spontaneous rhythmic labour springing mysteriously out of the inner working of some transparent, structureless tissue either gathered into individual masses called cells or spread in an unbroken layer over the surface of the body. In all we should see the movement exactly adapted to some special purpose, the position and form of the cilium, the direction, the force, the character, the time of the stroke all bearing towards the one end; that end being in most cases almost nothing to the cilium or ciliated cell itself, though perhaps all in all to the body of which they are the minutest and obscurest fragments. . . . All this we might see and

[23] Ibid., pp. 23-26.

[24] Ibid., pp. 31-33.

> learn and yet not be able, even with the highest powers of the most modern microscope, to catch so much as a glimpse of any structural machinery by which we might think ourselves able to explain the facts.[25]

Towards the end of this first lecture, Foster turned to the "amoeboid" movements of white blood cells and to the movements of the amoeba itself. Since the amoeba has no nervous system, indeed since it lacks any trace of specialized, differentiated structures of whatever sort, its movements are obviously "involuntary." But Foster also insisted that these movements have an "essentially vital" and "essentially muscular" character. Repeated attempts to trace amoeboid movements to mere physical agents had "signally failed." And the effects of external agents on the amoeba closely resembled their effects on cilia or ordinary muscles. In fact, said Foster, "one can hardly resist the conviction that all three [forms of movement] are but members of one class, bound together by the common possession of the same fundamental vital quality." He then asked vis-à-vis amoeboid movements the question he had already asked vis-à-vis ciliary motion: "what and where is the irregular stimulus which sets these movements going?" Chiefly because one form of presumably amoeboid motion (namely, the dispersion of chromatophore cells in the frog or chameleon) had recently been traced to nervous action,[26] Foster gave his answer in rather more qualified terms, but that answer remained essentially the same. In amoeboid movements, as in ciliary motion, the stimulus is "not from without but from within."[27]

In his second, climactic lecture, Foster focused on the heartbeat, "an involuntary movement of an undeniably muscular character." In fact, "the rhythmic pulsation of that important organ is independent, as far as its mere existence and continuance are concerned, of the whole central nervous system." One striking piece of evidence for this conclusion came from embryology, for the chick's heart began to beat during the second or third day of incubation, "while as yet the whole

[25] Ibid., pp. 26-34, indented quote on 33-34.

[26] Ibid., pp. 43-47. Milne-Edwards, *Leçons*, XIII (1878-1879), p. 260, n. 2, says that he had put forth a nervous explanation for chromatophore dispersion as early as 1834 and that his explanation has now been fully confirmed by Paul Bert's work of 1875. In the meantime, however, Joseph Lister had also produced important work leading to similar conclusions. See A. E. Best, "Lister's Microscopic Research on Chromatophores," *Proceedings of the Royal Microscopical Society (London)*, 2 (1967), 425 et seq. In his lectures of 1859, Foster gives no source for his perhaps grudging acceptance of a nervous explanation for chromatophore dispersion, but because Lister's work quite immediately preceded those lectures, and because Foster relied on his work in other contexts, it seems highly probable that he had Lister's results in mind as he spoke.

[27] Foster, *Three Lectures*, I, pp. 35-42.

nervous system is an unformed, almost shapeless thing, not even a rough sketch of what it is to be." For those who wished a more dramatic demonstration of the point, Foster directed attention to a tortoise laid out on a table in front of him. Although this tortoise was to all outward appearances dead and motionless, a straw lever attached to its heart rose and fell regularly, indicating not only that the heart remained alive, but indeed continued to beat "with well nigh the same strength and regularity as if its whole nervous system were present in the full swing of work." The same phenomenon could be seen in another tortoise heart which lay naked on the table, separated entirely from the rest of the body. To be sure, Foster continued, the heart of a bird or mammal would not sustain its beat in this sort of isolation, but the difficulties standing in the way were metabolic and mechanical rather than intrinsic or fundamental. Unlike the tortoise and other coldblooded animals which, as the phenomenon of hibernation showed, could accumulate a considerable store of nutritive materials, a warmblooded organism "lives up to its physiological income." Properly supplied and continually bathed with suitably nutritious blood, the mammalian heart too could be kept alive in isolation or after all signs of life had disappeared from the nervous system. In short:

> We may rest content that in all beating hearts the mechanism is the same, that they all beat for the same fundamental reasons, and if we take the cold-blooded heart as our lesson, it is because in it the wheels of life drag heavily and move slowly, giving us better hope of catching some glimpse of the wheels within wheels which will enlighten us as to how they move.[28]

That brought Foster face to face once more with the question, what stimulus could be responsible for this periodic motion. The blood, which periodically enters and leaves the heart, seemed a natural and plausible candidate. But, said Foster, the heart could be entirely emptied and cleansed of blood and yet continue to beat, so long as it was placed in a suitable medium. Nor would it do to respond that the blood must act not qua blood but qua fluid, so that any fluid that regularly enters and leaves the heart could serve as the stimulus to its beat. In fact, experiment shows that "no coming and going of blood or any fluid will solve the riddle of the heart." And so, declared Foster:

> I may say boldly and dogmatically at once that in none of the outward circumstances of the heart's existence can we find anything worthy of being regarded as a stimulus which comes and goes, which

[28] Ibid., II, pp. 1-6, indented quote on pp. 5-6.

> acts and ceases to act, and therefore [worthy] of being put forward as the cause why the heart beats and rests, rests and beats again: the cause of the heart's beat is somewhere in the substance of the heart itself.[29]

Thus far, of course, Foster had merely affirmed the independence of the heartbeat from the central nervous system and from fluids or other agents external to it. Within the heart itself, the issue remained whether the force responsible for its beat should be regarded as "diffused over the whole heart or fixed in some special centre or centres." Taking the frog's heart as his example, Foster then launched into a discussion which sounded at first like a standard brief on behalf of the neurogenic theory of the heartbeat. He reviewed the familiar consequences of various modes of slicing the frog's heart, emphasizing once again (as in his popular article of 1864) the quiescence produced in the ganglion-free lower part of the ventricle when it was divided "a little and only a little below its top." Turning to the vagus nerves, he described their remarkable inhibitory effect on the heartbeat and linked that effect with the pear-shaped nerve cells found in association with the vagus nerves. Such nerve cells, he said, enjoyed powers denied to ordinary nerves:

> Now all the results hitherto obtained in the physiology of the nervous system go to shew that while nerve fibres merely conduct, transmit or propagate nervous impulses, being wholly destitute of any power to originate them, nerve cells in addition to their capacity for simple conduction are able of themselves, out of their own inner molecular working, either to originate wholly new impulses or so to transform impulses which they receive that these issue from the cell as altogether different things from what they were when they entered it.

In short, whereas ordinary nerve fibers were "mere passive instruments," nerve cells were "active centres," which "can and do give out stimuli, set impulses going, though everything around them may be in a condition of most complete equilibrium."[30]

Foster then observed that the nerve cells in the frog's heart, "being in places scattered singly and in spots gathered together into little groups called *ganglia*," were particularly abundant in the area of the sinus venosus, in the auricular septum, and at the juncture between the auricles and ventricles. From the cluster of nerve cells at the top of the ventricle, fibers descended briefly into the substance of the ventricle itself. Below that point, no nerve cells or ganglia whatever

29 Ibid., pp. 7-9.

30 Ibid., pp. 9-14, indented quote on p. 13.

could be detected; above that point, they could be found in many parts of the heart. Now, said Foster, in the favorite refrain of the neurogenicists:

> It will not have escaped you that this structural feature of the frog's heart tallies remarkably with the results obtained touching the localisation of the power of spontaneous beat. Where nerve cells, where ganglia are present . . . there the spontaneous beat is present. Where ganglia are absent, in the lower part of the ventricle, in all the ventricle in fact except its top, there the spontaneous beat is absent too. The ventricle severed from the nerve cells which reside at its summit . . . has lost all power to give or keep up of itself a rhythmic beat. We infer therefore that these ganglia are in some way or other connected with the spontaneous beat.[31]

Somewhat later in this second lecture, Foster made another bow in the direction of the neurogenicists by stressing the extent to which the beating heart could be influenced by factors external to it. The excised heart responded to changes in temperature, to the quality of the ambient air, and especially to electric currents (which, depending on circumstances, might accelerate or decrease the rate of beat, augment or diminish its force, or even stop the beat entirely). Within the body, the intact heart could be affected dramatically by nervous action, by changes in blood pressure, and by "foreign" chemicals such as alcohol or poisons. Indeed, it was precisely because of its susceptibility to such influences that the heart offered "so sensitive, so true and so quick an index of the body's state," and that the taking of the pulse remained such a crucial part of clinical medicine.[32]

But if in these parts of his lecture Foster sounded for all the world like a convert to the neurogenic theory, he did manage elsewhere to convey his skepticism toward that theory. Thus, even as he described the striking coincidence between the location of ganglia in the frog's heart and the localization of its capacity for spontaneous pulsation, he sounded the following cautionary note:

> Now when a physiologist in his searching after the hidden cause of some secret motion finds a ganglion, he cries "Eureka!" and generally folds his hands as if his work were done. In the case of the heart, however, we may venture to go a little further and ask the question: in what way or by what means are these ganglia the cause of the heart's spontaneous beat? Is it that a stimulus or disturbance periodically arises in the substance of the potent active nerve cells

31 Ibid., pp. 15-16.

32 Ibid., pp. 20-25.

> and then hurries down to the muscular fibre as a nervous impulse causing it to contract? Or is it that the stimulus arises in the substance of the muscular fibre, or if you will, that like the cilia, the heart fibres periodically overflow with energy and from time to time burst out in action of their own accord, but that a conjunction with nerve cells is in some way or other necessary for the well being and perfect work of the muscle such as would ensure the periodical rise of a stimulus or overflow of energy.

Although most physiologists favored the first view, "which fits in most easily with our ordinary conceptions," Foster preferred the hypothesis that the ganglia exerted a continuous rather than intermittent influence.[33]

At this point, Foster offered just one piece of evidence for his choice. Describing the results contained in the brief paper published separately that same year, he insisted that the lower two-thirds of the frog's ventricle, though it lacks the power of spontaneous pulsation, is no ordinary muscle. Under the stimulus of an interrupted current, it is not (like ordinary muscle) seized with a "prolonged spasm of contraction lasting so long as the current continues to act." Rather, it begins to pulsate rhythmically and eventually with striking regularity, as if it possessed in itself "some mechanism . . . for the rhythmic beat, a mechanism which requires however to be set going and to be kept going by the galvanic current." Perhaps, then, in the heart as a whole, the cardiac ganglia normally played the role here played by the galvanic current. Perhaps they merely triggered or maintained a rhythmic mechanism actually located in the cardiac musculature itself.[34]

Toward the end of this second lecture, Foster took essentially the same position with respect to two other examples of involuntary movement in the vascular system—the beating of the so-called "lymph hearts" in the frog, and vasomotor action in the arterial systems of all vertebrates. The lymph hearts, of which every frog has two pair, one pair each at the upper and lower ends of the torso, resembled the ordinary blood heart in three ways. They too beat rhythmically, they too had nerve cells associated with them, and they too promoted the distribution of a fluid (in this case lymph) throughout the body. But the lymph hearts, "delicate, transparent little muscular sac[s]," each about "the size of a mustard seed," also differed from the blood heart in two important respects. For one thing, their nerve cells were not embedded in their substance, but were "removed to some distance from the muscular sacs and placed in the spinal cord." And, perhaps

[33] Ibid., pp. 16-18.

[34] Ibid., pp. 18-19.

for that very reason, they also had a much lower capacity for independent rhythmicity. In fact, the rhythmic beat of a lymph heart could be brought to a sudden halt by excising it from the body, by severing the nerve which connected it with the spinal cord, or by destroying a small but clearly defined portion of the spinal cord itself. Nonetheless, Foster cautioned, it remains uncertain even in the case of the lymph hearts how one is to distribute responsibility between the nerve cells and the muscular substance. Some physiologists insisted that the beat of a lymph heart, brought to a standstill in one of these three ways, did sometimes return several days later, even though its nerves and nerve cells remained nonfunctional. If so, Foster continued, "it would seem to indicate that the influence of the nerve cells here, as we saw it might be in the blood heart, is not in the way of rhythmic impulses, but of some continuous action which enables the muscular fibre to enjoy the high life of spontaneous rhythmic contraction."[35]

Finally, with respect to the arteries, Foster noted that "though in an ordinary way they have no active pulsation (what we call the pulse being but the stroke of the heart borne along their elastic walls)," they did nonetheless enjoy "an independent power of contraction and relaxation." This power could be traced to the ring of smooth muscular tissue of which the middle coat of the arterial walls was composed. Ordinarily, these arterial muscles were held, "so to speak, at half cock," being neither fully contracted nor completely relaxed. Under abnormal circumstances, however, this condition of moderate contraction (called tone or tonus) could be disturbed in either direction. Thus, to take the most obvious example, the muscles of the facial arteries could lose their usual tone either in the direction of increased contraction, which by constricting the arteries would cause the face to turn pale, or in the direction of increased relaxation, the tangible sign of which would be an "involuntary blush."

As usual, Foster's chief interest lay in the mechanism by which such phenomena might be explained. Many physiologists ascribed vascular tone to the action of peripheral ganglia in the so-called "sympathetic" nervous system. Others supposed that nerve cells in the spinal cord itself chiefly determined the process. While both hypotheses thus associated tone with the action of nerve cells, neither excluded the possibility that the responsible nerve cells did their work not by producing rhythmic impulses, but rather by virtue of "a mild continuous influence . . . borne along the sympathetic nerve to the muscles of the facial arteries, inducing in them a moderate contraction and thus keeping

[35] Ibid., pp. 32-37.

them in tone." If so, impulses traveling along appropriate neural paths might "break in upon the tonic nerve centres busy with their mild continuous labour," in such a way as either to arrest or increase their contractile influence, just as nervous impulses to the heart could either quicken its beat or bring it to a standstill, depending on the path by which they reached the cardiac ganglia.

On any view of vascular tone, one also needed to explain how it could be so indefinitely maintained. What kept the responsible muscles from losing their hold "out of mere weariness?" One hypothesis suggested that the apparent constancy of tone was in reality a somewhat illusory aggregate phenomenon, due to the combined action of a large group of individual muscle fibers. Any one of these fibers might alternately contract and relax without disturbing the overall tone if, while this fiber relaxed, "its neighbour fibres are busy at work and while it is at work they are at rest." On this hypothesis, vascular tone became the aggregate outward expression of a host of unseen individual pulsations, "and the influence proceeding from the nerve centre would be at bottom not constant but rhythmic." As an alternative, Foster offered an explanation more in keeping with the general thrust of his lectures to this point. Perhaps, "just as the nutrition of the heart is so pitched that its rhythmic pulsation goes on without tire, so the nutrition of these arterial muscles is specially adapted to meet the wants arising from their unbroken labour." If so, it could be further supposed that the capacity for contraction lay within the substance of the arterial muscles themselves, while the nervous system exerted a continuous influence, acting not so much as a stimulus in the usual sense but rather in a manner "analogous to what under one view we imagined to be the elevating and inspiriting influence of the nerve cells of the heart."[36]

Then, by way of summarizing and connecting his first two lectures, Foster added that whichever hypothesis one preferred for vascular tone and vasomotor action, and wherever one chose to locate the tonic center, "we have the same order of things that we met with in cilia, in protoplasm and in the hearts both of the blood stream and the lymph stream":

> In all cases we have a certain mechanism, a certain portion of vital material, either in the form of virgin protoplasm or differentiated . . . into muscular fibre, nerve fibre and nerve cell. This mechanism, this patch of living stuff, we find to be the seat of spontaneous movement, quite independent of our will, of movement slow or quick,

[36] Ibid., pp. 37-43.

regular or irregular, continuous or rhythmic, of movement touching the cause of which we can only say that it is the natural outcome of the molecular motions which constitute the nutrition of the tissues, that it is part of their very life.[37]

Somewhat earlier, while discussing the heartbeat, Foster had made the same point by insisting that "whether we imagine the cause of the rhythmic beat to be seated wholly in the ganglia or partly in the muscle, the cause itself is not any outward thing, but is . . . an outcome of that molecular travail of the heart which we call its nutrition." Emphasizing the complexity of this nutritive process, and hence "the deeply rooted and complex nature" of the heartbeat itself, he clearly distrusted attempts to explain the phenomena in simple chemical terms. The heartbeat could not be traced to "the heaping up of decomposable oxygen-needing substances, which in turn are decomposed, oxidized, or otherwise got rid of by the act of contraction." In fact, "all we can say at the present time . . . is that the heart grows [and] is nourished in such a way as to beat."[38]

At the very end of his second lecture, Foster returned briefly to the "final cause" of involuntary movements, saying that "both in the undisturbed accomplishment of the natural movements and, perhaps still more in the variations which circumstances call forth, we see . . . means delicately and adequately fitted to secure an end." He closed with an example of what would later be popularized as "homeostatic" mechanisms in Walter B. Cannon's *The Wisdom of the Body* (1932). This "beautiful mechanism, recently discovered and forming part of the general tonic system," allowed an overburdened heart to send impulses to the tonic centers, paralyzing them and thus releasing the blood vessels, so that "the heart with lightened labour does its work with ease." Foster promised to devote his third lecture to mechanisms of a similar sort.[39]

This third lecture, a generally prosaic discussion of reflex actions, focused on the labyrinthine structure of the central nervous system and on the movements of frogs deprived of relatively larger or smaller portions of their brains. These movements, along with such other reflexes as coughing, sneezing, blinking, crying, laughing, shivering, and salivating, shared two basic features with the sorts of involuntary movements Foster had thus far discussed. They ordinarily did or certainly could occur independently of conscious volition; and they had

[37] Ibid., pp. 43-44.
[38] Ibid., pp. 19-20.
[39] Ibid., pp. 44-46. The mechanism to which Foster refers (the depressor nerve reflex) was discovered by Carl Ludwig and Elie Cyon in 1866. See Rutherford, "Lectures," p. 842.

obvious utility for the organisms in which they appeared. Indeed, in most of them, Foster found "'purpose writ so large that . . . where intent is not visible, it is probably our fault that we cannot see it."[40] Otherwise, however, these reflex acts had little in common with ciliary motion, amoeboid movements, the rhythmic pulsations of blood and lymph hearts, or vascular tone and vasomotor action. Only rarely, and then crudely, did Foster manage to draw any analogies between the two different classes of involuntary movement. Only with evident strain did he manage to integrate this third lecture with the two that preceded it.

Any such efforts were doomed from the outset by the patent dependence of reflex actions on highly specialized structures and processes in the central nervous system. The very concept and definition of the tripartite "reflex arc" emphasized those neural structures and processes. Reflexes occurred when (1) sensory impulses traveled along incoming nerve fibers to the spinal cord, where (2) they were transformed into motor impulses and (3) sent out along outgoing nerve fibers to produce contraction in the muscles of the affected region. In his first two lectures Foster had discussed movements which did or conceivably could occur independently of specialized nervous structures, and sometimes even independently of any visible anatomical structure whatever. By contrast, the reflex actions depended so intimately upon structure, and the responsible nervous impulses picked their way "so delicately amid a labyrinth of paths," that "it becomes almost impossible to resist the conclusion that the intricate mesh work of the spinal cord [harbors] special bits of machinery for all these special ends, an anatomical groundwork for the physiological facts."[41]

Even in the case of breathing, the one partly reflex act that also displayed some of the automaticity of the heartbeat, Foster accepted the claim that its rhythmicity originated in a special "centre of rhythmic labour" in the medulla oblongata. To be sure, he did (like Paget) trace this rhythmicity ultimately to nutritive processes, saying that here (unlike the heart) one specific nutritive factor could be assigned primary responsibility—namely, the relative quantity of free oxygen in the ambient blood. And he did then try to suggest that both the heart and the respiratory center enjoyed a "natural intrinsic rhythmicity," which in both cases (if much less so in the heart) could be altered by a variety of nervous influences in order to meet "both the ordinary and extraordinary needs of the body economy."[42] But in the very attempt to draw this strained analogy, Foster threatened rather than

[40] Foster, *Three Lectures*, III, p. 18.
[41] Ibid., pp. 9-19.
[42] Ibid., pp. 12-14.

enhanced his case against the prevailing neurogenic theory of the heartbeat. Indeed, the neurogenicists themselves emphasized the analogy between respiratory and cardiac rhythmicity precisely because the former was so widely ascribed to periodic discharges from the medulla oblongata. Paget's own support of the neurogenic theory of the heartbeat depended mainly, if not solely, on its presumed analogy with respiratory movements.[43] Thus, to compare the heartbeat with respiratory movements, and then to ascribe the latter to a rhythmic nervous process, was to play directly into the hands of the neurogenically inclined. That Foster did so is all the more surprising because he knew of evidence that inactive respiratory muscles could be made to imitate rhythmic movements even after their nervous center had been mutilated or destroyed.[44]

Almost equally embarrassing to the myogenic cause were Foster's other attempts to forge some link between this third lecture and its predecessors. One such attempt, which came near the beginning of the last lecture, merely embellished to little end one of the vague notions set forth in his popular article of 1864. With the disclaimer that he offered "not exactly an hypothesis but [rather] a way of looking at the phenomena over which we have been travelling with the hope of giving to them some little bond of unity," Foster again asked his audience to think in terms of two basic groups of competing forces. The effect of one group would be to promote activity wherever they appeared—to produce contraction in a muscle or the propagation of an impulse in a nerve, for example. The second group of forces would act in such a way as to offer resistance to the first. Depending on circumstances, the two sets of forces would then interact to produce one of four conditions or results. First, the competing forces might be so equal in strength as to neutralize each other, leading to a state of rest. Second, the active forces might consistently overwhelm the resisting forces, leading to a continuous action such as that exerted (on one view) by the spinal cord on the arterial muscles or by the cardiac ganglia on the muscles of the heart. Third, the two sets of forces might be so adjusted that the active forces (though continuously at work) could only periodically overcome the resisting forces and could thus produce only intermittent action. Here, as in his article of 1864, Foster drew attention to the bubbles produced by driving a steady current of air through a column of water, suggesting that a similar process occurred "in cilia, in the beating heart [and] in rhythmic impulses flow ing from ganglia or from the spinal cord." Finally, said Foster, the

43 See Paget, "Rhythmic Motion," pp. 478-479.

44 E.g., ibid., p. 478, unnumbered footnote.

competing forces might be so delicately balanced that the slightest stimulus would decide the battle in favor of the active forces. This fourth condition, really only a special case of the first, could be seen not only in ordinary voluntary movements but also in reflex actions, which the rest of the third lecture then discussed at some length.[45]

Potentially more damaging was Foster's last, perhaps desperate, attempt to unify all involuntary movements by relating them in some fashion to ordinary voluntary actions. He virtually assumed the existence of special mechanisms by means of which voluntary acts were converted through habit and education into involuntary movements, and then (as if to be fair) proposed the existence of inverse mechanisms for the "voluntary" control or release of ordinarily "involuntary" movements. Ultimately, Foster so blurred the distinctions between muscular movements of whatever kind that he actually (if rhetorically) spoke of structureless infusoria as having volition and of "holding their cilia subject to their wills."[46]

The significance of the Royal Institution lectures: rhythmicity and the "physiological division of labour"

It should be clear by now that Foster's 1869 lectures to the Royal Institution have not earned this lengthy account by virtue of their cogency or consistency. Nor, for that matter, do they contain any strikingly original observations or insights. With one partial exception, to which we shall return, every fact that Foster transmitted was standard coin in the physiological world and could have been found in virtually any contemporary textbook or review article.[47] Even his lengthy discussion of ciliary motion, traceable no doubt to his tutelage under William Sharpey, contained nothing new. His crude attempt to conceptualize heart action in terms of competing forces had its antecedents.[48] His emphasis on rhythmic nutrition as the ultimate cause of rhythmic motion, though somewhat unusual, was almost certainly inspired by Paget's Croonian Lecture of 1857.[49] Some, including Paget, even surpassed Foster in the range of rhythmic movements they considered and in their interest in rhythmicity as a general biological phenomenon. And though the neurogenic theory certainly dominated

[45] Foster, *Three Lectures*, III, pp. 4-9 et seq.

[46] Ibid., pp. 46-50.

[47] E.g., Paget, "Rhythmic Motion," and Milne-Edwards, *Leçons*, IV (1859), pp. 120-168.

[48] See Milne-Edwards, ibid., IV (1859), p. 157.

[49] That the emphasis on rhythmic nutrition was both unusual and traceable to Paget seems clear from ibid., pp. 127-128 n. la. See also n. 73 below.

the field, Foster's skepticism toward it was shared by at least a few contemporaries, including William Sharpey himself, from whom Foster had quite probably acquired his own myogenic bias.[50]

Why, then, do Foster's lectures deserve resurrection? And why have they been examined at such length? The most obvious reason is that they provide a useful general account of the contradictory evidence and complex issues faced by any contemporary physiologist who wished to take a position on the problem of the heartbeat. That evidence and those issues served as the backdrop against which Foster and his students pursued the problem, and our account of that subsequent research can be materially abbreviated now that this background has been so fully laid out. From a methodological point of view, moreover, it seemed essential to examine the general context of Foster's lectures and thus to avoid the "whiggish" tendency to select for discussion only those events or ideas that seem to point toward the future. Although lengthy, the resulting account should make it clear that Foster's views on rhythmic motion in general or on the heartbeat in particular, as they stood in 1869, could not have led directly to the full-fledged myogenic resolution that the Cambridge School was ultimately to achieve. With that point in mind, we can now turn more comfortably to two themes in Foster's lectures that were to become of great significance in the cardiological research of the Cambridge School.

One of these themes was Foster's repeated insistence that ganglia (or other nervous mechanisms) might contribute to rhythmic motion only by exerting some sort of continuous influence, the real origin of rhythmicity lying within the muscular tissue itself. Although he seri-

[50] Sharpey's initial myogenic inclinations seem clear from the notes on his lectures taken by Joseph Lister in 1849-1850, just a couple of years before Foster himself took Sharpey's course, and just as most German physiologists were adopting the neuogenic theory. Lister's notes, covering some 429 folio pages, are deposited (as Medical Society of London MS 80) in the Wellcome Institute of the History of Medicine. I am grateful to the Medical Society of London for permission to examine this manuscript. It reveals that Sharpey (like Foster after him) compared the heartbeat to the movements of nerve-free cilia, sensitive plants, and amoeba, and that he preferred the "Hallerian doctrine" according to which contractility is a property belonging to the contractile substance itself independent of nerves (pp. 52-54, 94-95, 379-389).

By 1867, however, Sharpey had apparently abandoned or severely compromised his myogenic stance, at least insofar as one can judge from the sketchy notes of his lectures taken by G. D. Thane in 1867-1868. These notes are deposited (as Medical Sciences MH5c17) in the library of University College London. I am grateful to University College for permission to examine this manuscript. There Sharpey suggests that the regular motion of the heart probably depends on the cardiac ganglia, citing the well-known fact that in a divided ventricle only the upper part (containing the ganglia) will continue to beat (pp. 269-272).

ously compromised this position in his discussion of respiratory movements, Foster returned to it again at the very end of his last lecture. There he boldly drew "a curious analogy between the body at large and that pattern of simple life, the beating heart":

> If I touch the leg of a brainless frog, it kicks; if I touch a nerve, its muscle contracts; if I touch a dying motionless heart, it gives one beat and nothing more—we have in these cases instances of a simple equilibrium broken once and then restored. If I throw a brainless frog into water, it swims, the image of a natural living frog; when I take it out again, it resumes its normal unbroken quiet repose. If I subject the divided, motionless ventricle of the heart to an interrupted current, it begins to beat, the picture of a normal beating heart. When I withdraw the current, it lapses into its former quiet. In the one case, the mechanism of swimming—in the other, the mechanism of a rhythmic beat—is present; but the something to set them going and keep them going without need of stimulus is wanting. In the heart that want is supplied by the fuller life of the ganglia; in the frog, by the fuller life (whatever that may be) of the cerebral hemispheres.[51]

Here, for the second time in these lectures, Foster stressed the capacity of the ganglion-free lower part of the frog's ventricle to respond rhythmically to an interrupted current. This is the one fact of arguable originality in his three lectures. To be sure, Paget, in his paper of 1857, had already observed that the lower portion of a divided ventricle could be made to "imitate a rhythm by stimulating it at regular intervals, such as every two or three seconds,"[52] and the German physiologist Carl Eckhard had used an electric current to produce the same effect as early as 1858.[53] But Paget had also insisted that the contractile force of the ventricle was "much sooner . . . exhausted by the artificially excited actions, than by the apparently equal actions of the natural rhythm," and Eckhard's result (produced by a constant current rather than the interrupted current Foster used) had later been interpreted as merely a special case of the spasmodic or tetanic contraction produced by applying the constant current to an ordinary muscle-nerve.[54] Foster, in any case, considered his observa-

[51] Foster, *Three Lectures*, III, pp. 47-48. [52] Paget, "Rhythmic Motion," p. 474.

[53] Carl Eckhard, "Ein Beitrag zur Theorie der Ursachen der Herzbewegung," *Beiträge zur Anatomie und Physiologie, 1* (1858), 145-156.

[54] Paget, "Rhythmic Motion," p. 474; Rudolf Heidenhain, "Erörterungen über die Bewegungen des Froschherzens," *Müller's Archiv*, 1858, pp. 479-505; and Carl Eckhard, "Kritische Beleuchtung der über die Ursachen der Herzbewegung bekannten Thatsachen," *Beiträge zur Anatomie und Physiologie, 2* (1860), 123-157.

tion sufficiently novel or important to report it separately in the brief note published in 1869. There, too, he used the result in support of the notion that the cardiac ganglia exerted a continuous rather than a rhythmic influence. Neither there nor in his lectures did Foster attempt to specify the nature or function of this influence, but the issue persisted in his later work and he did eventually develop an hypothesis as to the role that might be played by continuous ganglionic discharges.

The second theme, which in retrospect seems far more important to Foster's myogenic conception of the heartbeat, was later to dominate his influential *Textbook of Physiology* (1877). There, in its general form, this theme was called "the physiological division of labour," a phrase that Foster probably borrowed from Charles Darwin[55] and certainly deployed on behalf of a broadly evolutionary vision of physiology. Foster's specific views on the problem of the heartbeat were part and parcel of this larger scheme.

Foster began his textbook by describing the lowest form of animal life, the unicellular amoeba. Now the amoeba, he noted, is nothing but a simple mass of undifferentiated protoplasm, and yet, since it does live, that undifferentiated protoplasm must exhibit and subserve all the functions essential to life. As for organisms higher in the evolutionary scale, all of their physiological actions are "the results of these fundamental qualities of protoplasm peculiarly associated together." Foster then moved to the core of the matter:

> the dominant principle of this association is the physiological division of labour corresponding to the morphological differentiation of structure. . . . In the evolution of living beings through past times, it has come about that in the higher animals (and plants) certain groups of the constituent amoebiform units or cells have, in company with a change in structure, been set apart for the manifestation of certain only of the fundamental properties of protoplasm, to the exclusion or at least to the complete subordination of the other properties.[56]

[55] Darwin, in turn, had adopted the phrase from Henri Milne-Edwards. On Darwin's use of Milne-Edwards' concept, see Camille Limoges, *La sélection naturelle: étude sur la première constitution d'un concept (1837-1859)* (Paris, 1970), pp. 135-136. That Foster's immediate debt was to Darwin seems likely from his general admiration for Darwin and from his imitation of Darwin's language rather than that of Milne-Edwards ("physiological division of labour" instead of "division of physiological labour"). When he introduced the phrase in his *Textbook* (p. 3), Foster in fact assigned credit neither to Milne-Edwards nor to Darwin.

[56] Michael Foster, *A Textbook of Physiology* (London, 1877), pp. 1-8, indented quote on 3.

Guided by this "dominant principle," Foster divided the rest of his textbook into sections each of which corresponds to a class of tissues that has evolved and become differentiated in such a way as mainly to embody one of the fundamental properties of protoplasm. In speaking of rhythmicity in particular, Foster argued that very little differentiation seems to be required for this function. The simple amoeba already possesses the capacity for spontaneous contractile movement, and ciliated cells demonstrate perfectly regular (or rhythmic) automatic activity.[57]

In his 1869 lectures, Foster never used the phrase "physiological division of labour," but the concept can be glimpsed in implicit form and is once or twice made quite explicit. At one point, comparing contractile waves in amoeba with those in ordinary skeletal muscle, Foster emphasized the following difference between them:

> Amoeboid waves are slow, muscular waves are quick; in this respect, muscle has the advantage over protoplasm. But there is a compensation. The amoeboid wave moves in all directions of space; the muscular wave is limited to one. Protoplasm is all-sided; muscle can do no more than bring its two ends together. What is gained in force and time is lost in character—the old tale writ large and often in the book of animal life.[58]

Later he included the heart in the comparison:

> In work [the heart] stands midway between protoplasm and [skeletal] muscle. The waves of its contractions move along its fibres in one direction only. It has lost the all-sidedness of protoplasm. But unlike ordinary muscle, it retains the spontaneity of protoplasm.

And if the heart thus stands physiologically midway between protoplasm and skeletal muscle, so too does it stand anatomically midway between them. Like ordinary muscle, the heart is composed of striated fibers, but its striations are less sharply marked. It lacks the elastic sheath (or sarcolemma) of ordinary muscle, and its fibers retain some of the cell-like and granular features of primordial undifferentiated protoplasm.[59]

To illustrate his point further, Foster than turned to embryology, more specifically to the early development of the chick's heart. Part of his aim here was to deny any "sharp demarcation between the protoplasmic crawl and the true rhythmic spasm." For in fact "the slow, irregular, crawling movements of the primordial protoplasm are [only

[57] Ibid., pp. 72-73, 77-78, 261ff.

[58] Foster, *Three Lectures*, I, pp. 41-42.

[59] Ibid., II, pp. 27-28.

very] gradually transformed and gathered up into the sharp short stroke of the heart's beat." It is therefore misleading to say that the chick's heart suddenly begins to beat during the second or third day of incubation. To use a musical metaphor, it is not the case that the cardiac cells gather as musicians do around a conductor, "fully equipped with powers of rhythmic pulsation, but quiet and inactive," waiting only for a wave of the baton to sound the first note in unison. Rather, it is as though these cells were a dispersed group of novice violinists, destined ultimately to come together, but each trying unsuccessfully to play the same tune as they come nearer to each other and to us:

> As we listen with stretched ear to them coming nearer and nearer, and as at each moment more and more performers fall into the one proper tune, the initial discordant noise as it gathers in intensity also gradually puts on a definite form and at last there comes a moment when we can say, "Now I hear them! Now they have the tune!" So it is with the growing heart. Looking at it earnestly with the microscope, we may fancy ourselves witnesses of how the cells, as they assemble together, little by little exchange the all-sided flow of protoplasm for the limited throb of a muscular contraction, gaining in force what they lose in form. And so there will come a moment when we say, "Now I can see it beat," though in reality it has been beating a long time before.[60]

Literary devices aside, Foster hoped to convey "the essential unity of the rhythmic beat of the heart and the amoeboid movement of protoplasm." In a passage that would echo repeatedly through the work of the Cambridge School, Foster then explicitly applied the physiological division of labor to the problem of the heartbeat:

> In the chick growing within the egg, the heart begins to beat very early, while as yet it is built up of nothing but protoplasmic cells. Many authors, overzealous as it seems to me for the prerogatives of nerve cells, find satisfaction in affirming that these constituent cells of the young heart, though apparently alike in structure, are various, some being potentially nerve cells, others potentially muscle. To my mind, each and every cell is not only potentially but actually both nerve and muscle. So long as they are still cells, that is, still tiny masses of untransformed protoplasm, each enjoys all the powers of life. What befalls them afterwards is not gain but limitation and loss. Some cells lose the power to move and so become

[60] Ibid., pp. 28-31, indented quote on 31.

nerve cells; other cells lose (to a great extent at least) the power to originate impulses and so become muscular.[61]

The full meaning and fecundity of this passage will emerge only through an extended study of subsequent research by Foster and his students. But, before turning at last to that task, it seems essential to insist on several points connected with the "physiological division of labour" and with Foster's use of that concept. First, Foster invented neither the concept nor the phrase. Credit for both is usually given to Henri Milne-Edwards (1800-1885), who certainly promoted the doctrine in its most developed form.[62] Second, while it was perfectly possible to accept the physiological division of labor without accepting Darwinian evolution (indeed, Milne-Edwards himself opposed Darwin's theory),[63] the two doctrines shared an idiom and scope that made them mutually reinforcing. Among those who enthusiastically adopted Milne-Edwards' concept and phrase was Charles Darwin himself. Third, though the broad comparative outlook already implicit in the physiological division of labor might itself engender doubts about the neurogenic theory,[64] an evolutionary perspective tended to quicken and solidify such skepticism. In the work of Walter Gaskell, in particular, it will be seen that evolutionary concepts could even provide a way out of mere skepticism toward the positive affirmation of myogenicity.

We shall also see a more general association between the myogenic theory and Darwinian thought, though this association was neither rigid nor logically necessary. In particular, the contribution to which Darwin himself probably attached most importance, the mechanism of natural selection, received no explicit attention, even from the most obviously "evolutionary" myogenicists, Foster and his students. Nonetheless, their campaign on behalf of myogenicity did draw

[61] Ibid., pp. 28-29.

[62] See Henri Milne-Edwards, *Introduction à la zoologie générale* (Paris, 1851); Milne-Edwards, *Leçons*, I (1857), pp. 16-27; ibid., XIV (1880-1881), pp. 280-281, n. 1; Camille Limoges, "Darwin, Milne-Edwards et le principe de divergence," *Actes du* XIIe *congrès internationale d'historie des sciences (Paris, 1968)*, vol. 8 (1971), pp. 111-115; and Limoges, *La sélection naturelle*, pp. 135-136. From the late 1850's on, Milne-Edwards could justifiably insist that the principle of the "division of physiological labor" had been adopted by nearly all zoologists. His attempts to claim sole priority for the principle carry rather less persuasion. In some respects at least, the concept had long been a standard assumption among naturalists, though Milne-Edwards did give it a widely imitated name and did pursue its consequences more systematically than his predecessors.

[63] And did so until his death in 1885. See, e.g., Milne-Edwards, *Leçons*, XIV (1880-1881), pp. 296-335.

[64] Again Milne-Edwards serves as an example, though a somewhat uncertain one. See Chapter 11, n. 44.

sustenance from other Darwinian concepts. Despite the superficially Aristotelian and teleological language of his lectures of 1869 (with his allusions to "final cause" and "purpose"), Foster was already drawn to the Darwinian camp. That position, as we shall see, helped him to develop a more consistent attack against the prevailing neurogenic theory of the heartbeat. But as a physiologist, especially as one eager to establish his distance from the anatomical bias of the English tradition, Foster was outwardly more concerned to provide experimental evidence for his position. Evolutionary concepts, though probably more fundamental to his myogenic beliefs, emerged only rarely in his published works on the problem.

Foster's later research on the problem of the heartbeat: the discovery of nerveless inhibition in the snail's heart

In 1872, in a paper published in German, Foster described the results of new experiments he had recently performed on the snail's heart.[65] Thus, after a lapse of more than a decade, he returned to the animal on which his first research had been done. This time, however, he used a different method directed toward a different end. Instead of cutting the heart into small pieces, he studied the effect exerted on the whole organ by interrupted currents like those he had earlier applied to the lower portion of the frog's ventricle. The effect was remarkable. Weak currents applied directly to the excised but still beating heart of a snail produced immediate diastolic standstill. In other words, the effect was an inhibition exactly like that produced in the frog's heart by stimulating its vagus nerve.

What made this inhibition in the snail so remarkable was its production solely and directly by stimulation of cardiac tissue. It could be referred to nervous action only on the hypothesis that ganglia (or some such local nervous mechanisms) were present in the excised snail's heart. But this hypothesis could claim no more support now than it had in 1859, when Foster reported the results of his first experiments on the snail's heart. He argued therefore that any explanation of how direct electrical stimulation could inhibit the snail's heartbeat must begin with the fundamental properties of its contractile tissue. He concluded with what he admitted to be a vague and undeveloped hypothesis pointed in this direction. Perhaps the interrupted current induced "molecular changes" in the cardiac tissue that interfered with and inhibited the usual sequence of events leading up to the heartbeat.

[65] Foster, "Hemmungswirkung" (see n. 13 above).

This paper of 1872 was the starting point for the extensive researches that Foster undertook between 1874 and 1876 in collaboration with A. G. Dew-Smith, his student, friend, and benefactor. These researches, in turn, mark the beginning of the brief but important period when Foster directed all of his energies as well as those of his students toward a solution to the problem of the heartbeat and its origin. The evolutionary foundations of Foster's position remained largely submerged in additional experimental evidence that was doubtless presented with at least one hopeful eye directed toward Continental physiologists.

On 1 February 1875, for example, the Royal Society received from Foster and Dew-Smith a paper describing the effects of electric currents on the heart in several species of mollusks, but principally in the common snail.[66] Their first experiments, like those in Foster's paper of 1872, involved the direct application of an interrupted current to the excised but still beating heart of a snail. These experiments confirmed in greater detail Foster's earlier results: in a heart beating spontaneously, induction shocks too weak to evoke contraction produced a prolongation of the diastole, and therefore inhibition, when repeated rapidly. When using constant instead of interrupted currents, Foster and Dew-Smith got similar results. The passage of a sufficiently weak current, no matter in what direction, tended to produce inhibition. They then returned to the problem from which they had begun, namely how direct stimulation of cardiac tissue could produce inhibition.

According to the "favourite theory," local nervous mechanisms were as responsible for inhibition as they were for the heartbeat. The leading assumption in this theory was that the heart contained separate nervous mechanisms for inhibition and for contraction.[67] Contraction, it was further assumed, could prevail over inhibition only because the inhibitory mechanisms were more susceptible to exhaustion. Contraction would occur, therefore, when inhibition was at its minimum. But in their experiments with interrupted currents, Foster and Dew-Smith found that inhibition *increased* with the strength of the current until finally "the action of the stimulus is suddenly reversed and a contraction or beat is caused instead of prevented."[68] Contraction, that is to say, succeeded *maximum* and not minimum

[66] Michael Foster and A. G. Dew-Smith, "On the Behaviour of the Hearts of Mollusks under the Influence of Electric Currents," *Proc. Roy. Soc., 23* (1875), 318-343.

[67] For an early statement of this assumption, see Stannius (n. 11), p. 92.

[68] Foster and Dew-Smith, "Hearts of Mollusks," p. 323.

inhibition, and so Foster and Dew-Smith dismissed the theory that inhibition, like contraction, was regulated by nervous mechanisms.

Their alternative theory focused instead on the effects of the current on the fundamental properties of contractile tissue, as indeed Foster had already insisted it must. They suggested that weak electrical currents might not by themselves provide enough energy to establish polarization (anelectrotonus and kathelectrotonus) in the ventricular tissue. Some, perhaps most, of the energy required for this polarization might be taken instead from the cardiac tissue itself. As a consequence, there would be less energy available for purposes of pulsation, and the heartbeat would be inhibited.[69] Foster's earlier vague appeal to "molecular changes" had merely become slightly more concrete.

In thus fulfilling their main goal—to explain Foster's 1872 discovery of electrically induced inhibition—Foster and Dew-Smith naturally focused on experiments that would yield inhibition, that is to say, on experiments in which currents were applied to the spontaneously beating heart. But they also described a series of experiments in which constant currents were applied to a previously *quiescent* snail's heart. Their major finding was that a quiescent heart could, under certain circumstances, be induced to rhythmic pulsations passing in the normal direction from auricle to ventricle, but never in the reverse direction.[70] Clearly, then, in the snail, as in the vertebrate, there existed some functional distinction between the upper and lower end of the ventricle.

In the vertebrate heart, this distinction was generally attributed to the ganglia found at the upper end of the ventricle, between it and the auricles. In the snail's heart, however, this hypothesis would not do. More impressive than certain physiological evidence against it was the simple fact that careful histological investigation failed to detect ganglia anywhere in the snail's heart. Therefore, argued Foster and Dew-Smith, nervous mechanisms could not explain why the induced beats began only at the auricular end of the heart. This result ought to be ascribed instead to the shape of the ventricle (the greater mass of contractile tissue lying toward the upper end) and to the fact that from the time the heart is first formed, "the nutrition . . . of each part is habituated and regulated to such conditions as are involved in the normal beat beginning at the auricular end."[71]

In a penultimate section entitled "General Considerations," Foster and Dew-Smith sought to relate their conception of the snail's heart-

[69] Ibid., p. 333. [70] Ibid., pp. 323-327. [71] Ibid., pp. 327-328.

beat to more fundamental biological and physiological processes. They began by emphasizing that the constituent fibers ("if fibres we may venture to call them") of the snail's heart were not isolated like those of an ordinary skeletal muscle, but were "physiologically continuous." The ventricle in its entirety might be compared to a single voluntary muscle fiber in the sense that "any change set up in any part of the ventricle can be propagated over the whole of it [just as] a contraction-wave set going at any point in a striated fibre is propagated along the whole." Both are masses of protoplasm acting together as a whole. The normal heartbeat, however, "is not a simple contraction-wave passing uniformly from one end to the other" but a complicated and coordinated act, "having for its object the ejection of fluid from the cavity in the best possible manner." For the sequence to be regular and orderly, changes in each part of the ventricle must be determined by what is going on in the other parts. There must exist "means of communications" between all parts of the ventricle so that each part not only "feels, if we may use the phrase," what is going on in every other part, but also communicates to those parts its own condition.

But precisely how is this coordination achieved? Just what are these "means of communication"? Since the snail's heart contained no ganglia or differentiated structures of any kind, this capacity for coordination and communication must reside in the undifferentiated protoplasm itself. Foster saw no difficulty in this apparently remarkable conclusion. The coordinated movements of cilia were just one example favoring the idea that there exists "a *consensus* among the several parts of any mass of protoplasm which acts together as a whole." Furthermore, Foster suggested,

> there must be in undifferentiated protoplasm the *rudiments* of all the fundamental functions present in the differentiated structures of higher animals. Thus the process by which [for example] the condition of the aortic region of the snail's ventricle is communicated to the auricular region seems to be the rudiment of the *muscular sense*.[72]

This conception led Foster to propose a new interpretation of the role played by ganglia, when present, in rhythmic and automatic movements. According to the prevailing conception, ganglia determined the rhythmic movements of the heart by discharging rhythmic motor impulses to the cardiac muscles, "the contractile elements being, so to speak, mere passive instruments of the nerve cells." Foster

[72] Ibid., p. 338.

argued instead that ganglia, when present, acted simply as coordinators, serving essentially the same function in higher organisms that "the rudiment of the muscular sense" performed in the undifferentiated protoplasm of the snail's heart. He illustrated his view with this discussion of the frog's heart:

> The contractile tissue of the frog's ventricle, arranged in bundles isolated by connective tissue, is not physiologically continuous . . . and the consensus we spoke of . . . cannot be effected by molecular communications. Hence the existence [in the frog] of differentiated organs, in the form of nerves and ganglia, by means of which indications of the condition of the isolated constituents of the ventricle . . . are carried, *as items of a muscular sense,* to a central organ, and thus the state of each part is made common to all. In this central organ the advent and character of each beat is determined as the expression of the nutritive condition, not of the nerve cells only, but of the contractile elements as well.[73]

Foster and Dew-Smith, 1876: the effects of electric currents on the frog's heart

It was doubtless to test this interpretation that Foster and Dew-Smith turned next to the effects of constant currents on the hearts of frogs and toads. They published in 1876 a series of experiments designed to show that the lower two-thirds of the frog's ventricle could be brought from a state of inactivity to one of rhythmic pulsation by direct application of appropriate constant currents.[74] That interrupted currents could do the same Foster had already claimed in his paper of 1869, but the view persisted that a ventricle deprived of its basal ganglia was unable to beat rhythmically. To controvert the established view, Foster and Dew-Smith had to do more than simply

[73] Ibid., p. 339. Here again Foster reveals his deep commitment to Paget's notion that all rhythmic movements depend ultimately on rhythmic nutrition. Indeed, he considered this bond of agreement more important than any differences between them—so much so that he could describe his own view as "simply a modification of Sir J. Paget's conception," even though he rejected Paget's more immediate conclusion that the vertebrate heartbeat was neurogenic.

Nearly fifty years after Paget delivered his Croonian Lecture, Foster tried to assess its influence. Though it did not, he wrote, "contain any striking original observations, it gave a careful account of the knowledge of the time, with many indications that the author had himself by experiment verified the statements of others." Of Paget's emphasis on rhythmic nutrition, Foster wrote that it had for its time "all the charm of an illumining idea," and that it had served as "the basis of many speculations, as well as the starting-point of many inquiries." See Stephen Paget, ed., *Memoirs and Letters of Sir James Paget* (London, 1902), p. 220.

[74] M. Foster and A. G. Dew-Smith, "The Effects of the Constant Current on the Heart," *J. Anat. Physiol., 10* (1876), 735-771.

demonstrate that their constant currents could evoke a series of beats. Carl Eckhard had already established this point without persuading the physiological community that the resulting beats were truly comparable to the rhythmic pulsations of the normal heartbeat. Foster and Dew-Smith set out in the hope of succeeding where Eckhard had failed.

Using currents at least strong enough to evoke a beat, Foster and Dew-Smith found a variety of effects depending on current intensity. When the intensity was gradually increased, the number of contractions produced was also gradually increased. Then, at a certain point a series of "distinct rhythmic pulsations" appeared, and though they lasted only as long as the current, these pulsations were "in all their characters" comparable to normal heartbeats. "Nothing like this," the authors insisted, "is ever witnessed in any ordinary muscle." No adjustment in the strength of the current applied could alter the frequency of its contractions. In an ordinary muscle, a current strong enough to produce any contraction at all produced a complete and rapid tetanus. In their experiments with the frog's ventricle, no comparable tetanus ever appeared.[75]

The artificially induced beats were, then, truly rhythmic. It remained to explain their origin. Referring them to local nervous mechanisms would not do. Ganglia certainly did not exist in this lower portion of the ventricle, and all attempts to detect distinct nervous elements of another kind had been so far unsuccessful.[76] To Foster and Dew-Smith the facts pointed instead to the conclusion that electrical stimulation was just one way to make manifest a latent rhythmicity in cardiac tissue and just one way of showing that the heartbeat was essentially independent of nervous mechanisms. The *direction* of the beats aroused by constant currents seemed equally independent of localized mechanisms and was more probably "connected with the shape of the ventricle and the consequent unequal distribution of the anodic and kathodic areas."[77]

Thus far Foster and Dew-Smith had considered only the isolated

[75] Ibid., pp. 737-739.

[76] In February 1876, just five months before Foster and Dew-Smith published this paper, a German histologist did describe "nets of nerve fibers" (Nervenfäsernetze) throughout the frog's ventricle. See Leo Gerlach, "Ueber die Nervendigungen in der Musculatur des Froschherzens," *Archiv für pathologische Anatomie und Physiologie, 66* (1876), 187-223. But then, in the nick of time, another German histologist denied Gerlach's claim. See Ernst Fischer, "Ueber die Endigung der Nerven im quergestreiften Muskel der Wirbelthiere," *Centralblatt, 14* (1876), 354-356. An expanded version of Fischer's paper appeared in *Archiv für mikroskopische Anatomie, 13* (1877), 365-390.

[77] Foster and Dew-Smith, "Constant Current," pp. 739-741.

lower portion of the frog's ventricle, but they found no reason to alter their basic position when they extended their experiments to include the ventricle as a whole or even the entire heart. They were especially impressed by the effects of the constant current on a heart rendered quiescent by "Stannius' experiment," that is by section of the boundary between the sinus venosus and the auricles. After this operation, which left intact the boundary between the auricles and ventricle, the dominant and most characteristic effect of applying constant currents was a ventricular beat followed by an auricular one. Under proper circumstances this initial reversal of the usual sequence (from auricle to ventricle) was followed by others until a new rhythm was established in the reverse direction.[78]

This result required emphasis because of the very different conclusions drawn from similar experiments by the German physiologist Julius Bernstein. He too had shown that the normal sequence of the frog's heartbeat could be reversed by the application of constant currents to a heart made quiescent by Stannius' experiment. But he had also claimed that the normal sequence would be established if the same currents were applied in the opposite direction. From his conclusion that the sequence was produced in the direction of the current, he developed an explanation based on the effects of the current on the nervous mechanisms located between the auricles and ventricle.[79] Admitting that the sequence could be produced in the normal as well as in the reverse direction, Foster and Dew-Smith insisted that this could be achieved only with strong currents. They concluded that the direction of the sequence depended not so much on the *direction* of the current as on its *strength*. And whereas current direction might exert its greatest effect on ganglia or other nervous mechanisms, current strength should exert its greatest effect on *muscular* tissue. Accordingly, the ventricle, because it is more muscular than the auricle, is also more responsive to external stimuli, and so the ventricular beat would generally precede the auricular upon stimulation of a heart rendered quiescent by Stannius' experiment.[80]

But if the more muscular ventricle really were more susceptible to external stimuli (such as the constant current) than the more nervous auricle,[81] why did the normal heartbeat always begin in the auricle?

[78] Ibid., pp. 753-759.

[79] Julius Bernstein, *Untersuchungen über den Erregungsvorgang im Nerven-und Muskelsysteme* (Heidelberg, 1871), esp. pp. 205, 225-227.

[80] Foster and Dew-Smith, "Constant Current," p. 763-764.

[81] An assertion that is, not so incidentally, exactly the inverse of Stannius' conclusion that the *auricle* is more susceptible to external stimuli. See Stannius (n. 11), pp. 88-89. In fact, by this assertion, Foster and Dew-Smith seem not only to have

For Foster and Dew-Smith, the fact that the normal direction of the frog's heartbeat could be reversed at all was sufficient proof that "the sequence is not the result of any fixed molecular constitution of the ganglia." It must depend instead on "a concurrence of circumstances," one being that "in the normal heart the nutrition of all parts is, so to speak, tuned for the production of a beat with the normal sequence." The basic determinant, though, was the general condition of the cardiac tissue in each part of the heart as it was somehow made known to every other part. On grounds no more conclusive than those given in their previous paper, Foster and Dew-Smith again suggested that the essential influence here was "a sort of muscular sense" and that the connecting nervous elements functioned merely as transmitters and coordinators.[82]

In the final section of this long and complex paper of 1876, Foster and Dew-Smith reported their most remarkable experimental result. All along they had been demonstrating that appropriate constant currents could bring cardiac tissue from a state of quiescence to one of rhythmic pulsation. *But they now showed that in a heart rendered quiescent by stimulation of the vagus nerve, constant currents, no matter how strong, were unable to evoke rhythmic beats.* So long as the vagus retained its efficacy, inhibition took place as usual and "very much as if no current were passing through the heart." The only observable effect was a postinhibitory "reaction" of such force that upon removing both the constant current and vagus stimulation the heart began to beat "with remarkable vigour and rapidity."

As Foster and Dew-Smith interpreted it, this striking result was just further evidence that nervous influences played a subordinate role in the movements of the heart. In the prevailing theory of inhibition, impulses were believed to descend the vagus nerve until they reached ganglionic centers, where by exalting certain "molecular inhibitory forces" they prevented the ganglia from discharging their usual rhythmic stimuli. According to this conception, as Foster and Dew-Smith so vividly expressed it, "the muscular fibres lie idle till the struggle in the ganglia is over." But if the muscular fibers really were isolated from ganglionic stimuli during inhibition, it followed that the constant current ought to be able, as usual, to arouse them to action. That it could not do so suggested that "stimulation of the va-

violated common physiological canons but also to have undermined their own suggestion that the ganglia could serve as "coordinating" elements precisely because nervous tissue is more easily excited than muscular tissue.

[82] Foster and Dew-Smith, "Constant Current," pp. 759-761.

gus . . . whatever effect it may have on the ganglia, has also an effect of such a kind that the irritability of the cardiac muscular tissue [itself] is impaired, and the production of rhythmic pulsations hampered in their muscular origin."[83]

Foster and Dew-Smith missed no opportunity to relate this work on the frog to their earlier work on the snail's heart. They were now able to emphasize the similarity between the inhibition produced in the frog's heart by stimulation of the vagus nerve and the inhibition Foster had produced in 1872 by applying electrical current to a spontaneously beating snail's heart. Both in the snail and the frog, inhibition resulted from the effects of external stimuli on previously active contractile tissue. Equally similar were the responses of the frog's heart and the snail's when external stimuli were appropriately applied to previously *quiescent* cardiac tissue; in both, the result was a series of rhythmic pulsations. In neither animal were ganglia necessary either for inhibition or for rhythmic pulsation.[84]

Foster and the effects of upas antiar on the frog's heart, 1876

Although the paper of 1876 with Dew-Smith represents Foster's last attempt to challenge the neurogenic theory through his own research, being in fact his last research paper in physiology,[85] he had already advanced still more evidence against it in a paper published earlier that same year. Here again the subject was the frog; but this time, instead of applying electrical currents to the heart, Foster studied the effects of the poison, upas antiar. He particularly emphasized the unusual effect of vagus stimulation on a heart poisoned with antiar. For a certain period after administration of the poison, vagus stimulation produced its usual inhibitory effect on the heartbeat. But this inhibitory phase soon gave way to an opposite, augmentative phase that increased as the influence of the poison increased. Eventually, when the antiar had brought the heart to a standstill, the sole effect of vagus stimulation was to produce a long series of vigorous beats.[86]

[83] Ibid., pp. 766-767. [84] Ibid., pp. 768-769.

[85] In 1877 Foster did publish (again with Dew-Smith) a brief paper on the heart of mollusks, but it involved no new research and was really only a reply to a critique of their joint paper of 1875. See Michael Foster and A. G. Dew-Smith, "Die Muskeln und Nerven des Herzens bei einigen Mollusken," *Archiv für mikroskopische Anatomie, 14* (1877), 317-321. For a full discussion of this paper, see Chapter 8 below, "Francis Darwin." Foster did of course publish many other papers after 1876 (notably on irises), but none of them grew out of new original research in physiology.

[86] Michael Foster, "Some Effects of Upas-Antiar on the Frog's Heart," *J. Anat. Physiol., 10* (1876), 586-594. Foster communicated the substance of this paper to

As usual, Foster displayed skepticism toward any nervous explanation of these phenomena. Others might have used his results to argue that the frog's vagus nerve contained not only inhibitory but also accelerator fibers, of which the former were gradually poisoned by antiar while the latter remained unaffected. Foster did not, even though accelerator fibers had recently been traced to the *mammalian* heart. A decade later, Foster's own favorite student, Walter Gaskell, was to establish the existence of accelerator fibers in the *frog's* cardiac nerves as well.[87] For now, however, Foster could refer to the hypothetical accelerator fibers in the frog as a "*Deus ex machina,*" and could offer at least one solid piece of evidence against their existence. In the mammalian heart, stimulation of its known accelerator fibers produced *weaker* if more frequent beats in a heart slowed by antiar. In the antiarized frog's heart, by contrast, stimulation of the vagus—and thus its presumed accelerator fibers—ultimately produced (in the augmentative phase) *stronger* as well as more frequent beats. Having thus cast doubt on the most probable explanation of his results by a nervous mechanism, Foster used his earlier work on electrically induced inhibition in the snail's heart to argue that his results with antiar were really just typical of inhibition in general. However achieved, cardiac inhibition was eventually followed by an augmentative phase.

In the excised and nerveless snail's heart, this postinhibitory augmentation must be due to processes taking place in the contractile tissue itself. Antiar was known to be a muscular poison, which produced its effects "chiefly at least by interfering with the functions of the muscular tissue of the heart." If only it were further admitted that vagus inhibition also resulted from a direct action on cardiac tissue, Foster thought he might be able to offer an explanation for the marked augmentation observed upon vagus stimulation of the antiarized frog's heart. Insofar as it can be divined from this often obscure paper, Foster's proposed explanation seemed to run something like the following. Antiar, we know, acts directly on the cardiac musculature; so, perhaps, does vagus inhibition. If both initially act in an "inhibitory" fashion, then perhaps both also participate in the opposed augmentative phase that typically follows inhibition. And if so, the augmentation produced by their combined action might be so marked as to yield beats at once stronger and more rapid than normal. Foster then expressed the hope, later fulfilled in his final paper with Dew-Smith, that he would soon be able to present "some

---

the Cambridge Philosophical Society on 15 November 1875. See *Proc. Camb. Phil. Soc.*, 2 (1876), 398-400.

[87] See Chapter 10 below, pp. 315-316.

facts which seem to me clearly to prove that the muscular mass of the frog's ventricle is affected during inhibition, in addition to whatever may be going on in the ganglia."[88]

The significance of Foster's work on the heart

Throughout these papers, then, the trend of Foster's thought is clear. The rhythmic heartbeat has its essential origin in the cardiac musculature. Moreover, whenever and whatever alteration appears in the heartbeat, whether inhibitory or excitatory, that alteration is produced primarily at least by an influence acting directly on contractile tissue. Nervous mechanisms of whatever sort play at most a secondary role. "The thesis," as Foster once put it, "of which all our experiments may be regarded as illustrations [is] that the causes of the rhythmic pulsations are to be sought for in the properties of contractile tissue."[89]

We now believe with Foster that the heartbeat is fundamentally myogenic. Heart transplant operations provide spectacular proof that even the highly specialized human heart can retain its capacity for rhythmic action after being severed from the central nervous system. Within the heart itself, the fibers in which the beat begins, as well as those by which it is transmitted from the auricle to the ventricle, are relatively undifferentiated. Cardiac fibers or cells, cultivated outside the body in tissue culture, show an inherent capacity for rhythmic contraction and give rise spontaneously to electrical action potentials, which can be recorded with microelectrodes even in individual cardiac cells. Perhaps now, as then, the most decisive evidence that the rhythmic heartbeat does not require specialized nervous structures is the fact that the fetal heart begins its rhythmic pulsations before it has received any nerves or ganglia. Even the otherwise anomalous heart of the invertebrate *Limulus* crab, which stops beating when isolated from its nervous system, is myogenic in the less differentiated larval stage. Moreover, cardiac inhibition and acceleration, though ordinarily the result of nervous action, arise more proximately from the effects of chemical substances (acetylcholine and epinephrine) released from nerve ends and acting directly on contractile cardiac tissue. Finally, though we could and would add specificity to Foster's concept of "rhythmic nutrition" (using such terms as "ionic balance"), we would scarcely alter his message in its essentials.[90]

88 Foster, "Upas-Antiar," pp. 590-594.

89 Foster and Dew-Smith, "Constant Current," p. 766.

90 See Arthur C. Giese, *Cell Physiology*, 4th ed. (Philadelphia, 1973), pp. 541, 610-615.

But, interesting as it may be to learn that Foster chose the "right" side in the myogenic-neurogenic debate, that is not why his research has been rescued here from its former near oblivion. Nor was his research by itself terribly influential in the debate. The Germans generally ignored his work, and the most significant and most spirited phase of the debate came after he had abandoned active research. In the first edition of his celebrated textbook, published one year after his final paper with Dew-Smith, Foster acknowledged that the neurogenic theory continued to prevail despite his efforts.[91] If Foster's research did not settle or even by itself greatly influence the debate over the origin of the heartbeat, it nonetheless deserves resurrection for at least three reasons. In the first place, it reveals an important but unfamiliar fact about Foster: that he was a competent, if unprolific, investigator with a capacity for original research. The quality of his work should not be undervalued simply because the German physiologists so largely ignored it. Having found little reason during the previous generation to pay close attention to English physiology, they probably did not bestow on Foster's work the consideration it would have received had it been published by a fellow German physiologist.[92]

In the second place, Foster's work on the heart provides a concrete demonstration of his biological and evolutionary approach toward physiology. For what especially distinguished his attempt to demonstrate the myogenicity of the heartbeat was the manner in which he appealed ultimately to fundamental biological considerations, to the basic properties of protoplasm and other contractile tissue, for example, and to evolutionary concepts. The evolutionary basis for his posi-

[91] Foster, *Textbook*, pp. 121-122.

[92] None of Foster's papers is cited in the highly documented discussion of the rhythmic heartbeat in Hermann's famous textbook. See Ludimar Hermann, ed., *Handbuch der Physiologie*, 6 vols. (Leipzig, 1879-1881), IV, pp. 349-374. In France, by contrast, Milne-Edwards found the last joint paper by Foster and Dew-Smith sufficiently meritorious to deserve citation, though he says nothing about their conclusions. See Milne-Edwards, *Leçons*, XIII (1878-1879), pp. 273-274, n. 1.

It is probably no accident that the only work by Foster to attract any discernible immediate attention in Germany was published not in English, but in German. That paper, on electrically-induced inhibition in the snail's heart, at least found its way into the leading German abstracting journal. See *Centralblatt, 10* (1872), 238. The tendency of the German physiological community to ignore or incompletely appreciate the English literature persisted to some extent throughout the rest of the nineteenth century, long after significant work had begun to appear in England. Walter Gaskell's epoch-making papers on the heart (see Chapter 9 below) received no attention in the *Centralblatt*, while his equally important work in 1886 on the involuntary nervous system was apparently unknown to the famous German histologist Albert von Koelliker nearly a decade after its publication. Cf. Sir Walter Langdon-Brown to Donal Sheehan, 11 March 1940, copy deposited in the obituary files of the Yale University Medical Historical Library.

tion emerges quite clearly in his final paper with Dew-Smith. There Foster and his co-author suggested that the "less differentiated cardiac muscular tissue" retained a capacity for spontaneous rhythm that in the course of evolution had been "almost entirely lost by those [more differentiated] muscles which are more completely under the dominion of the will." This capacity for rhythmic movement, they insisted, was a fundamental property of primitive protoplasm and traces of it could be found "in all protoplasm, from that of a bacterium or a vegetable cell upwards."[93] In a postscript to the same paper, they developed their argument further by means of an example that can be properly appreciated only in its full context.

In that postscript, Foster and Dew-Smith criticized a paper, recently published by Bernstein, in which the latter had argued that the isolated lower portion of the frog's ventricle was incapable of truly spontaneous beats. Bernstein's paper, in turn, was directed against the 1875 work of J. Merunowicz, who had conducted his experiments in Carl Ludwig's institute at Leipzig. In the most important of these experiments, Merunowicz had separated the lower two-thirds of the frog's ventricle from the ganglia located in the upper portion and had then supplied it with rabbit's serum or dilute rabbit's blood, whereupon rhythmic beats ensued.[94]

Bernstein did not dispute the experimental result, but rather the conclusion drawn from it. Merunowicz believed that his experiment could be explained only on the hypothesis that some "automatic motor apparatus" existed in the lower tip of the ventricle as well as in the auricles and in the auriculo-ventricular tissue. What most impressed Foster and Dew-Smith about Merunowicz's work was the support it might bring to their own work with the constant current. The essential feature in both cases was the production of rhythmic pulsations in a previously quiescent portion of the heart through application of a constant stimulus. Foster and Dew-Smith could therefore describe Merunowicz's paper "as corroborative of some at least of our own conclusions."[95] In their view, both investigations demonstrated the inherent rhythmicity of ventricular muscle and undermined the belief that ganglia were necessary for the production of the heartbeat.

Bernstein developed an explanation designed not so much to deny the rhythmic capacity of ventricular tissue as to assert that this capac-

[93] Foster and Dew-Smith, "Constant Current," pp. 739, 741.

[94] Julius Bernstein, "Ueber den Sitz der automatischen Erregung im Froschherzen," *Centralblatt, 14* (1876), 385-387. Cf. Bernstein, "Bermerkung zur Frage über die Automatie des Herzens," ibid., pp. 435-437.

[95] Foster and Dew-Smith, "Constant Current," pp. 735-736.

ity became manifest only under abnormal circumstances. He remained convinced that the normal heartbeat resulted from the rhythmic action of nervous ganglia. A new experiment seemed to him further proof of this view. When he pinched the frog's ventricle across its middle with a pair of forceps, the upper part (containing the ganglia) continued to beat while the lower part remained perfectly quiescent. Since the lower tip retained its anatomical connection with the rest of the heart and continued to receive a fresh supply of blood, Bernstein ascribed its quiescence solely to its physiological isolation from the ganglia known to exist in the upper part of the heart. He therefore insisted that the rhythmic beats Merunowicz had produced were not really spontaneous, but rather an artificial rhythm resulting from the action of foreign fluids (the rabbit's serum or blood) on "certain motor mechanisms in the cardiac muscles, . . . causing in them an intermittent discharge of energy."[96] Whatever meaning Bernstein may have attached to the phrase "motor mechanisms," his essential point was that the pulsations produced in the lower tip under the influence of Merunowicz's fluids, or of any other "foreign" stimulus (chemical or electrical), had nothing to do with the normal heartbeat.

Bernstein's work failed to convert Foster to the prevailing theory of the heartbeat. According to that theory, the impulses that caused the heartbeat, and determined the rate and character of its rhythm, "proceeded *in a rhythmic manner* from the ganglia in the sinus." The muscular tissue of the heart, it was assumed, "had no other task than to respond to those rhythmic impulses according to the measure of its irritability." But Bernstein now admitted that the ventricular tissue could respond rhythmically to a *constant* stimulus, to a stimulus "so constant in its nature and action as serum or blood." Since the beats thus produced differed in no observable way from normal spontaneous beats, and since in their own experiments with another constant stimulus (the constant current) equally spontaneous beats had been produced in this same portion of the frog's ventricle, Foster and Dew-Smith argued again that it was unnecessary to attribute rhythmic action to the ganglia. "Its presence would shew a wasteful want of economy," they wrote, moving on to more general biological considerations.

Foster had already emphasized the rhythmic capacity of the largely undifferentiated protoplasm of the snail's heart. So he very much doubted that differentiated nervous mechanisms were required to perform the same function in the vertebrate heart. To him, "looking at the matter from an evolution[ary] point of view, and seeing that

[96] See Bernstein (n. 94).

muscular or neuro-muscular tissue is anterior in evolution to strictly differentiated nervous tissue," it made no sense simultaneously to suppose that nervous mechanisms normally determined the rhythmic heartbeat and yet that cardiac muscle displayed under exceptional circumstances an inherent capacity for rhythmicity. If the rhythmicity of the vertebrate heartbeat really were normally determined by nervous mechanisms, then why, Foster wondered, should cardiac muscle retain an inherent capacity "which it could never, in actual life, have an opportunity of manifesting, since [this capacity] would ever afterwards be determined by its nervous master?" The way out of this difficulty was to admit that even in the vertebrate heart, despite all its nerves and ganglia, rhythmicity had its origin not in any nervous mechanisms but entirely in muscular tissue. On this view, the influence of any ganglionic stimuli, by analogy with the constant current and blood fluids, would be continuous and not rhythmic.[97]

Perhaps no single example better exemplifies Foster's evolutionary and essentially biological attitude toward physiology. He believed that to understand rhythmicity fully, it was as important to consider the evolution of that capacity as it was to consider its experimental manifestations. There was nothing explicitly or necessarily "Darwinian" about his position, insofar as "Darwinian" is used as a synonym for "natural selection," but Foster and his students did at least consistently address an issue that seemed never to occur to their Continental counterparts: why should more fully evolved (or more fully differentiated) nervous tissue be required for a function (namely, rhythmicity) that already could be performed by a more primitive (and thus, in terms of energy expenditure, less "expensive") tissue, namely, undifferentiated protoplasm. The more fully one appreciates the extent of Foster's evolutionary attitude, the better one understands not only his own research and his influence on the general development of the biological sciences at Cambridge, but also his great influence on the particular direction of physiological research there.

This brings us to the third and most important reason for our detailed examination of Foster's work on the heart: it was the starting point for so much of the research undertaken in the formative years of the Cambridge School of Physiology. The simple fact is that experimental physiology got its start at Cambridge when Foster directed the energies of his students toward a solution to the problem of the rhythmic heartbeat. His influence is even more manifest in the general approach his students took toward this and other problems than it is in the particular conclusions they reached.

[97] Foster and Dew-Smith, "Constant Current," pp. 770-771.

8. The Problem of the Heartbeat and the Rise of the Cambridge School

Every man of science . . . is strengthened and encouraged by the presence, at once stimulating and guiding, of his scholars. He feels that without these his mind would be in danger of getting entangled in cobwebs. The eager questions, the apt remarks, nay, even the silent faces of his pupils, serve to brush all cobwebs away; and the knowledge that, as he is working surrounded by learners, he is not only advancing science but forming a school which, even if it be small, will carry on his work after him, gives him a strength greater than his own.

Foster (1906)[1]

WHEN in the mid-1870's Foster concentrated his efforts on the problem of the heartbeat, he was far from alone. A. G. Dew-Smith is only the most obvious of his Cambridge collaborators. Another, much less important, was Arthur Sheridan Lea, who conducted a histological investigation of the snail's heart for Foster and Dew-Smith. Lea's task was to determine whether any distinctly specialized nervous structures could be detected in the snail's cardiac tissue. His conclusions—that no such structures could be found and, moreover, that no nerves entered or left the snail's heart—were never published separately, but were simply incorporated into Foster and Dew-Smith's own first paper.[2]

Francis Darwin

A more elaborate study of the same sort was made soon after by Charles Darwin's son, Francis, who received his bachelor's degree from Cambridge in 1870 and his M.B. in 1875. Darwin's study was thorough enough to merit independent publication.[3] In the first sentence of this short paper, he referred to the earlier work of Foster and Dew-Smith as an "important research" in which, however, "the anatomy of the organs studied was necessarily rather briefly dealt with." For that reason, Darwin continued, "Dr. Foster suggested that I should under-

[1] Michael Foster, "Art. IX. The British Museum (Natural History)," *Quarterly Review, 205* (1906), 491-510, on 509.

[2] See Michael Foster and A. G. Dew-Smith, "On the Behaviour of the Hearts of Mollusks under the Influence of Electric Currents," *Proc. Roy. Soc., 23* (1875), 318-343, on 320.

[3] Francis Darwin, "On the Structure of the Snail's Heart," *J. Anat. Physiol., 10* (1876), 506-510.

take the investigation of the histology of the snail's heart." So, fifteen years after he had dismissed the possibility (in his first work on the subject), Foster apparently remained concerned that someone might yet claim that intracardial nervous ganglia were responsible for the rhythmic pulsations he had produced in the excised snail's heart.

Darwin's investigation did much to remove this concern. He supported Foster and Dew-Smith in their major histological conclusion that the snail's heart contained no automatic nervous mechanisms and had no connection with the central nervous system. "I have," Darwin wrote, "searched carefully, but in vain, for ganglion cells, both by staining with chloride of gold, and also with picrocarminate of ammonia, which brings out the nuclei of the ganglia of the central nervous system very brightly."[4] Nor could he detect any nerve passing into the heart. Foster was much impressed by Darwin's work; and when in 1877 someone finally did dispute his conclusions about the myogenicity of the snail's heartbeat, Foster's reply depended primarily on Darwin's work.

This German-trained opponent, Johann Dogiel, admitted that his own research had been inspired by two of Foster's contributions, his 1872 paper on electrically induced inhibition and his 1875 paper (with Dew-Smith) on the effects of electrical currents on the hearts of mollusks. But Foster's conclusions seemed to Dogiel untenable on both histological and physiological grounds. For one thing, Dogiel insisted, not only could "apolar" nerve cells be detected in the hearts of snails and other mollusks, but these cells were in fact concentrated in the regions of greatest physiological significance, in the upper part of the auricle and in the tissue between the auricles and ventricles. Furthermore, he claimed that stimulation of certain extracardiac nervous structures could produce true inhibition in the molluskan ventricle, while direct electrical stimulation of the cardiac tissue simply brought the ventricle instead to a state of tetanic (systolic) standstill. Dogiel concluded that "the heart contractions in the mollusks studied are influenced by the nervous system, and . . . apolar nerve cells are imbedded in the striated cardiac musculature of these animals."[5]

In their reply, Foster and Dew-Smith criticized Dogiel's experiments on the basis of their own earlier work. They suggested, for example, that his failure to produce true inhibition in the snail's ventricle by direct electrical stimulation was due simply to his use of currents too strong to yield the phenomenon. As for the inhibition he had allegedly

[4] Ibid., p. 506.

[5] Johann Dogiel, "Die Muskeln und Nerven des Herzens bei einigen Mollusken," *Archiv für mikroskopische Anatomie, 14* (1877), 59-65, quote on 65.

produced by stimulating extracardiac nerves, it should be ascribed not to nervous influences but to the purely mechanical changes accompanying such experiments (e.g., a sudden drop in the blood pressure). But their most important criticism had to do with Dogiel's histological conclusions, and their most serious charge against him was that he had ignored Darwin's careful work on the histology of the snail's heart. On the basis of that work, Foster and Dew-Smith considered it very likely that the cells Dogiel called "apolar nerve cells" were in fact nothing but the pyriform connective tissue cells described by Darwin. Certainly nothing in Dogiel's paper persuaded them to abandon the essential conclusion of their work on the snail's heart: that a spontaneously contracting organ could be inhibited without the intervention of any special, differentiated nervous mechanisms.[6]

But if Darwin's study of the snail's heart thus contributed to the Cambridge campaign against the neurogenic theory, his earlier work on inflammation most decidedly did not. Indeed, that work reminds us forcefully that the general alliance between myogenicity and evolution was neither inevitable nor universal. For the fact is that Charles Darwin's own loyal son and disciple had no apparent difficulty reconciling his commitment to evolutionary theory with a basically neurogenic point of view. His acceptance of the neurogenic position, foreshadowed in a paper of 1874, emerges explicitly in his M.B. thesis of 1875, the chief object of which was to explain the vascular dilation that accompanies inflammation. The connection between it and the problem of the heartbeat lay in the general question of inhibition.

To explain vascular dilation of any kind, it was necessary to explain how the normal state of mild contraction or tonus in the blood vessels could be overcome. For cases of vascular dilation accompanying inflammation, the German pathologist Julius Cohnheim had recently proposed the hypothesis that inflammatory irritants directly paralyzed the muscular walls of the arteries, thereby destroying what he took to be their inherent power of tonic contraction. He also denied the existence of peripheral ganglia such as might account for the arterial dilation produced in the frog's tongue by applying irritants directly to it. In his paper of 1874, Darwin disputed the latter claim by insisting that in at least one organ, the mammalian bladder, the arterial system is not only accompanied in its ramifications by close-lying ganglia, but is also supplied with nerve fibers proceeding from those ganglia.[7] Then,

[6] Michael Foster and A. G. Dew-Smith, "Die Muskeln und Nerven des Herzens bei einigen Mollusken," ibid., *14* (1877), 317-321.

[7] Francis Darwin, "Contributions to the Anatomy of the Sympathetic Ganglia of the Bladder in their Relation to the Vascular System," *Quarterly Journal of the Microscopical Sciences, 14* (1874), 109-114.

in his M.B. thesis, Darwin argued that if Cohnheim's theory of inflammatory dilation were true, local irritants should always produce vascular dilation, whereas he had found that they sometimes produced *constriction* instead, particularly in the web of the frog's foot.[8] Whatever the effect of local irritants, whether dilation or constriction, it seemed to be the same as that produced by stimulating the vasomotor nerves in the region to which the irritants had been applied. Darwin therefore concluded that inflammatory irritants exerted their effects not by paralyzing the muscular walls of the arteries but rather by stimulating the endings of nearby vasomotor nerves.

But the usual result of stimulating a motor nerve was muscular contraction, which in the case of the arterial walls would produce constriction. To explain how stimulation of a vasomotor nerve could sometimes produce vascular dilation instead, Darwin adopted the standard view that dilation resulted from a nervous "inhibition" of the vasoconstrictor forces responsible for the normal tonus in the blood vessels. According to this view, the blood vessels were ordinarily kept in a state of mild contraction by nervous forces proceeding from local ganglia, and dilation ensued when impulses reaching these ganglia by way of "vasodilator" fibers prevented their usual action. The arterial walls then relaxed and yielded to the pressure of the blood that flowed into and distended them. For this form of inhibitory dilation to take place, nervous ganglia were considered essential.[9]

But the distinguishing feature of the inhibition Foster and Dew-Smith had produced in the snail's heart was its occurrence in the absence of any such ganglia. Darwin met the difficulty by drawing a sharp distinction between the causes of inhibition in simple as opposed to complex organisms. He approached the problem obliquely, discussing first the role of ganglia in the production of the rhythmic heartbeat. Conceding Foster and Dew-Smith's claim that the snail's heart was capable of rhythmic pulsation in the absence of ganglia, he disputed their extension of this conclusion to the fully developed vertebrate heart. He denied their assumption that the ganglia in the frog's heart acted merely to coordinate the rhythmic movements of its separate cavities, accepting instead the more common view that the ganglia in fact determined the rhythmic pulsations themselves. Now, if the rhythmic beats could thus be traced to different mechanisms in the snail and the frog, so too might inhibitory processes have very different causes in the two organisms. That rhythmicity and inhibition

[8] Francis Darwin, "On the Primary Vascular Dilation in Acute Inflammation," *J. Anat. Physiol., 10* (1876), 1-16, esp. 5-7.

[9] Ibid., p. 7.

could arise independently of ganglia in the simple snail by no means implied that these processes must also occur independently of such nervous mechanisms in the more complex frog:

> We have then a simple contractile organ (snail's heart) apparently capable of contracting rhythmically without the help of ganglion cells, and we have a complex contractile organ (frog's heart) in which ganglia . . . are apparently essential to the production of rhythm. In the same way, is it not possible that ganglia may be necessary for the *inhibition* of a complex organ [i.e., the frog's heart or arterial system] in spite of the possibility of inhibiting a *simple* organ [the snail's heart] where no such ganglia exist?[10]

Against this background, it should come as no great surprise to learn that Darwin was not Foster's student in any formal sense. He had graduated B.A. from Cambridge just before Foster arrived there, and his work on inflammation was, by his own testimony, "mainly conducted in the Laboratory of the Brown Institution [in London], under the supervision of [German-born pathologist] Dr. [Edward] Klein."[11] In some of its histological aspects at least, this work was well underway before Foster and Dew-Smith even began their joint investigations. By the time he published the results of his work for the M.B. thesis, Darwin had become impressed by the connection between it and the recent work of Foster and Dew-Smith, but he discussed their work on the snail only to deny its extension to vertebrates.

What probably most interested Foster about these two early papers by Darwin was their demonstrated command of the histological techniques used to detect ganglia. The search for ganglia in the snail's heart might therefore be entrusted to him with full confidence. That he believed ganglia to be essential for the vertebrate heartbeat, and that his early work had been directed by someone else, only made him a more attractive candidate. For if he did confirm Foster and Dew-Smith's view that the snail's heart was destitute of ganglia, the confirmation would gain in credibility by his independence from their own work. In fact, after Darwin had confirmed their view, Foster and Dew-Smith used precisely this line of reasoning in their brief debate with Dogiel. If, they asked, ganglia actually did exist in the snail's heart, as Dogiel believed, how could they have escaped so careful an observer as Darwin, who had no a priori desire to confirm their own position?[12]

[10] Ibid., p. 8.

[11] Ibid., p. 1. For a biographical sketch of Klein, see *Journal of Pathology and Bacteriology, 28* (1925), 684-697.

[12] Foster and Dew-Smith, "Nerven des Herzens," p. 318.

But even as Darwin did confirm Foster and Dew-Smith's conclusions with respect to the snail's heart, he displayed no hint of conversion to their general position on the myogenic-neurogenic debate.[13] If Foster harbored any hopes that Darwin might yet undergo such a conversion and join fully in the Cambridge effort against the neurogenic theory, those hopes went unfulfilled. Between 1875 and 1882, Darwin spent most of his time in Kent, where he served as secretary to his famous father and joined him in studies of the motile and sensitive plants. All of his papers after 1876 are devoted to botanical topics, and his scientific reputation rests chiefly on his contributions to physiological botany. After his father's death in 1882, Francis returned to Cambridge, where he became university lecturer in botany in 1884 and reader in 1888.[14]

In these capacities, Darwin participated in the later stages of the more general "Fosterian" revolution in Cambridge biology, but by then the Cambridge School of Physiology had already attained maturity. His earlier work on inflammation and vasomotor action is of interest here chiefly because it concerned problems strikingly similar to those pursued almost simultaneously by Foster's most loyal disciple, Walter Gaskell, whose related research is discussed at the very end of this chapter. By comparing Gaskell's approach to these problems with Francis Darwin's, we may be led to appreciate how much Foster was himself responsible for forging and promoting the alliance between myogenicity and evolution. Indeed, we may even go so far as to suggest that had the young Francis Darwin worked longer and more intimately with Foster, he might have played a crucial role in the campaign for myogenicity and in the rise of the Cambridge School of Physiology. In the event, he fell only briefly under Foster's influence and contributed only fleetingly and equivocally to either end. Fortunately for Foster, other young Cambridge graduates were eager to join the small group then at work on the problem of the heartbeat. Prominent among them was John Newport Langley, who had just received his B.A. but was already assisting Foster in his teaching and was soon to be appointed his demonstrator.

13 In fact, Darwin very cautiously limited his support for Foster and Dew-Smith to the immediate *histological* question that Foster had asked him to investigate; insofar as he refers to the *physiological* issues, his remarks are either ambiguous or vaguely opposed to the myogenic theory. See Darwin, "Snail's Heart," esp. p. 506, n. 4 and p. 510.

14 On Francis Darwin's career, see J. Reynolds Green, *A History of Botany in the United Kingdom from the Earliest Times to the End of the Nineteenth Century* (London, 1914), pp. 545-553; and A. C. Seward and F. F. Blackman, "Francis Darwin," *Proc. Roy. Soc., 110B* (1932), i-xxi.

John Newport Langley

Of Foster's direct and immediate influence on Langley's initial research there can be no doubt. In his first publication, in February 1875, Langley began with the following sentence: "Dr. Foster, having received from Dr. [Sydney] Ringer . . . a small quantity of the alcoholic extract of jaborandi, placed the drug in my hands, and requested me to observe its physiological action."[15] Langley then described briefly the general effects of jaborandi on the nervous system, the circulation, and the secretory organs of the frog, rabbit, and dog. By April of the same year, he was concentrating primarily on the effects of jaborandi on the circulation,[16] and by October, when his first full length paper was published, his focus had narrowed to the effects of the drug on the heart.[17] Langley's basic approach here was almost identical to that used by Foster in his own work on the effects of upas antiar on the frog's heart.[18] In both, the focus was on the behavior of the poisoned frog's heart when the vagus nerve was electrically stimulated. In both, atropia (or atropin) was used as an antagonist to the drug of primary concern (jaborandi and antiar, respectively). Finally, both lent support to Foster's claim that the essential seat of the heartbeat is the muscular tissue itself.

Langley's immediate aim was to determine the mechanism responsible for the inhibitory effect of jaborandi on the heartbeat. The French pharmacologist E. F. A. Vulpian supposed that jaborandi exerted this effect by stimulating the peripheral fibers of the vagus nerve. He based this claim on experiments suggesting that jaborandi lost its usual inhibitory powers if the vagus nerve had been previously "paralyzed" by administration of urari. But when Langley administered urari gradually and moderately, an injection of jaborandi produced a slowing of the heartbeat even after direct electrical stimulation of the vagus no longer did so. Jaborandi therefore could not produce inhibition by acting on vagus nerve fibers.[19]

Langley went much further. Showing that the sinus, auricle, or ventricle could be separately inhibited by jaborandi, he argued that

15 J. N. Langley, "Preliminary Notice of Experiments on the Physiological Action of Jaborandi," *Brit. Med. J.*, 1875 (1), 241-242.

16 J N. Langley, "On the Physiological Action of Jaborandi," *Proc. Camb. Phil. Soc.*, *2* (1876), 402-404.

17 J. N. Langley, "The Action of Jaborandi on the Heart," *J. Anat. Physiol.*, *10* (1876), 187-201.

18 Michael Foster, "Some Effects of Upas-Antiar on the Frog's Heart," ibid., pp. 586-594. For a detailed discussion of this paper, see Chapter 7 above, "Foster and the effects of upas antiar."

19 Langley, "Jaborandi on the Heart," pp. 187-190.

inhibitory nervous mechanisms could be involved only if they were so widely scattered throughout the heart that each part had its own mechanism. From the opposite capacity of atropia independently to remove or prevent this inhibition in each part of the heart, it followed that the "excitatory" nervous mechanisms on which it allegedly acted must also be scattered throughout the heart. Coupled with well-known histological evidence to the contrary, the supposition that both kinds of nervous mechanisms were present throughout the heart involved a number of physiological difficulties. Rather than resort to a bewildering set of subsidiary hypotheses, Langley found it "a far easier way out of the difficulty to suppose that jaborandi (and therefore of course atropia also) acts directly on the whole neuro-muscular cardiac tissue."[20] Langley's position, in short, was that jaborandi (and atropin) alter heart action not by acting on differentiated nervous elements but rather by acting directly on the cardiac tissue itself. Foster's view of upas antiar was, of course, precisely the same, and the essentially myogenic orientation of Langley's work should now be clear.

Like Dew-Smith, Arthur Lea, and Francis Darwin, Langley too was quickly distracted from the problem of the heartbeat. One of the most striking effects of jaborandi was to evoke copious secretion from the submaxillary gland. Recognizing in this phenomenon an opportunity to clarify the nature of salivary secretion and secretion in general,[21] Langley undertook a systematic investigation of the secretory organs, which occupied him for the next fifteen years. Toward the end of his career, he turned to a wide range of problems connected with the structure and function of the involuntary nervous system. Even though secretion was one of the phenomena sometimes linked with ganglionic action, and though the involuntary nervous system constitutes the most obvious realm of interplay between the nervous and cardiovascular systems, it would be misleading to suggest that Foster exerted a powerful lasting influence on the research of the man who was to succeed him as professor of physiology. In the case of secretion, the internal dynamic of Langley's own work led him quite logically from one focused investigation to another until he had elucidated the general problem "with a precision which had never before been approached."[22] Langley's later and even more famous work on the involuntary nervous system may also seem to reflect certain "Fosterian"

20 Ibid., pp. 195-198, quote on 198.

21 See J. N. Langley, "The Action of Pilocarpin on the Sub-maxillary Gland of the Dog," *J. Anat. Physiol., 11* (1877), 173-180, on 173. Pilocarpin is the alkaloid of jaborandi.

22 C. S. Sherrington, "John Newport Langley," *DNB, 1922-1930*, pp. 478-481, quote on 478.

themes, but Foster probably played no immediately active role in it either.[23] Nonetheless, the fact remains that Langley was launched on his distinguished career in physiological research when he joined the band of workers who, under Foster's direct influence, sought to resolve the problem of the heartbeat.

George John Romanes and the problem of rhythmic motion

Although better known for his later writings on general evolutionary issues, George John Romanes was yet another of Foster's students who did important work bearing on the problem of the heartbeat. Indeed Romanes, whose pertinent research focused on the contractile movements in the jellyfish umbrella, played a vastly more influential role in the resolution of the problem than did Dew-Smith, Lea, Darwin, or Langley. To judge solely from the impact of their published work, he played a more important role than Foster himself. Behind the scenes, however, Foster exerted an influence on Romanes' work remarkably similar to and scarcely less profound than his influence on Walter Gaskell's.

When Foster arrived at Cambridge in 1870, Romanes was among the first students attracted to him. Romanes had entered Gonville and Caius College in 1867 and had obtained a second class in the Natural Sciences Tripos just a few months before Foster's arrival. Clearly, then, Foster was not responsible for his original interest in science. On the future direction of that interest, however, Foster was a strong influence. Until Foster arrived, Romanes had read none of Charles Darwin's works.[24] Soon thereafter he began to study evolutionary theory, and by 1873 had written a note to *Nature* (on color variation in fish) which attracted Darwin's attention and opened the way to a mutual correspondence and friendship lasting until Darwin's death in 1882.

Between 1870 and 1874 Romanes read Darwin, studied medicine briefly (withdrawing because of poor health), and worked in the Cambridge physiological laboratory under Foster. By 1872 or 1873, he had begun to spend vacations at his family's summer home on the Scottish coast, where he constructed a small seaside laboratory. Already, it seems, he had begun to collect and study jellyfish. For two or three years after taking his M.A. from Cambridge in 1874, Romanes worked sporadically in the physiological laboratory at University College un-

[23] See Chapter 10 below, "Gaskell, Langley, and the autonomic nervous system."

[24] See Ethel Romanes, *The Life and Letters of George John Romanes* (London, 1896), p. 8.

der John Scott Burdon Sanderson. His first paper on the medusae was published during this period. Although he gave public and university lectures, Romanes depended for his income on family money. His later years were devoted primarily to the broader themes of evolutionary biology and psychology. Shortly before his premature death in 1894, he gathered the results of this work into a three-volume treatise called *Darwin and after Darwin*.[25]

Romanes undertook his detailed investigations of the jellyfish (or medusae) in an attempt to decide whether or not they possessed nerve tissue. The question was of great evolutionary significance because nerve tissue had been found in all animals higher in the evolutionary scale. Romanes failed in his initial attempts to tease recognizable nerve tissue out of the delicate gelatinous tissue of the jellyfish. He therefore deferred the histological portion of his inquiry and turned instead to the physiology of the jellyfish locomotor system. In the autumn of 1874 he made the "fundamental observation" that he could paralyze the medusa swimming bell either by removing the strip of differentiated marginal tissue that ran around the rim of the bell or by removing all the marginal bodies (or lithocysts) from the strip. From this observation, he concluded that the lithocysts controlled the normal spontaneous movements of the swimming bell.[26]

Romanes described these experiments in the Croonian Lecture for 1876, along with others in which he showed that the swimming bell was almost incredibly indifferent to mutilation by section. So long as at least one lithocyst remained intact, the bell would continue its normal contractions despite extensive spiral section and interdigitating cuts. Eventually, however, when progressive section produced a sufficiently narrow band of tissue, the contractile waves proceeding from the margin could be interrupted or "blocked." The phenomenon was remarkably consistent: "*At whatever point in a strip that is being progressively elongated by section the contractile wave becomes blocked, the blocking is sure to take place completely and exclusively at that point.*" Pressure, rather than section, could also produce such blocking.[27]

As we shall see, these experiments and Romanes' use of the term "block" were to have a clear echo in Gaskell's later work on the heart.

25 On Romanes' life and work, see ibid.; and John Lesch, "George John Romanes," *DSB, XI* (1975), 516-520.

26 G. J. Romanes, "Locomotion of Medusidae," *Nature, 11* (1875), 29.

27 G. J. Romanes, "Preliminary Observations on the Locomotor System of Medusae," *Phil. Trans., 166* (1876), 269-313, quote on p. 293. For official verification that Romanes was Croonian Lecturer for 1876 see *Record of the Royal Society of London for the Promotion of Natural Knowledge*, 4th ed. (London, 1940), p. 360.

More generally, as Richard D. French has emphasized in two important and suggestive papers, Romanes' work on the medusae had a remarkably broad impact on the development of English physiology. In one of these papers, French shows that Romanes' work played a prominent part in the genesis of the neurone theory and in the elucidation of its physiological consequences. In the summer of 1877, Romanes asked E. A. Schäfer (yet another of Foster's former students) to join him in his studies of the medusae and, more particularly, to seek out those nervous elements that had thus far escaped Romanes. Schäfer, a master of histological technique, managed to find what Romanes had not: a distinct, albeit primitive and unusual, nervous system in the medusa swimming bell. As a result, Schäfer produced a paper containing perhaps the first clear statement of the neurone theory, for he described the medusa nervous system as one "formed of separate nerve units without anatomical continuity." The work of Romanes and Schäfer apparently led in turn to the rapid and immensely fertile development of the synapse concept by Charles Sherrington.[28]

In his other paper, which is more pertinent here, French explicitly links Romanes' research on the medusae with the work of Foster and especially Gaskell on the problem of the heartbeat.[29] To underscore Romanes' influence on Gaskell, French quotes from the following passage of 1935 by Sherrington, who studied under both Foster and Gaskell:

> [Romanes was engaged in] a study of the contractile movement of the swimming-bell of the Medusa. His observations carried out with simple means were novel and fundamental. The questions which he put to the swimming-bell and answered from it, led, it is not too much to say, to the development of modern cardiology. Medusa swims by the beat of its bell, and Romanes examining it discovered there and analysed the two phenomena now recognized world-over in the physiology of the heart, and there spoken of as the "pace-maker" and "conduction-block." Romanes' work, published in the *Philosophical Transactions* of 1877, had directly inspired Gaskell's on the heart. . . .[30]

[28] Richard D. French, "Some Concepts of Nerve Structure and Function in Britain, 1875-1885: Background to Sir Charles Sherrington and the Synapse Concept," *Med. Hist., 14* (1970), 154-165.

[29] Richard D. French, "Darwin and the Physiologists, or the Medusae and Modern Cardiology," *J. Hist. Biol., 3* (1970), 253-274. On the relationship between French's work and mine, see Acknowledgments above.

[30] C. S. Sherrington, "Sir Edward Sharpey-Schafer and His Contributions to Neurology," *Edinburgh Medical Journal*, n.s. *92* (1935), 393-406, on 397. Cf. French, "Medusae," p. 268.

Two decades earlier, in an obituary note on Gaskell, William Maddox Bayliss had already suggested that Gaskell's contributions to cardiac physiology "arose from those of Romanes upon the contractile umbrella of jelly-fishes."[31] When we examine Gaskell's work on the heart, we will find that it was indeed linked with Romanes' study of the medusae; nonetheless, it seems an exaggeration to say that Gaskell's work "arose from" or was "directly inspired" by Romanes'. In the chain connecting them, the crucial link was Foster himself.

In the first place, of course, both Romanes and Gaskell were taught by Foster. More importantly, there exists decisive evidence that Foster exerted a powerful influence on Romanes' own work from the outset. Thus, in June 1875, Romanes wrote as follows to Schäfer:

> You no doubt remember that in the paper we heard [Dr. Foster] read, he said that the snail's heart had no nerves or ganglia, but nevertheless behaved like nervous tissue in responding to electrical stimulation. He hence concluded that in undifferentiated tissue of this kind, nerve and muscle were, so to speak, amalgamated. Now it was principally with the view of testing this idea about 'physiological continuity' that I tried the mode of spiral and other sections mentioned in my last letter.
>
> The result of these sections, it seems to me, is to preclude on the one hand, the supposition that the muscular tissue of Medusae is merely *muscular* (for no muscle would respond to local stimulus throughout its substance when so severely cut), and, on the other hand, the supposition of a nervous plexus (for this would require to be so very intricate, and the hypothesis of scattered cells is without microscopical evidence here or elsewhere). I think, therefore, that we are driven to conclude that the muscular tissue of Medusae, though more differentiated into *fibres* than is the contractile tissue of the snail's heart is, as much as the latter, an instance of 'physiological continuity.' . . . Dr. Foster fully agrees with me in this deduction from my experiments, and is very pleased about the latter, thus affording additional support to his views.[32]

In this letter, Romanes seems to give the phrase "physiological continuity" a slightly different meaning from that which Foster gave it. Ordinarily at least, Foster used the phrase to express his belief that the snail's cardiac fibers were not isolated from one another, but acted

[31] W. M. B[ayliss], "Walter Holbrook Gaskell," *Heart, 6* (1915), 1-2.

[32] Romanes, *Life and Letters*, pp. 30-31. French cites but does not quote this letter in his "Medusae," p. 266, n. 51. The paper to which Romanes obviously refers is Foster and Dew-Smith's "Heart of Mollusks," received by the Royal Society on 1 February 1875. For an extensive discussion of this paper, see Chapter 7 above, "Foster's later research on the problem of the heartbeat."

together as a whole. He did sometimes also suppose that "in undifferentiated tissue of this kind, nerve and muscle were, so to speak, amalgamated," but that was not his main point. What he really sought to emphasize was the capacity of the snail's heart to beat rhythmically in the absence of distinct, differentiated nervous structures. Now since Romanes at this point attributed the rhythmic movements of the medusae swimming bell to the differentiated lithocysts rather than to the undifferentiated contractile tissue of the bell, one wonders why Foster should have been so cheered by his conclusions.[33] Perhaps he was pleased that Romanes at least denied that these lithocysts gave rise to a plexus of nerves scattered throughout the rhythmically contracting bell.

The essential point, in any case, is that Romanes' work was influenced from the beginning by Foster's work on the heart. The influence seems even stronger and more direct in Romanes' later papers on the medusae, after Schäfer and others had established that the swimming bell did in fact contain histologically distinguishable nerve tissue. In the light of this discovery, Romanes gradually abandoned his earlier notion that medusae might contain the structurally undefined "lines of discharge" or "incipient nerves" about which Herbert Spencer had speculated. Instead Romanes began to explore the question of the inherent rhythmicity of medusa tissue. In his first paper he had scarcely considered the possibility that the medusa swimming bell could respond rhythmically to stimulation after its lithocysts were removed.[34] But in his second major paper (1877), Romanes explicitly referred to the lithocysts as ganglia and began to study more carefully the responses of the "deganglionated" swimming bell to stimuli. Having discovered that the bells of covered-eye medusae could respond rhythmically to mechanical or chemical stimuli in the absence of lithocysts or ganglia, he announced that he was developing a new theory of ganglionic action to explain these and other results.[35]

[33] To put the same point another way, Romanes' early work on the medusae more nearly *supported* than challenged the prevailing neurogenic theory. Thus Milne-Edwards, who had once been skeptical of the neurogenic theory, cited Romanes' Croonian Lecture of 1876 as one piece of mounting evidence that rhythmic motion depended on "nervous ganglia." Romanes' results, wrote Milne-Edwards, "leave no doubt as to the nervous character of the parts in question [i.e., the lithocysts]." See Henry Milne-Edwards, *Leçons sur la physiologie et l'anatomie comparée de l'homme et des animaux*, 14 vols. (Paris, 1857-1881), XIII, p. 199. Not until Romanes turned his attention to "deganglionated" medusae and applied Foster's theory of ganglionic action to the jellyfish did his work become susceptible to a fully myogenic interpretation.

[34] Cf. Romanes, "Preliminary Observations," pp. 281-282.

[35] See George J. Romanes, "Further Observations on the Locomotor System of Medusae," *Phil. Trans., 167* (1877), 659-752, esp. 681, 747, and the n. on 730. Note

This theory, presented in full in Romanes' third major paper on the medusae (1880), was almost identical to the one Foster had already advanced for the vertebrate heart. At the base of Romanes' "new theory" lay the suggestion that ganglia might normally release a continuous impulse to which the muscular tissue of the medusa swimming bell could then respond rhythmically. As evidence for the general applicability of this theory, he pointed out how many examples could be given of rhythmic motion in the absence of nervous ganglia. He cited the contractile vacuoles of unicellular organisms, the rhythmic movement of cilia, Foster and Dew-Smith's experiments on the hearts of snails and frogs, Burdon Sanderson's experiments on the insectivorous plant *Dionaea*, and his own unpublished experiments on the frog's tongue.[36]

What all of this suggests is that Romanes' original concern with the evolution of nervous tissue in medusae was transformed largely under Foster's influence into an interest in the general problem of rhythmic motion. Almost simultaneously, Foster was directing Gaskell's work on the vasomotor nerves into the same channels. In the event, there was little need for Romanes to inspire Gaskell's study of the heart; Foster was already doing so.

Walter Holbrook Gaskell and vasomotor action in skeletal muscle, 1874-1878

In 1874, just as the problem of the heartbeat became the focus of physiological activity at Cambridge, Walter Gaskell went off to Leipzig to work in Carl Ludwig's physiological institute. He almost certainly did so at the suggestion of Foster, who had just convinced him to try his hand at physiology. If the choice seems merely obvious, Ludwig's institute being then the most active and lavishly equipped center for physiological research in the world, it should be remembered that Foster very soon after sent Arthur Lea and John Langley to Heidelberg instead, to study physiological chemistry and glandular secretion under Willy Kühne.[37] Moreover, Foster clearly had private reservations about Ludwig and his laboratory. In July 1873, in a letter to his friend and former student, E. A. Schäfer, Foster wrote that he was growing a bit tired of the increasing "monotony" of Ludwig's

---

also that in this paper, unlike his first, Romanes devotes a separate section to rhythmicity.

[36] G. J. Romanes, "Concluding Observations on the Locomotor System of Medusae," *Phil. Trans.*, *171* (1880), 161-202, esp. 171, 173, 180-181, 184-189, 196-198.

[37] See Chapter 6 above, "Dramatis personnae."

labors. The Leipzig institute, he felt, was in need of "something a little 'bahnbrechend.' "[38]

Why, then, should Foster send Gaskell there? Forty years after the event, Gaskell recalled that he went to Leipzig "with the intention of working at problems connected with the sympathetic nervous system."[39] If so, Foster would certainly have recognized that Ludwig's laboratory offered special advantages for a man of Gaskell's inclinations. Part of Ludwig's own reputation depended on his demonstration in 1851 that stimulation of the chorda tympani fibers in the "sympathetic" trunk of the lingual nerve caused secretion in the maxillary gland.[40] And though the cardiovascular system had become the focus of attention in Ludwig's institute, the amount of interplay between that system and the sympathetic or involuntary nervous system encouraged a strong secondary interest in the latter. At Leipzig, just three years before Gaskell went there, Oswald Schmiedeberg had traced the accelerator nerves to the hearts of warmblooded animals and had found that these nerves passed to the heart directly from three separate ganglia of the sympathetic system.[41] A decade after returning from Leipzig, Gaskell would establish the existence of similar cardiac accelerator fibers in the sympathetic system of the frog.[42]

In the event, however, Gaskell spent his time at Leipzig studying vasomotor action in the skeletal muscles, notably in the quadriceps extensor muscles of the dog.[43] He did so, he says, "at the suggestion and with the help of Professor Ludwig."[44] Especially in view of Ludwig's legendary influence over the research done in his laboratory,[45] there is little basis for doubting the literal accuracy of Gaskell's

[38] Foster to Schäfer, 21 July [1873], Sharpey-Schafer Collection, Wellcome Institute of the History of Medicine, London.

[39] W. H. Gaskell, *The Involuntary Nervous System*, new [2nd] ed. (London, 1920), p. vii.

[40] Carl Ludwig, "Neue Versuche über die Beihülfe der Nerven zu der Speichelsecretion," *Zeitschrift für rationelle Medizin*, n.s. *1* (1851), 254-277.

[41] Oswald Schmiedeberg, "Ueber die Innervations verhältnisse des Hundeherzens," *Arbeiten aus der physiologischen Anstalt zu Leipzig*, *5* (1871), 34-56.

[42] See Chapter 10 below, pp. 315-316.

[43] See W. H. Gaskell, "Ueber die Aenderungen des Blutstroms in den Muskeln durch die Reizung ihrer Nerven," *Arbeiten aus der physiologischen Anstalt zu Leipzig*, *11* (1877), 45-88. For an English version, to which the footnotes below refer, see W. H. Gaskell, "On the Changes of the Blood-stream in Muscles through Stimulation of their Nerves," *J. Anat. Physiol.*, *11* (1877), 360-402.

[44] On ibid., p. 360, Gaskell says he undertook this investigation "at the suggestion of Prof. Ludwig." In a subsequent paper (n. 52 below, p. 439), he added the words "and with the help of."

[45] On Ludwig as research director, see Heinz Schroer, *Carl Ludwig, Begründer der messenden Experimentalphysiologie, 1816-1895*, Bd. 33 in Grosse Naturforscher (Stuttgart, 1967), passim. Cf. also H. P. Bowditch, *Nature*, *3* (1870), 142-143; and

testimony. Nonetheless, it is entirely possible that he went to Leipzig already determined to study vasomotor action, which constitutes one of the most obvious realms of interplay between the cardiovascular and sympathetic nervous systems. Certainly he could already have acquired from Foster a special interest in the vasomotor nerves. Their existence and importance had been established in the first place by the work of Claude Bernard, and it was he, more than any other Continental physiologist, whom Foster had long most admired. Two months after Bernard's death in February 1878, Foster spoke of his work and influence in an address to the advanced students in physiology at Cambridge. He emphasized especially Bernard's work on the vasomotor nerves and insisted that there was good reason why the students heard so much at Cambridge of vasomotor nerves and vasomotor action. "If," he said, "we were to mark out the history of physiology into epochs, the present might fairly be called the vaso-motor epoch."[46]

By this time, Gaskell had returned from Leipzig to Cambridge, where his continuing work on the vasomotor nerves of skeletal muscle may have added local interest and special force to Foster's remarks. But the real occasion for the address was Bernard's death, and Foster's interest in vasomotor action certainly antedated Gaskell's departure for Leipzig. If that interest had already been transmitted to Gaskell, Ludwig's influence may have consisted of proposing and directing a specific topic of research within the general area of vasomotor action. This possibility, which may now seem somewhat strained, should gain in credibility as we become more familiar with Gaskell's research on circulation in skeletal muscles.

Gaskell's first published paper, which appeared in Ludwig's *Arbeiten aus der physiologischen Anstalt zu Leipzig*, concerned vasomotor action in the quadriceps extensor muscles of the dog. Having severed a vein in the dog's extensor muscles, Gaskell devised a means of connecting its proximal end with a kymograph so as to obtain a continuous record of variations in blood flow from the open vein. Assuming that variations in this venous flow correlated with variations in the arterial diameters, he examined first the effects produced by cutting

J. N. Langley, "Walter Holbrook Gaskell," *Proc. Roy. Soc., 88B* (1915), xxvii-xxxvi, on xxvii-xxviii, where Gaskell is said to have testified that during his own stay at Leipzig, Ludwig retained the practice of directly assisting the workers in his laboratory and yet encouraging them to publish under their own names.

[46] Michael Foster, "Claude Bernard," *Brit. Med. J.*, 1878 (1), 519-521, 559-560, quote on 521. Cf. also Michael Foster, "Vivisection," *Macmillan's Magazine, 29* (1874), 367-376, on 374. The latter source establishes Foster's deep interest in vasomotor action even before Gaskell went to Leipzig.

the crural nerve (the motor nerve of the extensor muscles). He found that this operation resulted in a large temporary spurt from the open vein, followed by a more prolonged but less dramatic increase in the normal venous flow. Then, applying an electrical current to the peripheral end of the severed nerve, he observed another spurt of blood from the vein, followed this time by a temporary decrease or even cessation of flow. This decrease, in turn, gave way to yet another large outflow and then to a series of more moderate ebbs and flows until the normal volume of flow eventually returned.[47]

In his attempt to explain these results, Gaskell revealed a consistent skepticism toward the standard notion that such phenomena resulted from the action and removal of vasoconstrictor forces. Thus, he ascribed the initial spurt of blood upon section of the dog's crural nerve not to the removal of previous vasoconstrictor action, but chiefly to a temporary stimulation of vasodilator fibers during the operation itself. Albeit less confidently, he offered a similar explanation for the secondary but more prolonged increase in blood flow following section of the nerve. His argument here ran as follows. The act of section was for him only a prelude to experiments in which he applied electrical currents to the peripheral end of the severed nerve. In preparing for these electrical experiments, he had, soon after cutting the nerve, inserted its peripheral end into electrodes. Current was not applied until later, but Gaskell suggested that the act of insertion might itself produce a secondary stimulation of vasodilator fibers sufficient to yield the observed increase in blood flow.[48] Finally, to account for the phenomena produced by electrical stimulation of the severed nerve, he proposed an hypothesis based on the fact that the dog's extensor muscles contracted when the crural nerve was stimulated. Perhaps, he suggested, the initial spurt upon stimulation of the nerve resulted from the pressure of the contracting extensor muscles on the blood vessels leading to the open vein, thereby forcing blood to spurt out of it. The compression of these vessels would then produce the observed temporary decrease or cessation of flow until the obstructed spaces could be gradually refilled as the contracted extensor muscles relaxed. If the crural nerve did in fact contain any vasoconstrictor fibers, they might play a secondary role in these phenomena.[49]

Apart from a brief paper of no consequence on the capillary walls of the lymph system,[50] Gaskell did no further research under Ludwig's

[47] Gaskell, "Blood-stream in Muscles."

[48] Ibid., pp. 370-374. [49] Ibid., pp. 391-396.

[50] W. H. Gaskell, "Ueber die Wand der Lymphcapillaren," *Arbeiten aus der physiologischen Anstalt zu Leipzig, II* (1877), 143-146.

direction. He returned to England in the summer of 1875 and was soon back at work in Cambridge. Especially since Foster's laboratory then consisted of a mere two rooms,[51] Gaskell must have found himself virtually surrounded by men at work on beating hearts. But he was not yet tempted to join them, and pursued instead his research on circulation in skeletal muscles. By autumn 1876 he was ready to report the main results of this second phase of his investigation. In October he sent a brief note to the leading abstracting journal in Germany, and in November Foster communicated a longer abstract to the Royal Society.[52] Gaskell published the paper in full in July 1877.[53]

In this paper, Gaskell sought to clarify some of the more doubtful results of his previous work on circulation in the dog's extensor muscles. To do so, he adopted a different experimental approach and used a different experimental animal. At Leipzig he had used a method which, in its general features, clearly reflects the influence of Ludwig, who was chiefly responsible for the invention and physiological exploitation of the kymograph.[54] In fact, Gaskell's method there closely resembled one already used in 1869 by another of Ludwig's students, though valuable improvements had since been made in the apparatus employed.[55] At Cambridge, as if in imitation of his smaller and less elaborate surroundings, Gaskell used a smaller organism and a simpler and more direct method of measuring arterial phenomena. He chose now to experiment on the mylohyoid muscle of the frog. In its simpler tissues, the arteries were more distinct; and since their diameters could be measured directly with a micrometer eyepiece, there was no need to resort to the indirect method of measuring the blood flow from an open vein. In the frog, moreover, the skeletal muscles could be selectively paralyzed by injection of curare, allowing the effects

[51] See Chapter 6 above, "Enrollments, bricks, and mortar."

[52] W. H. Gaskell, "Beobachtungen über den Blutstrom im Muskel," *Centralblatt, 14* (1876), 557-558; and W. H. Gaskell, "Preliminary Notice of Investigations on the Action of the Vaso-motor Nerves of Striated Muscle," *Proc. Roy. Soc., 25* (1877), 439-445.

[53] W. H. Gaskell, "On the Vaso-motor Nerves of Striated Muscles," *J. Anat. Physiol., 11* (1877), 720-753.

[54] On the development of the kymograph, see Hebbel E. Hoff and L. A. Geddes, "Graphical Registration before Ludwig: the Antecedents of the Kymograph," *Isis, 50* (1959), 5-21; Hoff and Geddes, "The Technological Background of Physiological Discovery: Ballistics and the Graphic Method," *J. Hist. Med., 15* (1960), 345-363.

[55] See W. Sadler, "Ueber den Blutstrom in den ruhenden, verkürzten und ermüdeten Muskeln den lebenden Thieres," *Arbeiten aus der physiologischen Anstalt zu Leipzig, 4* (1869), 77-100. For a detailed description of Gaskell's method, and of the differences between it and Sadler's, see Gaskell, "Blood-stream in Muscles," pp. 360-367.

of nervous action to be separated from any effects produced by the contraction of the mylohyoid muscle itself.[56] In the dog, by contrast, this distinction seemed impossible because curare eliminated the power of the crural nerve to produce arterial dilation at the same time as it prevented contraction of the extensor muscles.[57] No direct evidence exists that Foster had anything to do with Gaskell's switch in method, but it is at least a suggestive coincidence that Foster was just then using the frog in his own experiments.[58]

In any case, Gaskell showed by his new method that section of the motor nerve to the frog's mylohyoid muscle produced a result precisely like that produced in the dog by section of its crural nerve: an immediate marked dilation of the arteries in the muscle, followed by a more prolonged but less pronounced dilation. On the other hand, his new experiments convinced him that this secondary dilation was too prolonged to be explained solely by any transitory stimulation of dilator fibers upon their insertion into electrodes. But Gaskell continued to insist that the result could not be ascribed simply to the removal of previous vasoconstrictor action either. If such action is removed by section, it ought to be restored by stimulating the peripheral end of the severed nerve; but this procedure failed to return the arteries fully to their presection state of tone.[59]

As a matter of fact, Gaskell now developed an explanation for this and other results that emphasized the role of vasodilator fibers while making vasoconstrictor fibers virtually superfluous. A crucial observation allowed him to do so with considerable confidence. Thanks to his new method of measuring the arterial diameters directly, he could report that stimulation of the severed mylohyoid nerve produced *in the arteries* not constriction but rather a steady dilation. *Only in the veins did constriction occur, and then only when stimulation of the severed nerve also caused contraction of the mylohyoid muscle itself.* In these cases, the restoration of venous flow occurred when the increased flow through the dilated arteries produced enough internal pressure in the veins to push away the contracted muscle. But if muscular contraction were prevented by large doses of curare, no vascular constriction appeared anywhere. Thus constriction, when it did occur, must be due entirely to the pressure of the contracted muscle on the veins, with nervous action making no contribution to

[56] Gaskell, "On the Vaso-motor Nerves of Striated Muscles," pp. 720-725.

[57] See Gaskell, "Blood-stream in Muscles," p. 375.

[58] See Chapter 7 above, "Foster and Dew-Smith, 1876" and "Foster and the effects of upas antiar."

[59] Gaskell, "On the Vaso-motor Nerves of Striated Muscles," pp. 727-728.

it. Gaskell therefore concluded that the vasomotor fibers in the frog's mylohyoid nerve must either be exclusively vasodilator fibers, or else that any constrictor fibers present in the nerve were always overpowered by dilator fibers when both kinds were simultaneously stimulated.[60]

Gaskell then pointed out, as he had in his initial paper on the dog, that any circulatory effects produced in a skeletal muscle by its own contraction ought to occur whenever that muscle contracted, whether through artificial, reflex, or voluntary stimulation of its nerve. At the time of his first paper, he had carried out no experiments to test this supposition, and he could only say from "chance observation" that the blood-flow through the dog's extensor muscles seemed to quicken strongly and more or less permanently every time the experimental animal moved those muscles.[61] In his work on the frog, Gaskell sought a more rigorous answer to this question.

Since the frog could not announce which of its movements might be "voluntary," Gaskell examined instead the effect of reflex contractions induced by stimulating sensory nerves. So long as the stimulated nerves produced no contraction of the mylohyoid muscle itself, its arterial circulation seemed little affected. If any effect were observed in these cases, it was always a slight constriction, suggesting perhaps that vasoconstrictor fibers were activated to some extent. When, however, stimulation of a sensory nerve evoked a strong localized contraction in the neighborhood of the mylohyoid muscle, then some reflex dilation of its arteries seemed also to occur. This led Gaskell to propose "what may prove to be a general law: namely, that stimulation of a sensory nerve causes dilation of the smaller arteries of the part with which it is in functional relation, and constriction more or less marked of the remaining smaller arteries of the body."[62] To show that this "law" applied to mammals as well as to the frog, Gaskell projected a full investigation of the effects of reflex contraction on arterial circulation in mammalian muscles. As the mammal to be studied he chose the dog, and he used again the method employed in his initial researches at Leipzig—that of measuring the outflow of blood from a severed vein in the dog's extensor muscles.

But before Gaskell had completed and published the results of these new experiments, Rudolf Heidenhain of Breslau published a paper in which he challenged some of Gaskell's earlier experimental

[60] Ibid., pp. 729-737.

[61] Gaskell, "Blood-stream in Muscles," pp. 401-402.

[62] Gaskell, "On the Vaso-motor Nerves of Striated Muscles," pp. 739-743, quote on 742.

results and, pari passu, some of the conclusions he had drawn from them.[63] To state the issue crudely, Heidenhain was far more inclined than Gaskell to emphasize the importance of vasoconstrictor action and rather less inclined to admit the importance of vasodilator action. Heidenhain had also conducted a series of experiments to decide the very question underlying Gaskell's own projected research—namely, the effect of reflex action on the circulation in skeletal muscles—and he disputed in advance the applicability to mammals of Gaskell's proposed law. In all of this work, Heidenhain used a method different both from Gaskell's "Leipzig method" of measuring blood flow from an open vein and from his "Cambridge method" of measuring arterial diameters directly with a micrometer eyepiece. Heidenhain measured instead the *temperature* changes which nervous action caused in the arteries of the gastrocnemius muscle of the dog. For him a rise in temperature implied a coextensive arterial dilation, while a fall in temperature implied arterial constriction.

Heidenhain's paper forced Gaskell to reconsider his earlier conclusions and greatly to expand the scope and objective of the experiments he had already begun. The problem of reflex action became just one part of a general reinvestigation of the effect of nervous action on the circulation in muscle arteries. Gaskell reported the results in his third major paper on vasomotor nerves.[64] In this work, Gaskell first repeated his earlier Leipzig experiments and confirmed their results in every case. The effects of severing the muscle nerve deserved special attention because Heidenhain had asserted that this operation produced in the muscle arteries not a more or less temporary dilation, as Gaskell claimed, but a more or less *permanent* dilation (lasting hours rather than minutes). On the basis of new experiments, Gaskell reasserted his earlier claim. Next, because Heidenhain had reported that section of the abdominal sympathetic nerve also produced a permanent dilation of the muscle arteries, Gaskell performed the same operation and found that it produced the same effect as section of the muscle nerve—a more or less *temporary* increase of flow lasting only a few minutes. How was the discrepancy between their results to be explained? Gaskell thought the explanation lay in the inadequacy of Heidenhain's method to answer the question at hand. The method

[63] R. Heidenhain et al., "Beiträge zur Kenntnisse der Gefässinnervation, I, II. Ueber die Innervation der Muskelgefässe," *Archiv für die gesammte Physiologie des Menschen und der Thiere, 16* (1878), 1-46. Cf. W. H. Gaskell, "Preliminary Note of Further Investigations upon the Vaso-motor Nerves of Striated Muscle," *J. Physiol., 1* (1878), 108.

[64] W. H. Gaskell, "Further Researches on the Vaso-motor Nerves of Ordinary Muscles," *J. Physiol., 1* (1878), 262-302.

of measuring temperature changes was simply too indirect to give an accurate indication of the state of the vessels at any given moment. For whatever reason, the rise in temperature was much more permanent than the increase in flow and was probably not due to that increased flow alone.[65]

The discrepancy between Heidenhain's results and those of Gaskell was more important than it may sound at first, since the period of dilation carried implications for the manner in which it were achieved. If permanent, dilation must be due primarily to the removal of a previous vasoconstrictor action; if temporary, it must instead be due primarily to the stimulation of vasodilator fibers during the act of section. Heidenhain, of course, favored the former; Gaskell, the latter. This difference persisted in the interpretation that each gave to the effects of *stimulating* the peripheral end of the severed muscle nerve or abdominal sympathetic, even though their experimental results did not directly conflict in this case.

Gaskell, that is to say, did not dispute Heidenhain's assertion that stimulation of these two nerves produced a slight constriction of the arteries in the skeletal muscles of a *curarized* dog. But he did claim that, under ordinary circumstances, the opposite effect would occur. The only reason Heidenhain had failed to produce dilation by stimulation, he argued, was that his use of large doses of curare paralyzed the dilator fibers. When Gaskell used just enough curare to prevent muscular contraction and its attendant complications, but not so much as to paralyze the dilator fibers, he always observed an increase of flow after stimulation. He concluded therefore "that in the dog and in the frog, the vasomotor system for the muscles consists essentially of dilator nerve fibres . . . and of constrictor fibres, which are insignificant in comparison to the dilator, but which are manifested when the dilator fibres are put out of play by the action of curare."[66]

When Gaskell finally discussed the question his experiments had originally been designed to answer—namely, the effect of reflex action on the circulation in skeletal muscles—he claimed that his new work supported his previously proposed law. In the dog, as in the frog, stimulation of a sensory nerve which caused reflex contraction in a skeletal muscle also increased the blood flow through the vessels of that muscle. Stimulation of a sensory nerve which produced no contraction in a skeletal muscle also failed to increase the blood flow

[65] Ibid., pp. 265-270, 294-297. To what, besides the increased flow, the observed rise in temperature might be due, Gaskell was unsure. He could find no evidence in support of Claude Bernard's hypothetical "thermic nerves." See ibid., p. 295.

[66] Ibid., pp. 271-282, quote on 281.

through its vessels. Because Heidenhain had found a rise in temperature (implying vasodilation) in certain skeletal muscles upon stimulating a nerve (the vagus) that caused no reflex contraction in those muscles, Gaskell repeated the experiments but did not observe the rise in temperature. The only possible explanation he could offer for this discrepancy was the fact that Heidenhain's dogs had been injected with curare alone, while the new English antivivisection law required him to use anaesthetics in addition. Whether this were the real explanation, "I must," Gaskell wrote, "leave to others to decide, as I am not able to settle the question in England."[67]

Foster, Gaskell, and the analogy between vasodilation and cardiac inhibition

As a matter of fact, Gaskell's work on vasomotor action in skeletal muscle left a host of other issues equally unresolved. Even his Cambridge colleague, J. N. Langley, who believed that Gaskell's experiments on the frog's mylohyoid muscle had provided "the most decisive instance known at the time of [vasodilator] action in a purely muscular structure," nonetheless disputed the more general significance of this work. In particular, Langley emphasized that the mylohyoid muscle more nearly resembled the frog's own tongue muscles than it did ordinary vertebrate skeletal muscle, and he therefore doubted the validity of generalizations from the former to the latter. In the end, Langley suggested, other contemporary physiologists, including Heidenhain, had contributed far more profoundly than Gaskell to our knowledge of the existence and action of vasodilator fibers in skeletal muscle.[68]

However that may be, Gaskell's first excursion into physiological research probably had a significance not to be found in any discussion of the issues with which he was directly concerned. For there exists impressive indirect evidence that this work belonged, perhaps even from the outset, to the Cambridge effort to resolve the problem of the heartbeat. The long-standing analogy between vasodilation and cardiac inhibition makes the suggestion at the very least plausible. This analogy, to which Foster's own mentor William Sharpey had drawn attention as early as 1862,[69] had become commonplace even in England by

[67] Ibid., pp. 283-289, 297, quote on 297.

[68] See Langley, "Gaskell," p. xxviii; and J. N. Langley, "The Sympathetic and other Related Systems of Nerves," in *Textbook of Physiology*, ed. E. A. Schäfer, 2 vols. (Edinburgh, 1898-1900), II, pp. 616-696, on 641.

[69] William Sharpey, "Address in Physiology," *Brit. Med. J.*, 1862 (2), 162-171, on 168.

the 1870's.[70] It was only natural therefore that Francis Darwin in 1875 should have recognized a connection between his work on inflammatory vasodilation and that of Foster and Dew-Smith on electrically induced inhibition in the snail's heart.[71]

Most importantly, of course, Foster had himself frequently compared vasodilation with cardiac inhibition. Indeed, he did so even in his Royal Institution lectures of 1869, albeit in a way that reflected the more general timidity and uncertainty of his campaign for myogenicity at that time.[72] By the end of his research career, he was drawing the analogy in such a way as to leave no doubt that his commitment to the myogenic theory of the heartbeat implied an almost equally strong commitment to a myogenic theory of vasomotor action. In neither case could he deny the power of nervous action to influence muscular events, but he did consistently seek to minimize the role of local nervous ganglia and to draw attention instead to the inherent properties of relatively primitive contractile tissue.

Thus, in his final paper with Dew-Smith, Foster claimed that the vagus, like any other motor nerve, acted "directly on the muscular tissue with which it is connected." He insisted that its peculiar inhibitory action had nothing to do with any effect it might exert on inhibitory nervous mechanisms in the vertebrate heart, but depended rather on the simple fact that "while ordinary nerves are connected with muscles ordinarily at rest, the [vagus] is connected with a muscle [i.e., the heart] in a state of rhythmic pulsation." Scarcely different, in Foster's view, was the mode of action of those "other marked inhibitory nerves, the vaso-dilator nerves." The muscles innervated by them were "normally in a state of activity (tonic action) which is more closely allied to rhythmic pulsation than to any other form of muscular activity; indeed in many cases . . . the two merge into each other, and it seems difficult to regard the tonic contraction of blood-vessels in any other light than that of an obscure rhythmic pulsation." On grounds such as these, Foster suggested, a sense of identity could be established between ordinary vagus inhibition and the apparently extraordinary inhibition he had produced in the snail's heart by direct application of an interrupted current: the essential element in both cases is that "stimulation is brought to bear on a spontaneously active tissue." Certain objections could be brought against the absolute identity of the

[70] See, e.g., T. Lauder Brunton, "On Inhibition, Peripheral and Central," *West Riding Lunatic Asylum Medical Reports, 4* (1874), 179-222, esp. 181-182.

[71] See Chapter 8, "Francis Darwin."

[72] See Chapter 7, "The Royal Institution lectures of 1869." For his 1869 version of the analogy between vasodilation and cardiac inhibition, see Foster, *Three Lectures*, II, pp. 41-44.

two forms of inhibition, but Foster looked hopefully toward further research on the vasodilator and vasoconstrictor nerves "to afford a solution of the difficulties."[73]

Against this background, it seems reasonable to suggest that Gaskell may have undertaken his work on vasomotor action at Foster's suggestion and with the analogy between vasodilation and cardiac inhibition firmly in mind. If so, it must be admitted that Gaskell kept his ultimate object remarkably well concealed. Throughout his three major papers on circulation in skeletal muscles, he focused entirely on the problem of vasomotor action itself. In none of them did he relate his work explicitly to vagus inhibition or to the more general problem of the heartbeat. Even his effort to establish the existence of vasodilator nerves in skeletal muscles did not by itself do much to advance the myogenic cause. After all, as we have seen in Francis Darwin's work, widespread agreement as to the existence of vasodilator fibers in other tissues had posed no serious threat to the ganglionic theory of vasodilation. For it could easily be imagined that these vasodilator nerves ran to central nerve cells or peripheral ganglia and "inhibited" their usual tonic action, just as it was supposed that the vagus nerve inhibited the heartbeat by virtue of its action on intracardiac ganglia. Only if one challenged the usual theory of their action did vasodilator fibers represent a challenge to the neurogenic theory.

In his initial paper on the circulation in the dog's extensor muscles, Gaskell did not even raise the possibility that an alternative theory of vasodilator action might exist. In his third major paper, his main concern was simply to defend his earlier experimental conclusions against Heidenhain's recent criticisms, and he once again made no attempt to specify the modus operandi of vasodilator fibers. But at the end of his second major paper, on vasomotor action in the frog's mylohyoid muscle, Gaskell devoted a separate section to an extensive discussion of the "two chief rival theories" of vasodilator action. One was, of course, the prevailing "inhibitory" theory, according to which dilation resulted from the inhibition of vasoconstrictor forces and the amount of arterial dilation depended simply on the blood pressure within the artery. Gaskell now challenged this theory on the grounds that he had produced dilation of the arteries in the frog's mylohyoid muscle even after cutting off their blood supply (and thus presumably eliminating the blood pressure within them). He favored instead the less popular "active" theory of vasomotor action, according to which both vasoconstriction and vasodilation resulted from the active con-

[73] Michael Foster and A. G. Dew-Smith, "The Effects of the Constant Current on the Heart," *J. Anat. Physiol., 10* (1876), 735-771, on 768.

traction of the smooth muscle of the arterial walls, the difference between them consisting in the direction of contraction. Whereas vasoconstriction was presumed to arise from a contraction in the direction of the long axis of the muscular wall (a transverse contraction), vasodilation was supposed to result from a longitudinal contraction. On this theory, the stimulation of a given vasomotor nerve might produce either constriction or dilation, depending on the manner in which that nerve terminated in the arterial wall. Dilation could thus be explained without resort to inhibitory nervous forces acting on vasoconstrictor ganglia. This theory required, instead, the existence of vasodilator fibers acting directly on the arterial walls.[74]

What is most interesting about Gaskell's position here is that by favoring the "active" over the "inhibitory" theory of dilation, he was also choosing the theory better suited to lend analogical support to Foster's view of vagus inhibition and heart action. Just as Foster insisted that vagus inhibition of the frog's heart and electrical inhibition of the snail's heart should be ascribed to the direct effect of stimulation on previously active contractile tissue, so Gaskell now suggested that vasodilation could be produced without the participation of local ganglia by the direct action of dilator fibers on the previously tonic muscular walls of the arteries.

Ultimately, to be sure, Gaskell failed to sustain this version of the theory of active dilation. At the very end of this 1877 paper on the frog, he admitted the existence of one crucial piece of evidence against it—namely, that *outside of the frog's body*, no dilation could be produced in the arteries of the mylohyoid muscle by stimulation of its nerve. This ugly fact forced Gaskell to concede that vasodilation might ordinarily depend after all on the blood pressure within the dilating vessels. In those cases where he had produced dilation after cutting off the blood supply to the arteries, the blood pressure in the rest of the vascular system must somehow have been communicated to the empty artery. Perhaps, indeed, this extrinsic blood pressure was even sufficient to account for the observed dilation solely on the hypothesis of inhibitory action.[75]

But if this reluctant concession sounds like a total capitulation to the inhibitory theory, Gaskell's final sentence does not:

> Although therefore for the present I do not deem the evidence strong enough to overthrow the accepted theory of inhibitory action, yet it might possibly be worth while to investigate, whether between

[74] Gaskell, "On the Vaso-motor Nerves of Striated Muscles," pp. 745-753.
[75] Ibid., p. 753.

> the simple unspecialized contractile protoplasm of the amoeba, capable of contraction in all directions, and the highly specialized striated muscle fibre, intermediate forms of contractile tissue may not be found, in which the power of contraction is limited along certain axes; and whether the smooth unstriped muscular fibre may not represent one of these intermediate stages.[76]

In its broadly biological and evolutionary attitude toward physiology, this sentence bears the unmistakable stamp of Foster's influence; and if Gaskell ultimately felt obliged to qualify his support of active dilation, his vigorous efforts on its behalf suggest that he was receptive to any idea that might lend support to Foster's conception of heart action. Whether or not Gaskell actually undertook his study of vasomotor action at Foster's suggestion, his mentor's influence is evident enough in the 1877 paper on the frog. Foster was obviously following Gaskell's work closely and was clearly alert for any opportunity to point that work more directly toward the problem of the heartbeat. By 1879 he had found that opportunity. With Foster leading the way, Gaskell turned from the vasomotor nerves to the heart itself by way of a transitional investigation on the tonicity of the blood vessels as well as of the heart.

Gaskell's study of cardiovascular tonicity, 1880

In March 1880 Gaskell sent the Royal Society a brief preliminary account of his work on cardiovascular tonicity.[77] In this account, communicated to the Society by Foster, Gaskell reported only his experimental results, saying nothing of their general significance. When a full discussion did appear,[78] Foster's clear and dominant influence was impressed on every page.

Gaskell situated his work within the general context of the problem of rhythmic motion. He began with this long sentence:

> The tendency of physiologists of the present time to attribute to nerve cells the chief agency in the causation of rhythmical phenomena has led to the generally received hypothesis that the beat of the heart is brought about by the action of certain ganglion cells situated in the heart itself, while the cardiac muscular tissue is credited with the purely subordinate *rôle* of responding to the impulses gen-

[76] Ibid.

[77] W. H. Gaskell, "On the Tonicity of the Heart and Arteries," *Proc. Roy. Soc., 30* (1880), 225-227.

[78] W. H. Gaskell, "On the Tonicity of the Heart and Blood-vessels," *J. Physiol., 3* (1880-1882), 48-75.

> erated in those nerve cells; and, further, that the muscles of the smaller arteries are kept in a state of semi-contraction, that is, of tonicity, by the influence of the nerve cells of the vaso-motor centre.

Faced with the fact that the arteries eventually regained their tone after separation from the vasomotor center in the spinal cord, physiologists had merely invoked the subsidiary hypothesis that "a local peripherally situated ganglionic vasomotor mechanism . . . was able to take on the functions previously held by the centrally situated nerve cells." As might be expected, the most widely accepted explanations of vagus inhibition and vasodilation were precisely those that best conformed to this general mode of thought. Crucial to these explanations was the idea that both the vagus and the vasodilator nerves exerted their effects by interfering with ganglionic processes in such a way that the result was, on the one hand, a suspension of the rhythmic heartbeat and, on the other, a suspension of vascular tonicity.

With this as background, Gaskell drew attention to a fundamental defect in the foundation upon which this elaborate superstructure was built. Like Foster before him, he asked how ganglia could be necessary for the normal heartbeat if rhythmic pulsations could be produced in their absence, as in the lower two-thirds of the frog's ventricle. Since, moreover, no ganglia had been found in the muscular tissue of the arterioles either, Gaskell thought it reasonable to consider "whether the phenomena in question may not be explained by the properties of the muscular tissue *per se*." He proposed to examine this possibility by studying "the alterations of the character of the heart's action, and of the tone of the arteries, produced by alterations in the nutritive condition of the muscular substance."[79]

In its perception that both vasomotor and heart action belong to the more general problem of rhythmic motion; in its skepticism toward the idea that either form of rhythmicity depends on nervous ganglia; in its receptiveness toward the alternative view that rhythmicity depends in both cases on the properties of the muscular tissue itself; and in its attention to the nutritive conditions of the muscular tissue—in all these ways, Gaskell's approach bears the distinctive mark of Foster's thought. If any doubt remains that Foster was at this point exerting an active and immediate influence on Gaskell's work, it can be entirely dispelled by considering the first of the six sections in which Gaskell reported his experimental results. In this section, Gaskell returned to the problem of the frog's heartbeat precisely at the point where Foster had left it in his final research paper, published in 1876.

[79] Ibid., pp. 48-49.

Foster had ended that paper by criticizing the conclusions drawn by Julius Bernstein from his demonstration that the lower tip of the frog's ventricle remained indefinitely quiescent when isolated (by pinching) from the ganglia in the rest of the heart. Foster was particularly reluctant to accept Bernstein's conclusion that this experiment established the dependency of the normal rhythmic heartbeat on the action of ganglia. Nor was he satisfied by Bernstein's suggestion that an abnormal chemical action of foreign fluids was responsible for the rhythmic pulsations produced in the lower tip by Merunowicz's method.[80] What Gaskell's 1880 paper on tonicity now clearly demonstrates is that Foster had not in the meantime totally abandoned his cardiological research, that he had in fact passed on to Gaskell the results of his continued investigation into the influences affecting rhythmicity in the ganglion-free lower tip of the frog's heart. This work had convinced Foster (and through him Gaskell) that Bernstein's famous experiment required a totally new interpretation. Along with his earlier theoretical and primarily evolutionary objections to Bernstein's interpretation, Foster had now found direct experimental evidence of its inadequacy. A modification of Bernstein's own procedure provided this new evidence, which Gaskell described in these words: "As Dr. Foster has shown me, and as I have often seen for myself, if, after the ventricle has been nipped, and the apex [lower tip] thus reduced to quiescence, the aortic branches be clamped, and the arterial flow from the heart thereby prevented, then by the continuous beating of the upper part of the heart, the pressure in the apex is increased, and the apex, though only supplied with its own blood, will beat spontaneously."[81] To Gaskell this experiment suggested that the crucial factor in the apex beats produced by Merunowicz's method was the pressure in the apex cavity, rather than, as Bernstein believed, the chemical action of foreign fluids. Of the broader implications of this experiment more will be said later. It is introduced now only to show that Foster was at this point directly influencing the direction of Gaskell's research.

Foster's influence also underlies the rest of Gaskell's paper on tonicity. Still using the frog as subject, the basic experimental procedure was simply to apply dilute acid and alkaline solutions to the heart and arteries (thereby producing "alterations in the nutritive condition of the muscular substance"), and then to observe and compare the effects. Gaskell's results and conclusions were all that Foster could have wished. His major finding was that these solutions produced the same effect on the muscles of the smaller arteries as on the cardiac muscle:

[80] See Chapter 7 above, pp. 233-235.
[81] Gaskell, "On the Tonicity of the Heart and Blood-vessels," p. 51.

in each case lactic acid solutions caused muscular relaxation (diastolic standstill in the heart and dilation of the arteries), while alkaline solutions produced muscular contraction (systolic standstill in the heart and constriction of the arteries).[82] Gaskell argued that these effects were produced by the action of the solutions on the muscular tissue itself, as were the cardiological effects of the most common poisons, some of which acted like alkalies, other like acids. Muscarin, for example, acted like an acid, producing diastolic standstill—in other words, an inhibition indistinguishable from that produced by stimulating the vagus nerve. From his own experiments and others, Gaskell felt certain that muscarin produced this effect not by exciting any nervous apparatus, but by its direct action on muscular tissue.[83]

What all of this proved to Gaskell was that the influence of nervous mechanisms, both on the heart and on the arteries, had been greatly overemphasized, to the neglect of the possible role of the nutritive fluid surrounding those organs. By focusing on this factor, Gaskell developed what was, in effect, a new mechanism for active (as opposed to inhibitory) vasodilation. He began by suggesting that the structure of the arterioles was such that the "naked muscular fibres" of their walls "must be continually bathed by the lymph fluid of the tissue they supply." Now in a muscle at rest, this lymph fluid was alkaline and would therefore contribute to vasoconstriction. But it was well known that muscular contraction was accompanied by the production of lactic acid. When as a consequence the surrounding lymph fluid became acidic instead of alkaline, the muscular fibers of the arterioles would relax and the end result would be vasodilation.[84]

This proposed mechanism was in perfect accord with Gaskell's earlier conclusion that dilation occurred in the arteries of a muscle whenever that muscle contracted, whether the contraction was produced directly by stimulation of the muscle nerve, or reflexly by stimulation of appropriate sensory nerves.[85] And compared to Gaskell's one previous attempt to offer a mechanism for active dilation, this new and more original proposal had several advantages: there was no crucial piece of evidence against it; it was based exclusively on phenomena taking place in muscular tissue, without requiring even vasodilator nervous action; and it was capable of support by analogy with other cases (particularly the gland cells) in which the activity

[82] Ibid., pp. 53-57, 62-66. [83] Ibid., pp. 58-61. [84] Ibid., esp. p. 69.

[85] On the other hand, the new mechanism failed to explain how stimulation of these same nerves could in some cases produce dilation even when contraction was prevented by curare. Cf. Gaskell, "On the Vaso-motor Nerves of Striated Muscles," pp. 731, 735, 743; and Gaskell, "Further Researches," pp. 287-289.

of an organ was accompanied by the production of lactic or carbonic acid as well as by vasodilation.[86] In terms of its meaning for heart action, however, this mechanism remained incomplete; for though it could explain vasodilation in exclusively muscular terms, how could it account in the same way for the presumably analogous phenomenon of vagus inhibition? Was it perhaps by exerting some unknown acidic effect on the metabolism of cardiac muscle that the vagus nerve produced diastolic or "atonic" standstill in the heart?

Gaskell's work on vasomotor action had thus now brought him face to face with the problem of the heartbeat. If he had been an indirect participant from the outset in the Cambridge effort to resolve that problem, his direct attack upon it did not come until Foster's other students had effectively abandoned it. A. S. Lea and J. N. Langley had moved on other problems, while A. G. Dew-Smith had left scientific research entirely. Francis Darwin had ended his fleeting association with the Cambridge School and was now in Kent with his father. Even George Romanes, whose work on the medusae drew so much attention to the problem of rhythmic motion, had begun to move out of active research in the area.[87]

To be sure, yet another of Foster's students, H. Newell Martin, was just now beginning work that would show that even the highly complex mammalian heart could be sustained in isolation from its nervous system. This work, described in part in the Croonian Lecture for 1883,[88] offered further suggestive evidence on behalf of the campaign for myogenicity. But Martin conducted this research at Johns Hopkins University, where he had become professor of biology in 1876, and it is difficult to establish any direct inspiration from Foster or from the Cambridge setting Martin had left to come to Baltimore.[89]

Certainly at Cambridge, where the physiological laboratory now

[86] Gaskell, "On the Tonicity of the Heart and Blood-vessels," pp. 67-70.

[87] To be sure, Romanes did later publish at least two more works on the medusae. See G. J. Romanes and J. C. Ewart, "Observations on the Locomotor System of Medusae," *Phil. Trans., 172* (1882), 829-885; and G. J. Romanes, *Jelly-fish, Star-fish and Sea-urchins: Being a Research on Primitive Nervous Systems* (London, 1885). But the paper with Ewart, which was in fact the Croonian Lecture for 1881, contained nothing new vis-à-vis the problem of rhythmic motion, and the book of 1885 merely brought Romanes' earlier results together in one place. Beginning about 1880 Romanes had begun to gather materials for his influential writings on "animal intelligence."

[88] H. N. Martin, "On the Direct Influence of Gradual Variations of Temperature upon the Rate of Beat of the Dog's Heart," *Phil. Trans., 174* (1883), 663-688. Cf. *Record of the Royal Society of London for the Promotion of Natural Knowledge*, 4th ed. (London, 1940), p. 361.

[89] On Martin's research at the Hopkins, see C. S. Breathnach, "Henry Newell Martin (1848-1893): A Pioneer Physiologist," *Med. Hist., 13* (1969), 271-279.

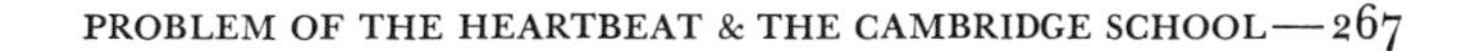

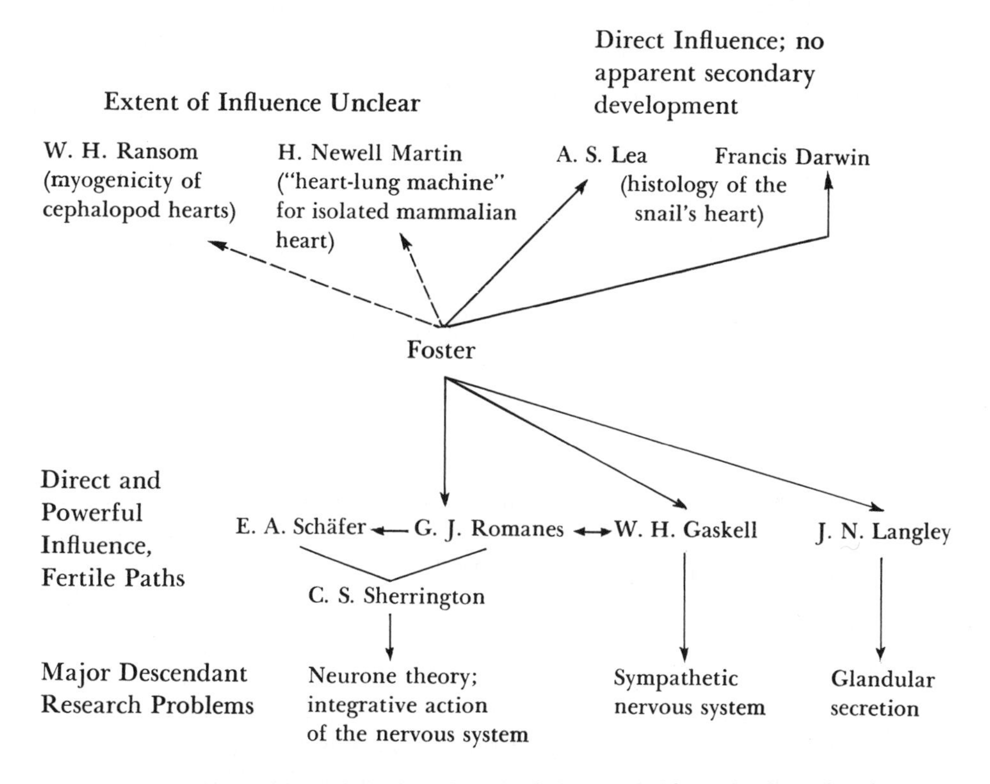

Figure 1. The problem of the heartbeat and the Cambridge School: paths of influence and descendant problems.

occupied an entire floor of the new biology building,[90] Gaskell was no longer surrounded by men at work on beating hearts. The research tradition established there by focusing on the heart had spawned an increasingly wide range of descendant research problems. The new building had thus been built partly to accommodate young physiologists who pursued research increasingly remote from Foster's own immediate interests and influence. But it was also in this new setting that Gaskell achieved his impressive resolution of the problem of the heartbeat. That resolution differed significantly from and went far beyond any of Foster's efforts, but it also provided a stunning vindication of his myogenic and evolutionary intuitions. And as Gaskell worked with astonishing speed toward that resolution, Foster was never far from his side.

[90] See Chapter 6 above, pp. 169-170.

9. The Maturation of the Cambridge School: Gaskell's Resolution of the Problem of the Heartbeat, 1881-1883

Some important features of the social aspects of scientific activity can be illuminated by the concept of style. Since the personal style of a matured scientist is influenced by his earlier experience, we can speak of the transmission of style from a master to his pupils. In this way one can construct intellectual genealogies; and to understand a man's work it may be relevant to know who was his teacher's teacher. This is clearly an important element in the creation of scientific "schools." . . . [But] it is impossible for a personal style of scientific work to be transmitted perfectly; for no two people, especially two of different generations, can have experiences so similar as to provide bases for nearly identical styles. . . . And in spite of a complete personal allegiance and an attempted copying of the master's style, [the disciple] cannot re-live his experience, and so cannot be a replica of him.

J. R. Ravetz (1971)[1]

At the end of his 1880 paper on cardiovascular tonicity, Gaskell insisted that vasodilation could result from the direct effect of acidic metabolites on the muscular walls of the arteries. On this view, it was natural to wonder whether the vagus nerve might produce cardiac inhibition by releasing some unknown acidic metabolite which then acted directly on the cardiac musculature. And it was therefore natural that Gaskell should next undertake a thorough study of the action of the vagus nerve on the frog's heart. In March 1881, he spoke of his ongoing research before the Cambridge Philosophical Society. He described briefly a new method he had devised for studying heart action, as well as several of the experimental results he had so far reached with that method. He also suggested, on the basis of these and other results, that "a possible explanation of the action of the vagus might be found on the hypothesis that the vagus is the trophic [i.e., metabolic] nerve of the cardiac muscle." But he admitted that he could not yet give "any definite explanation" of his experimental results, and he projected further work on the problem.[2]

[1] Jerome R. Ravetz, *Scientific Knowledge and Its Social Problems* (Oxford, 1971), pp. 105-106.

[2] W. H. Gaskell, "On the Action of the Vagus Nerve upon the Frog's Heart," *Proc. Camb. Phil. Soc.*, *4* (1883), 75-76.

The Croonian Lecture for 1882: Gaskell on vagus action

By December of the same year, Gaskell's investigation was more complete and he sent a long paper to the Royal Society, where Foster had just become biological secretary and where Gaskell's paper became the Croonian Lecture for 1882.[3] This paper was divided into two parts, the second and longer of which dealt with the action of the vagus nerve. Gaskell's discussion here seems to conform throughout to the general position Foster had so long been urging on the question of vagus action. Without denying that the vagus exerted an effect on the cardiac ganglia, Gaskell insisted that the results of vagus action depended more importantly on events taking place in the cardiac musculature. The precise effect of vagus stimulation seemed to be determined primarily by the nutritional condition of the heart at the time.[4] The in situ heart was more readily brought to standstill than the excised heart, and the heart of a given organism was differentially susceptible to standstill at different times of the year.[5] But exactly how could the vagus nerve interact with nutritional events in the cardiac muscle to produce the manifold and sometime contradictory results of vagus stimulation?

The particular mechanism Gaskell proposed is notable less for its cogency than for its bold attempt to draw an analogy between the heartbeat and glandular secretion, and to base that analogy on a presumed similarity in the metabolic processes taking place in undifferentiated protoplasm. With special reference to the work that Langley had been doing at Cambridge on glandular secretion, Gaskell suggested that cardiac contractions depended on the presence of an "explosive substance," which was formed in several stages from muscle protoplasm, just as the ultimate products of secretion were formed in several stages from the protoplasm of the gland cell. Assuming, further, that the vagus nerve simply increased the rate of these formative processes in the cardiac muscle, one could then explain both its excitatory effect on flagging or inactive hearts and its quieting effect on previously active hearts. In the former case, vagus action must promote the formation of the explosive substance where previously there had been none; in the latter, it must push the process beyond its explosive stage toward a condition of temporary quiescence. If the vagus

[3] W. H. Gaskell, "On the Rhythm of the Heart of the Frog, and on the Nature of the Action of the Vagus Nerve," *Phil. Trans., 173* (1882), 993-1033. For official verification that this was the Croonian Lecture for 1882, see *Record of the Royal Society of London for the Promotion of Natural Knowledge*, 4th ed. (London, 1940), p. 361.

[4] Gaskell, ibid., p. 1027.

[5] Ibid., p. 1011.

also acted on the ganglia, it probably did so in a similar fashion. "The vagus," Gaskell suggested, "is the trophic nerve of both the muscular tissue and the motor ganglia, meaning thereby that it increases the activity of the various formative processes going on in both these kinds of tissue, and it produces all its effects by virtue of this quality."[6]

Gaskell also believed that his conception of vagus action explained one of the most remarkable results of Foster and Dew-Smith's experiments with the constant current. During and only during vagus standstill, they had found, did the constant current lose its usual capacity for inducing rhythmic pulsation in previously quiescent cardiac tissue.[7] That result, Gaskell argued, found a natural explanation on his hypothesis that vagus standstill arose from an induced reduction of explosive substance in the cardiac tissue. For unless the explosive substance were present, all stimuli—including the constant current—would be equally ineffectual in evoking contraction.[8]

At the end of his lecture, Gaskell returned to the question with which he had probably begun his study of the vagus: could the effects of vagus stimulation be reconciled with the effects of alkaline and acid solutions on cardiovascular tonicity? His answer was atypically cautious: "whether the trophic action of the vagus nerve is connected with some such action of acids and alkalies I cannot say, and only put forward the resemblance in the hope that a possible clue may thus be found to the chemical action of the vagus upon the muscular substance of the heart."[9]

There is much here that suggests the influence of Foster and the Cambridge setting. In a way, Gaskell's entire discussion of vagus action seems little more than another attempt to add specificity to Foster's suggestion (now a decade old) that the vagus exerted its effects by producing molecular changes in the cardiac muscle.[10] The emphasis on the nutritional condition of cardiac tissue, as well as the suggestion that the vagus exerted a chemical action perhaps connected in some way with the formation of acidic and alkaline metabolites, is entirely in keeping with Foster's basic position and with the spirit of Gaskell's own work on tonicity.

Gaskell's defection: continuous vs. discontinous ganglionic discharges

It comes as a great surprise, therefore, to discover that in the first and shorter part of this Croonian Lecture, Gaskell adopted a con-

6 Ibid., pp. 1028-1029.

7 See Chapter 7 above, "Foster and Dew-Smith, 1876."

8 Gaskell, "Heart of the Frog," p. 1029.

9 Ibid., p. 1031.

10 See Chapter 7 above, pp. 221-223.

ception of the heartbeat differing fundamentally from Foster's in what was to the latter the most crucial point: namely, the cause of the rhythmicity of the normal heartbeat. In this section on the rhythm of the heart, Gaskell became the first of Foster's students to suggest that the cardiac ganglia directly determined the rhythmicity of the normal vertebrate heartbeat.[11] His defection seems all the more remarkable because it came at a time when most German physiologists had actually begun to turn away from the original version of the neurogenic theory. Without recognizing it, since Foster's work remained largely ignored, many German physiologists were moving toward his position. A decade after Foster had first insisted upon it, the view was now becoming general that the frog's ventricular muscle possessed a natural capacity for rhythmic response. To understand why Gaskell deserted his mentor at this very moment of apparent vindication, it is necessary first to recognize that the myogenic-neurogenic debate had recently undergone a subtle and complex transformation. The nature of this transformation can be appreciated only by outlining the developments that had persuaded many German physiologists to adopt a position strikingly similar to Foster's.

During the 1870's, and especially in the wake of the important experiments by Merunowicz and Bernstein, the myogenic-neurogenic debate had become increasingly murky. Merunowicz's work acquired much of its importance from the fact that it was just one particularly notable example in a series of repeated demonstrations that the ganglion-free lower tip of the frog's ventricle could respond rhythmically to a variety of chemical and nutrient fluids. Several of these demonstrations, including Merunowicz's, carried with them the authority of Ludwig and the Leipzig school. As this evidence continued to accumulate, the view that ganglia determined the rhythmicity of the heartbeat began increasingly to be questioned. Perhaps, it was argued, the isolated ventricular tip remained quiescent not because of its separation from ganglia but rather because the methods of isolation prevented it from receiving its normal supply of fresh nourishing blood.

At the same time, however, the neurogenicists found in Bernstein's famous experiment a means of explaining away all of this evidence. For Bernstein's method of isolating the ventricular tip from the ganglia by pinching rather than section allowed a continuous supply of the frog's own fresh blood to reach it. Still the tip remained quiescent. Because this seemed to Bernstein the most natural of possible

[11] Although, as we have seen, Francis Darwin certainly accepted the standard neurogenic theory before he came briefly under Foster's influence. See Chapter 8 above, "Francis Darwin." Thereafter, to my knowledge, Darwin took no position on the issue.

circumstances, he argued that the pulsations produced by Merunowicz's method were due to the influence of "foreign" chemical stimuli supplied by the rabbit's blood he had used. This argument could easily be applied to similar cases, and it seemed to make little difference to the debate when in 1878 two other members of Ludwig's school extended Merunowicz's experiments with nutrient fluids.[12]

About 1880, however, the debate took on a new dimension when a German team and Gaskell independently reported that the ventricular tip isolated according to Bernstein's own method could be brought to rhythmic pulsations merely by raising the blood pressure in its cavity. This was of course the meaning of the experiment Foster had devised and shown to Gaskell, who had then repeated it himself before describing it in his paper on cardiovascular tonicity.[13] In his biographical sketch of Gaskell, Walter Langdon-Brown placed great emphasis on this experiment:

> Bernstein had shown that if the ventricle of the frog's heart were "physiologically disconnected" by crushing the auriculo-ventricular junction with a fine pair of wire forceps, it remained quiescent, while the rest, which contained ganglion cells, continued to beat. But Gaskell simply raised the intracardiac pressure by ligaturing the aortae, and saw the ventricle begin to beat rhythmically once more. To my mind at that moment and by that experiment modern cardiology was born.[14]

Langdon-Brown here ignores the place of the German team in the same story, as well as the role played by Foster in Gaskell's own version of the experiment. More seriously, he seems to imply that Gaskell established the myogenicity of the rhythmic heartbeat through this one experiment, so that all of his subsequent work amounts to a confirmation and extension of a point already established. In fact, the story is much more complicated.

As Foster had emphasized as early as 1869, in his lectures at the Royal Institution, it is easy to confuse two issues about the origin of the rhythmic heartbeat: (1) are ganglia normally involved in its pro-

[12] See Dr. Stiénon, "Die Betheilung der einzelnen Stoffe des Serums an der Erzeugung des Herzschlags," *Archiv für Anatomie und Physiologie (Physiologische Abtheilung)*, no vol. number, 1878, pp. 262-290; and McGuire, "Ueber die Speisung des Froschherzens," ibid., pp. 321-322.

[13] See Chapter 8 above, pp. 263-264.

[14] Walter Langdon-Brown, "W. H. Gaskell and the Cambridge Medical School," *Proceedings of the Royal Society of Medicine, 33* (1939), Section of the History of Medicine, pp. 1-12, on 5.

duction? and (2) even if they are, do they determine the rhythm by giving off *discontinuous* discharges at regular intervals? In 1876, in his final paper with Dew-Smith, Foster raised this point again and argued that nothing Bernstein or anyone else had done eliminated the possibility that the ganglia discharged impulses *continuously* rather than discontinuously. If this were the case, then the actual rhythmicity of the heartbeat depended not on the ganglia per se but on the unique and inherent capacity of cardiac muscle to give a rhythmic response to a constant stimulus.[15]

In the same year, Bernstein simply reaffirmed his adherence to the old assumption of rhythmic ganglionic discharges.[16] In 1878, H. P. Bowditch repeated and extended Bernstein's experiments, agreeing with him that the auriculo-ventricular ganglia were essential to the normal rhythmic beats in the frog's ventricular tip, though without committing himself as to whether the ganglionic discharges were continuous or discontinuous.[17] But as evidence accumulated that the ganglion-free tip responded rhythmically under a wide variety of stimuli, the consensus shifted toward the view that the ganglia discharged their impulses continuously. It is in this context that the work of Gaskell and the German team on intracardiac blood pressure properly belongs. For their work provided yet another example of the capacity of cardiac muscle to respond rhythmically to a constant stimulus. This capacity had already been demonstrated under the influence of constant chemical and electrical stimuli, but they now insisted that the same capacity became manifest under the influence of a constant *mechanical* stimulus no more "foreign" than the heart's own blood under pressure.

In the midst of great disagreement about the precise meaning of the experimental work of the 1870's, the view became general that ventricular muscle could respond rhythmically to *any* constant stimulus of appropriate strength. Many adopted S. Ritter von Basch's suggestion that cardiac muscle possessed the unique ability to sum up the

[15] See Chapter 7 above, "The significance of Foster's work on the heart."

[16] See Julius Bernstein, "Bemerkung zur Frage über die Automatie des Herzens," *Centralblatt, 14* (1876), 435-437. Although Bernstein does not explicitly insist here on rhythmic ganglionic discharges, he does announce his "complete agreement" with the theory set forth by Isidor (Julius) Rosenthal in *Bemerkungen über Thätigkeit der automatischer Nervecentra insbesondere über die Athembewegungen* (Erlangen, 1875). In this influential work Rosenthal repeated and extended the old analogy between the rhythmic heartbeat and the respiratory movements, ascribing both to the rhythmic action of nerve cells or ganglia.

[17] H. P. Bowditch, "Does the Apex of the Heart Contract Automatically," *J. Physiol., 1* (1878), 104-107.

effects of individual stimuli in such a way as to convert a continuous impulse into a discontinuous, rhythmical pulsation.[18] These developments encouraged the rise of a radically different conception of the normal vertebrate heartbeat. No longer was it so widely believed that the cardiac ganglia themselves determined the normal rhythmic beats of the individual cavities of the heart. Many assigned this role instead to the cardiac musculature, on the assumption that it responded rhythmically to a continuous impulse sent out by the ganglia. But even then, the nerves and ganglia remained crucial as well. Because nervous structures are more easily excited than muscle, they were believed to provide a more responsive mechanism for the overall regulation of the heartbeat. Accordingly, nerves and ganglia were made responsible for the genesis of the continuous impulse in the sinus, for its conduction to the auricle and ventricle, and for the sequence of ventricular upon auricular beat.[19] According to Gaskell, this was the most widely held conception of the normal vertebrate heartbeat at the time of his Croonian Lecture of 1882:

> Motor ganglia in the sinus send out a continuous impulse along nerve fibres to auricle and ventricle; such impulse produces a discontinuous rhythmical result, owing to the nature of the muscular tissue; the sequence of the contraction of the ventricle upon that of the auricle is due to a delay in the impulse, owing to its passage through the ganglia at the auriculo-ventricular junction.[20]

Whether this conception should be called myogenic or neurogenic is perhaps mainly a semantic issue. To the extent that the ganglia were now assumed to discharge a continuous impulse, with rhythmicity per se ascribed to the cardiac musculature, Foster could have claimed that his long-standing position had at last been vindicated, even though his own work continued to be ignored in Germany.[21] On the other hand, it had not yet been accepted that the cardiac musculature could

[18] S. Ritter von Basch, "Ueber die Summation von Reizen durch das Herz," *Sitzungsberichte der Mathematisch-Naturwissenschaftlichen Classe der kaiserlichen Akademie der Wissenschaften (Wien)*, *79* (1879), 37-75.

[19] For a summary of the debate as it stood in 1880, with extensive references, see H. Aubert, "Die Innervation der Kreislaufsorgane," *Handbuch der Physiologie*, ed. Ludimar Hermann, 6 vols. (Leipzig, 1879-1881), IV, pp. 345-374, esp. 369ff. Cf. also M. Lowit, "Beiträge zur Kenntniss der Innervation des Froschherzens, Part II," *Archiv für die gesammte Physiologie des Menschen und der Thiere*, 25 (1881), 399-496, esp. 416-466.

[20] W. H. Gaskell, "The Contraction of Cardiac Muscle," in *Textbook of Physiology*, ed. E. A. Schäfer, 2 vols. (Edinburgh, 1898-1900), II, pp. 169-227, on 171. Cf. Gaskell, "Heart of the Frog," p. 993.

[21] Foster's work received no attention whatever from Aubert or Lowit (n. 19). Cf. Chapter 7 above, p. 232, n. 92.

generate its own rhythm spontaneously, independently of ganglionic stimuli. Its rhythm was conceived as a rhythm of *response*. Confusion and disagreement continued to prevail on all these issues.

It was against this background that Gaskell began the research leading to his Croonian Lecture for 1882. On the assumption (which even Foster shared) that ganglia were somehow involved in the normal vertebrate heartbeat, Gaskell decided first to focus on the question whether the ganglionic discharges were continuous or discontinuous. And it was in the course of this investigation, described in the first part of his Croonian Lecture, that he felt obliged to differ from Foster and to return to the more purely neurogenic view that the ganglia determined the rhythm by discharging discontinuous impulses.

The evidence that caused Gaskell to adopt this independent stance had its origin in the new "suspension method" he had devised for studying heart action. Besides allowing the auricular and ventricular beats to be measured and recorded separately, this new method was distinguished by its deployment of a micrometer screw clamp. This clamp, wrote Gaskell, "enables us to study the effects of compression at different points much more delicately than the old plan . . . for it is possible either just to hold the tissue so as not to injure it physiologically or to compress it up to any required amount by the simple movement of the micrometer screw." Further refinements made it possible to confine various stimuli to the cardiac tissue on one side of the clamp only. Since the frog's heart could be most conveniently fixed in position by placing the clamp in the groove of tissue between the auricles and ventricle, the vast majority of experiments concerned the relationship between the auricular and ventricular beats.[22]

Gaskell later gave this brief account of the reasoning behind the experiments described in the first part of his Croonian Lecture:

> It seemed to me that the question whether continuous or discontinuous impulses passed from the sinus to auricle and ventricle, could be easily answered by increasing the excitability of ventricle or auricle respectively. A more excitable ventricle must respond to a constant stimulus with contractions at a more rapid rate than a less excitable one; while, on the other hand, the rate would remain unaltered if the impulses from the sinus were discontinuous, and the excitability of the ventricle only were altered.[23]

To increase the excitability of the frog's auricle or ventricle respectively, he applied a moderate amount of heat, first above the clamp and

22 Gaskell, "Heart of the Frog," pp. 994-995.
23 Gaskell, "Cardiac Muscle," pp. 171-172.

then below it. He observed that heat applied to the sinus and auricles alone (and thus presumably to the ganglia found there) resulted in a marked and synchronous increase in the rate both of the auricular and ventricular beats, while heating the ganglion-free ventricular tip alone produced no change in the rate of beats in either cavity.

"This experiment," wrote Gaskell, "seems to me positive proof that in the whole heart the rhythm is due to discrete impulses proceeding from certain motor ganglia [in the sinus and auricles] to the muscular tissue, each of which impulses causes a contraction of that tissue." This conclusion seemed even more certain from the effects of stimuli that *decreased* the excitability of auricle or ventricle respectively. When applied to the sinus and auricles alone, atropin, muscarin, and coolants all slowed the auricular and ventricular beats synchronously, while none of them affected the rhythm, either of ventricle or auricles, when applied to the ventricular tip alone.[24] All of these experiments demonstrated that the rate of ventricular beats depended directly on the rate of auricular beats (and thus presumably on the rate of ganglionic discharge). Any of the stimuli applied to the ventricular tip alone (where no ganglia could be found) produced no effect on the rhythm.

Gaskell's compression experiments only further strengthened his new conviction that the ganglia directly determined the normal rhythm of the heartbeat. When the micrometer screw on the clamp was loose, so as just to hold the tissue in the auriculo-ventricular groove, the normal sequence of ventricular upon auricular beat continued undisturbed. By tightening the screw, however, this normal sequence could be altered. As the compression increased, the ventricular beats became less and less frequent, while the auricular rhythm remained unaffected. Eventually, at a point when the compression was very severe, the sequence was entirely destroyed and the ventricle became quiescent. Up to this point, however, the altered ventricular rhythm still depended in a definite way on the unchanged auricular rhythm. Instead of beating once for each auricular beat, the ventricle merely beat once for every second, third, fourth, or more auricular beats as the compression increased. To Gaskell these results demonstrated that "direct compression of the tissue between the motor ganglia and the ventricle causes a certain number of the [ganglionic] impulses to be ineffective as far as the causation of a ventricular beat is concerned, although, as shown by the [unaltered] auricular beats, that compression has not interfered with the rate of discharge from the ganglia."[25]

Gaskell drew similar conclusions from experiments in which heat

[24] Gaskell, "Heart of the Frog," p. 997.
[25] Ibid., p. 1000.

and poisons were applied to the heart in such a way as to produce a want of sequence between the auricular and ventricular beats. In every case in which the ventricle beat at a slower rate than the auricles, the force of its contractions, when they did occur, was at least as great as under normal circumstances. Thus, the excitability of the ventricular muscle must not have been decreased. Rather, the efficacy of the ganglionic stimuli must somehow have been impaired. No longer was each ganglionic discharge able to evoke a ventricular beat. Only because the ventricular muscle possessed the power of summing up their effects could these weaker impulses eventually produce a ventricular beat, but the essential point was that *separate* impulses were being cumulated.[26]

It may seem astonishing that here, in this section of his Croonian Lecture, Gaskell could return to the old neurogenic view of the rhythmic heartbeat, while throughout the second and longer part he insisted with Foster that the vagus acted primarily on cardiac muscle rather than on ganglia. The two positions did not necessarily clash, however, for Gaskell's experiments on vagus stimulation demonstrated that its most important effect was to alter the *force* of cardiac contractions and not necessarily to alter their *rhythm*.[27] Nor did Gaskell find it necessary to deny the capacity of cardiac muscle to respond rhythmically to a constant artificial stimulus. He admitted that such a stimulus could eventually make the ventricular tip so excitable that "no separate stimulus is required: the muscle, in fact, can be spoken of as capable of spontaneous contraction." But he argued that this had nothing to do with the normal heartbeat, where the much lower level of muscular excitability required a separate ganglionic stimulus for each contraction.[28]

So even though Gaskell still granted the cardiac musculature a latent capacity for rhythmicity, he had nonetheless become a neurogenicist in the most fundamental sense. For he now believed that *under normal conditions* the rhythm of the heartbeat was a direct expression of the rate of ganglionic discharge. This conception went directly against Foster's evolutionary arguments for myogenicity. Foster might well have asked Gaskell, as he had already asked Bernstein: if the rhythm of the vertebrate heartbeat is normally determined by nervous mechanisms, then why should cardiac muscle retain a capacity for rhythmicity which it could never use in actual life? If this question occurred to Gaskell, he responded obliquely at most, and merely by

[26] Ibid., pp. 997-1002.
[27] Ibid., pp. 1006-1009 et seq., esp. 1026.
[28] Ibid., pp. 997-1005.

asserting that the rhythmic capacity of cardiac muscle was "a special safe guard for the maintenance of . . . rhythmical action."[29] He was more impressed by the apparent meaning of the experiments his new method allowed.

Thus, while this Croonian Lecture reveals Foster's continuing influence on Gaskell, it also reveals more dramatically the latter's full emergence as an independent investigator. Foster's influence, though still very real, had become more diffuse and indirect; and particularly in terms of experimental method, Gaskell had by now surpassed his former mentor. Nor was this independence entirely ill-advised. Foster's evolutionary perspective allowed him an almost intuitive insight into the cause of rhythmic motion, but his conception of the heartbeat carried with it very real difficulties. Most seriously, he had never offered a satisfactory explanation of how the vertebrate heartbeat normally acted as a coordinated whole, with the auricle and ventricle beating in perfect sequence. In the nerve-free snail's heart, he simply ascribed this coordinating faculty to a rudimentary "muscular sense" by which the condition of each part of the heart was made known to every other part. In the frog's heart, Foster believed, *each separate cavity* also acted together as a whole by virtue of a "muscular sense" like that operating throughout the snail's heart. And since, in isolation, each cavity of the frog's heart could carry on "a rhythmic pulsation of long duration and wholly like that of the entire heart," he supposed that in the intact heart too each cavity possessed "a rhythm of its own dependent on its own circumstances"—in other words a myogenic rhythm. But Foster could perceive no muscular continuity between the separate cavities of the frog's heart, and so assumed that the ganglia and other nervous structures found there must be responsible for the fact that these cavities ordinarily beat in a regular and coordinated sequence.[30]

If he was undisturbed by the vagueness of this conception, Foster might have been suspicious of it on the basis of his own evolutionary criteria. For if the coordinated sequence of the snail's heart could be made a property of its relatively primitive contractile tissue, then why in the vertebrate heart should differentiated nervous mechanisms be required to perform the same function? Perhaps Foster did feel a little uncomfortable about his position, since he did consider, before rejecting, the possibility that muscular events (in the form of a contractile wave) might be as responsible for coordinating the vertebrate

[29] Ibid., p. 1004.

[30] See Chapter 7 above, p. 225. Cf. J. N. Langley, "Sir Michael Foster—In Memoriam," *J. Physiol.*, *35* (1907), 233-246, on 240.

heartbeat as they were for the rhythmicity of the beats in each of its separate cavities. He felt compelled to reject this possibility for several reasons, but especially because "in that case, the systole of the ventricle would provoke a systole of the auricles, as well as the systole of the auricles a systole of the ventricle, and a rhythmic pulsation with long pauses between the whole beat (of both auricles and ventricle) would be impossible."[31]

Thus, even though Foster had long insisted that the cardiac ganglia played a subordinate part in the heart's action, they retained an important "coordinating" role in his own scheme. And, whether the discharges from these ganglia were presumed to be continuous or discontinuous, his scheme proved vulnerable even in its most fundamental feature—the insistence that each cavity of the heart enjoyed a spontaneous rhythmicity. For if the ganglia released discontinuous impulses, it was natural to assume that the rhythmic beats of each separate cavity depended more importantly on the rate of ganglionic discharge than on the special properties of cardiac muscle. Doubtless for this reason, Foster favored the view that the ganglia discharged a continuous impulse to which the cardiac tissue could then respond rhythmically. But, as Gaskell's new clamping experiments clearly showed, this view faced serious obstacles of another sort.

Gaskell's defection is revealing, then, not only because it illustrates the complexities of the problem, but also because it reminds us that Foster had not entirely resolved the issues even in terms of his own evolutionary principles. In particular, there was much to be learned from a careful investigation of the coordinated heartbeat as a whole. Foster clearly recognized the significance of the sequential character of the normal heartbeat, saying at one point that no theory of the heartbeat could be fully adequate unless it was "at the same time a solution of the difficult problem, why in a normal heart beat the sequence of constituent contractions is always such as it is."[32] But he concentrated his own research on the isolated ganglion-free ventricular tip and never really probed the problem of the coordinated vertebrate heartbeat as a whole. In his Croonian Lecture, Gaskell had begun to do just that, and as he now continued his work on the heart, he was prepared to pay special attention to the problem. When, in the end, he returned to the myogenic theory of the heartbeat, he did so in a way that, dependent though it was on what he had learned from Foster, owed even more to his own skill as an investigator. Largely as a result of Gaskell's work,

[31] Michael Foster and A. G. Dew-Smith, "The Effects of the Constant Current on the Heart," *J. Anat. Physiol., 10* (1876), 735-771, on 760-761.

[32] Ibid., p. 759.

the problem of the heartbeat was resolved in favor of the myogenic theory far more completely and persuasively than Foster had ever considered possible.

Gaskell's classic paper of 1883: a fully myogenic resolution of the problem of the vertebrate heartbeat

On 3 July 1882, Gaskell added a postscript to his Croonian Lecture.[33] Recent experiments on the hearts of tortoises (rather than frogs) had produced a radical change in his conceptions. His new views depended crucially on this switch in experimental animal, chiefly because the anatomy of the tortoise heart differed from that of the frog in such a way as to permit a much better appraisal of the effects of nervous action on the heart. Also vital to Gaskell's new conceptions was another methodological innovation. By slicing the tortoise's auricle in different directions and to different depths he had produced clear and unique evidence for the myogenic theory of the heartbeat. In the postscript and in three separate papers published during the next several months,[34] he developed his views with increasing confidence and persuasion. The results were gathered together in a long paper, really a monograph, published in 1883.[35]

Gaskell emphasized that he had turned to the tortoise's heart out of conviction that "the study of the evolution of function . . . is the true method by which the complex problems of the mammalian heart will receive their final solution."[36] Views of heart action had been based too exclusively on experiments with the frog, and it was therefore desirable to make "a corresponding elaborate series of observations upon the hearts of a large number of other animal types, and in this way to trace the evolution of function in the same way as the morphologist tracks that of structure."[37] Although a few other physiologists had already studied the tortoise's heart, and Foster himself had discussed it incidentally in his Royal Institution lectures of 1869,[38] Gaskell recognized

[33] Gaskell, "Heart of the Frog," pp. 1031-1032.

[34] See W. H. Gaskell, "Preliminary Observations on the Innervation of the Heart of the Tortoise," *J. Physiol., 3* (1880-1882), 369-379; Gaskell, "Observations on the Innervation of the Heart: On the Sequence of the Contractions of the Different Portions of the Heart," *Brit. Med. J.*, 1882 (2), 572-573; and Gaskell, "On Certain Points in the Function of the Cardiac Muscle," *Proc. Camb. Phil. Soc., 4* (1883), 277-286.

[35] W. H. Gaskell. "On the Innervation of the Heart, with Especial Reference to the Heart of the Tortoise," *J. Physiol., 4* (1883), 43-127.

[36] Ibid., p. 48.

[37] Ibid., p. 43.

[38] See Chapter 7 above, p. 205.

and exploited to an unprecedented degree its special advantages for yielding basic insights into the problem of the heartbeat.

In the first section of his classic paper of 1883, Gaskell emphasized the purely physiological differences between the tortoise's heart and the frog's. His basic experimental procedure was the same "suspension method" used in his earlier work on the frog, with one crucial difference in detail. In the frog, he had ordinarily used a clamp between the auricles and ventricles to fix the heart in position; in the tortoise, he found it easier simply to place the aortic trunk in a suitable clasp. The tortoise heart when suspended according to this method gave the following appearance:

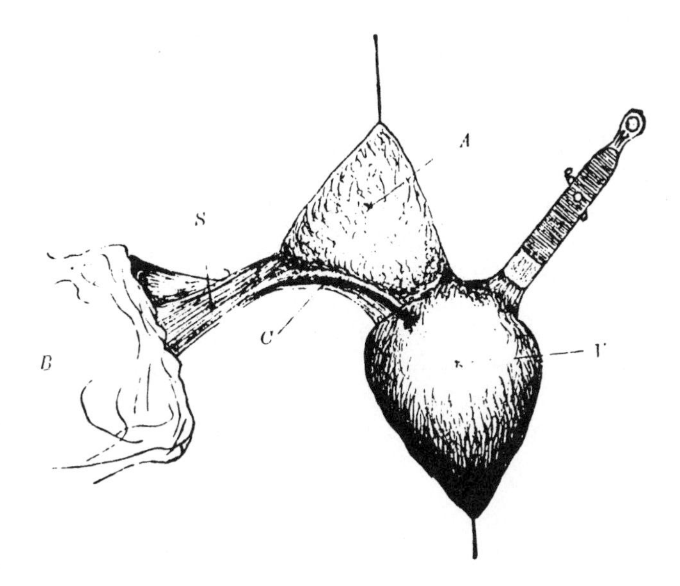

Figure 2. (B) upper part of body; (S) sinus; (C) cardiac nerve; (A) auricles; (V) ventricles. *Journal of Physiology, 4* (1883), 49.

Although the use of the aortic clasp was only a modification in detail of the earlier method, it had important consequences. In fact, it led to a new procedure for isolating the different parts of the heart from each other. In the tortoise, Gaskell ordinarily isolated the auricle from the ventricle by slicing the auricular tissue and not by clamping as in the frog.[39]

Gaskell was struck immediately by the difference in behavior be-

[39] Gaskell, "Heart of the Tortoise," pp. 48-50.

tween the isolated ventricle in the tortoise and the frog. In the tortoise, the isolated ventricle developed rhythmic pulsations "with as great a certainty as the isolated auricle," and it did so without the assistance of any external stimulus such as was required to produce rhythmic beats in the isolated frog's ventricle. Despite differences of degree, the automatic rhythm of the isolated tortoise's ventricle was of the same kind as that of the isolated auricle (and as that of the sinus), so that all three automatic rhythms must have a common origin, whether that origin be in the cardiac ganglia or in some inherent rhythmical property of cardiac muscle.[40]

To decide between these two possibilities, Gaskell focused on a strip of muscular tissue cut from the lower tip of the tortoise's ventricle. Upon the application of an interrupted current, rhythmical pulsations eventually appeared in the strip. "These contractions," Gaskell emphasized, "are clearly both myogonic and automatic":

> They are myogonic because no special nerve structures are to be found here, and when the rhythm is once well established the strip can be cut into small pieces, each of which will still continue its rhythm for a long time; the rhythmical power then is not specialized in any portion of the strip, but is distributed over the whole of it. . . . That the rhythm is not only muscular in origin, but also due to some quality inherent in the muscle itself, i.e., is automatic, is clearly shown by the method of its development. It is evidently not a "rhythm of excitation" due to the direct stimulation of the muscle by the single induction shocks or the interrupted current, for *it continues at least 30 hours after the discontinuance of all stimulation*. . . .

Furthermore, rhythmical pulsations scarcely less vigorous than those induced by electrical currents would appear if the muscular strip were merely left suspended in a moist chamber, undisturbed by any stimulus or foreign fluid.[41]

Since Gaskell had argued that the rhythm of the ventricular tip was of the same kind and therefore of the same origin as that of the sinus, it followed that the normal automatic rhythm of the sinus must also be muscular in origin. But this conclusion was so radical as to require careful demonstration. Only detailed and persuasive evidence would establish that ganglion-free ventricular muscle could develop rhythm at a rate truly comparable to that of the sinus. Gaskell showed that in the tortoise the isolated ventricle could in fact develop a rate of rhythm equal to that of the suspended heart as a whole (and thus equal to

[40] Ibid., pp. 50-51.

[41] Ibid., pp. 51-56, indented quote on 53.

that of its "leader," the sinus) if only its coronary blood supply and pressure were maintained at a suitable level. He established this point through an imaginative experiment which was possible only because the tortoise's ventricle (unlike the frog's) possessed a well developed coronary system. He also showed that even an isolated strip from the tortoise's ventricular tip could be "taught" (by an interrupted current) to beat spontaneously at a rate equal to that of the sinus. Since, therefore, no clear distinction could be drawn between the rates of the automatic rhythms in the ventricular tip and in the sinus, Gaskell rejected the idea that they could have different origins. Instead he reaffirmed his conclusion that "the rhythm of the sinus and therefore of the whole heart depends upon the rhythmical properties of the sinus [itself], and not upon any special rhythmical nervous apparatus."[42]

Gaskell had yet to explain why and how the normal heartbeat begins in the sinus, passes after a brief pause to the auricle, and then, after another pause at the auriculo-ventricular junction, ends in the ventricle. It was in fact by his investigation of this problem that he so dramatically confirmed the spirit of Foster's evolutionary approach toward the problem of the heartbeat at the same time that he transcended the limitations and difficulties of his mentor's particular attempt to resolve it. In an understandable reaction to the dominant neurogenicists, Foster had focused nearly all of his attention on the ganglion-free tip and had never carefully investigated the problem of the coordinated sequence of the vertebrate heartbeat as a whole. Meanwhile the neurogenicists, by focusing so exclusively on the ganglia-rich regions of the heart, fell victim to an equally pronounced myopia. At this point, further analysis of the separate cavities of the heart would be less fruitful than a new synthetic look at the heart as a whole and in different types of animals. More than anyone else, Gaskell supplied this enlarged perspective, and it was mainly by virtue of his synthetic approach that he made his most fundamental contributions to cardiovascular physiology.

Before the discovery of intracardiac ganglia in the 1840's, physiologists had generally conceived of the heartbeat as a peristaltic wave passing in regular order from one end of the heart to the other. By the time Gaskell began to study the problem, however, this view had been replaced by the belief that the coordinated sequence of the cardiac beats, if not their rhythm, certainly depended on the ganglia. This new conception seemed justified by the following evidence. From T. W. Engelmann's measurement of the rate of contractile propagation in the

[42] Ibid., pp. 56-60, quote on 56.

ventricle, it seemed clear that the normal heartbeat passed too slowly from the sinus to the ventricle to be considered a simple peristaltic wave of contraction. If, moreover, the heartbeat were followed in its course through the cavities of the heart, distinct pauses could be observed at the junctions between the separate cavities. And since it was precisely here, in these junctions, that ganglia were most abundant, it seemed reasonable to conclude (as indeed Foster himself had) that they must play some important role in producing the pauses and thus in regulating the sequence. Finally, the contribution of ganglia to the normal sequence of ventricular upon auricular beat seemed to be demonstrated beyond doubt by experiments in which the sequence was destroyed by extirpation of the auriculo-ventricular ganglia from the frog's heart. But despite all of this evidence, Gaskell had now become convinced that the older view carried more persuasion, and that the heartbeat should in fact be viewed as a simple peristaltic wave of contraction.[43]

Gaskell's work on the coordinated sequence of the heartbeat, even more than his work on rhythmicity, owed much of its value to his decision to study the tortoise's heart instead of the frog's. In this case especially, it was the anatomical arrangement of the tortoise's heart that made it so useful for Gaskell's purposes. In the tortoise's heart, unlike the frog's, the intracardiac nerve trunks and their accompanying ganglia are located outside of the heart itself, lying in or just beneath the pericardium. Among other things, this meant that "all experiments involving the removal of the different ganglionic masses or section of the large intra-cardiac nerve trunks can be performed on the tortoise with much greater ease and certainty than on the frog."[44] What Gaskell found upon performing such experiments was that "section of all the nerve trunks between the sinus and ventricle or even absolute removal of them with their accompanying ganglia does not in the very slightest degree affect the due sequence of ventricular upon auricular beat."[45] On this basis, he dismissed the hypothesis that the sequence depended on nervous stimuli passing from the sinus to the auricles and ventricle through the cardiac nerves and ganglia.

As a first step toward constructing an alternative explanation, Gaskell now reported the results of experiments in which he sliced deeper and deeper into the auricular tissue. With the tortoise's heart suspended as before (see Fig. 2), he made a slit through the auricular partition and continued the incision in such a way that the only connection between the sinus and the ventricle consisted of two bands

[43] Cf. ibid., pp. 61-62. [44] Ibid., p. 47. [45] Ibid., p. 64.

of auricular tissue (A_s and A_v) joined at the upper ligature by a narrow bridge of tissue. The suspended heart now appeared as follows:

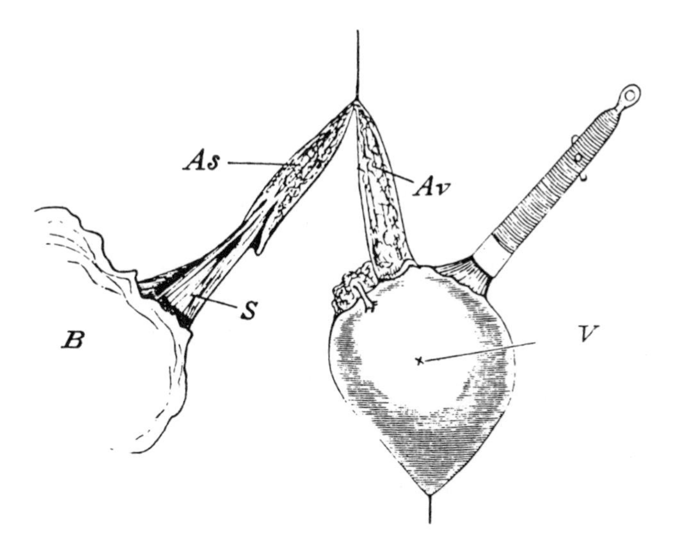

Figure 3. Junction wall and auricle slit up so as to separate auricle into two parts A_s, A_v. *Journal of Physiology, 4* (1883), 65.

None of this slicing disturbed the sequence of ventricular upon auricular beat, and the heart behaved just as if no section had been made. Gaskell continued his sections, slicing more and more tissue away from the auricle. For some time this procedure produced no observable result, but "at last . . . when the bridge is quite thin, it is apparent that *a wave of contraction* passes from the sinus up A_s and then after a slight pause down A_v to the sulcus where another pause occurs and then the ventricle contracts."

Still the sequence remained unaltered. But as the section became more and more severe, so that the bridge of tissue between A_s and A_v became increasingly narrow, eventually a point was reached when only every second wave passing up A_s was able to pass over the bridge and down A_v. At this point, the usual sequence was altered, with the ventricle now beating only once for every two auricular contractions. By still further slicing, the number of A_s waves passing over the bridge to A_v could be further reduced and the sequence correspondingly modified. Every time a wave of contraction did succeed in passing over the bridge, a ventricular beat resulted; and every time a wave was interrupted at the bridge, the ventricle remained quiescent. Finally, when the bridge was made so narrow that it seemed another slice would sever it completely, *none* of the waves passing up A_s was able to pass down A_v. In other words, a complete "block" had been established, and any ventricular beats now observed were absolutely independent of the contractions of the sinus and auricles.[46]

To Gaskell, the meaning of these results was crystal clear: "the ventricle contracts in due sequence with the auricle because a wave of contraction passes along the auricular muscle and induces a ventricular contraction when it reaches the auriculo-ventricular groove."[47] From the results of these experiments on the tortoise's heart, he now realized (though he did not openly admit) that he had misinterpreted the clamping experiments described in his Croonian Lecture on the frog's heart. Compression between the auricles and ventricle altered the sequence not because it made ganglionic impulses ineffective but because it created an obstacle to the advance of the contractile wave.[48]

With these points established, Gaskell now developed in detail his conception of the heartbeat as a peristaltic wave of contraction beginning at the sinus and passing from one end of the heart to the other, insisting that this model could be fully reconciled with the fact that the separate cavities of the intact heart beat in a regular and coordinated sequence. Bringing forth a wealth of histological and experimental evidence in favor of his view, Gaskell developed it in a distinctly evolutionary context. He focused first on the ring of tissue in the auriculo-ventricular groove and showed that the portion of the ring richest in nerve fibers and ganglia could be cut away without disturbing the correlation between auricular and ventricular beats. He then described in detail the histological appearance of the tortoise's cardiac tissue. Its most striking feature was the existence of a well-defined circular ring of parallel muscular fibers at the sino-auricular junction and again at the auriculo-ventricular junction. To Gaskell this arrangement was "very suggestive of the primitive origin of the heart from a tube like that of an artery or of the ureter, where the fibres are throughout arranged in a circular manner." He continued:

> The existence of these two muscular rings connecting the three muscular cavities of the heart is amply sufficient to account for the passage of the contraction from the sinus to the auricle, and from the auricle to the ventricle, without the necessity of invoking the presence of ganglion cells, and leads directly to the view originally held by physiologists that the cavities of the heart contract in reg-

[46] Ibid., pp. 64-66. [47] Ibid., p. 64. [48] Ibid., p. 66.

ular sequence, because a peristaltic wave of contraction commencing at the sinus passes from one end of the heart to the other.[49]

Two pressing questions remained unanswered: (1) How could one explain the pauses occurring at the sino-auricular and auriculo-ventricular junctions? And (2) why did the normal heartbeat begin in the sinus? In response to the first question, Gaskell argued that the junctions themselves presented a natural obstacle to the advance of the contractile wave, not only because the ring of muscular tissue at each junction is narrow, "but essentially because the structure of the muscular fibres here is different from those of the auricle or ventricle." The fibers in the junctions were less distinctly striated and their general appearance suggested that they belonged to an earlier stage of development than the fibers in the auricle and ventricle. This difference provided a clue not only to the reason for the pauses at the junctions but also to the reason for different rhythmical capacities in various parts of the heart. At this point, Gaskell's answer to the first question became intimately bound up with his answer to the second. He began by positing an antagonism between the capacity for rhythmicity and the capacity for the rapid conduction of contractile waves. Like Foster before him, he used the concept of the physiological division of labor to suggest that the capacity for rhythmicity was greatest in undifferentiated contractile tissue, while the capacity for rapid conduction increased pari passu with differentiation and specialization. Unlike Foster, however, Gaskell used this principle to account for the entire range of phenomena observed in the coordinated and sequential heartbeat as a whole.

Since the least striated and least developed muscular fibers were found in the sinus, Gaskell argued that it must also possess the greatest capacity for rhythmicity. It was only natural, then, that the heartbeat should begin there. The more highly developed auricular and ventricular fibers, on the other hand, were especially adapted to conduct the contractile wave rapidly. But the muscular rings connecting the heart cavities consisted of relatively undifferentiated tissue, and so the contractile wave naturally passed more slowly through them. This feature of the muscular rings, rather than ganglionic action, explained the pauses observed at the sino-auricular and auriculo-ventricular junctions. At each junction, the wave of contraction passed through a muscular region of enhanced rhythmicity but diminished conductivity. It was, then, erroneous to assume that a contractile wave would travel at

[49] Ibid., pp. 62-71, indented quote on 70-71.

the same rate through all parts of the heart. In fact, it would pass most rapidly through the ventricle, where the muscular tissue was most fully striated and differentiated. And so Englemann's measurement of the rate of propagation there could not properly be used as an argument against the peristaltic, purely muscular conception of the heartbeat.

Gaskell even managed to refute those seemingly decisive experiments in which the sequence between auricular and ventricular beats had been destroyed by extirpation of the auriculo-ventricular ganglia in the frog's heart. The real reason the sequence had been destroyed in such experiments, he argued, was that the muscular ring in the auriculo-ventricular groove had been simultaneously cut away or damaged. If great care were taken to remove the ganglia without damaging the muscular ring, the sequence remained undisturbed.[50] Since Gaskell had now denied the ganglia any role whatever in the ordinary coordinated vertebrate heartbeat, it was reasonable to ask why they existed at all. As Gaskell later put it, that was a question he was "ready to answer, and to answer with confidence." The cardiac ganglia, he asserted, "are cells connected only with the inhibitory fibres of the vagus, and as such are simply part and parcel of the mechanism of inhibition."[51]

Toward the end of his paper of 1883, Gaskell returned to his theme that the heart in its undeveloped state was a simple tube with muscular walls. The adult heart should then be considered "a piece of artery or vein, the muscular walls of which have developed in a special manner."[52] On the assumption that different parts of the original tube underwent development at different rates, the auricles and ventricles of the adult heart could be viewed as highly developed cavities which had bulged out from the original tube, while the remnant of that tube could be recognized in the undeveloped fibers of the sinus, the sino-auricular ring, the auricular partition, and the auriculo-ventricular ring. As a result of this differential mode of development, different sections of the adult heart possessed differing capacities for rhythmicity and conductivity. And so:

> the peristaltic wave of contraction which originally passed smoothly from end to end, passes finally along a tube of irregular calibre, the muscular walls of which have become so modified in their rates of contraction and conduction, as well as in the arrangement of their fibres, as to form out of a simple peristaltically contracting tube such an efficient muscular tube as is represented by the adult heart.[53]

[50] Ibid., pp. 71-77.
[51] Gaskell, "Cardiac Muscle," p. 197.
[52] Gaskell, "Heart of the Tortoise," p. 116.
[53] Ibid., p. 78.

In support of this conception, Gaskell appealed to experiments on the hearts of animals simpler than the tortoise. In the basically tubular and undifferentiated ascidian heart, "the rhythmical power of the two ends of the heart is so nearly equal, that the contractions start with equal, or nearly equal, facility from either end." In the slightly more developed heart of the skate, where the peristaltic nature of the beat was dramatically visible, a slight mechanical stimulus at the conus arteriosus was sufficient to reverse the normal direction of the rhythmic contractions. Since no ganglia could be found in this conus, its high capacity for rhythmicity must be due to the undeveloped state of its muscular fibers. Engelmann's work on the highly rhythmic but ganglion-free bulbus in the frog's heart led to the same conclusion. In every really important respect, in its rhythmicity and in its sequence, the heartbeat depended not on nervous influences but on the properties of the cardiac musculature.[54]

In a long section on the cardiac nerves, Gaskell sought to explain their action in terms that conformed to this general view. He argued, as he had in his Croonian Lecture on the frog, that the bewildering range and variety of the effects of vagus stimulation could all be explained on the assumption that the vagus was the trophic or anabolic nerve of the heart and particularly of the cardiac musculature. If the idea remained a little vague, it had gained in cogency from Gaskell's new perception that the fundamental properties of cardiac muscle were as responsible for the rhythm and sequence of the heartbeat as they were for the *force* of cardiac contractions. He was now able to suggest that many of the most important effects of vagus stimulation resulted from its effects on the *conductivity* of the muscular tissue in various parts of the heart.[55]

Finally, with specific reference to Foster and Dew-Smith's experiments, Gaskell also showed that the cardiac musculature responded to vagus stimulation in the same way that it responded to the direct application of an interrupted current too weak to cause muscular contractions. He suggested that the quieting effect of both types of stimuli on previously active hearts might belong to a more general law, namely, that direct stimulation of a rhythmically contracting tissue inhibits the rhythm of that tissue.[56] This proposed law is virtually identical to a suggestion Foster had made in 1876: that the essential factor linking vasodilator action, ordinary vagus inhibition, and the inhibition produced in a snail's heart by electrical currents was that in each case "stimulation is brought to bear on a spontaneously active tissue."[57]

[54] Ibid., pp. 78-81. [55] Ibid., pp. 81-103. [56] Ibid., pp. 103-111, esp. 110-111.
[57] Foster and Dew-Smith, "Constant Current," p. 768.

Gaskell's cardiological research and the Cambridge setting: toward a comparative evaluation of the contributions of Foster, Romanes, and Gaskell

Gaskell's paper of 1883 marks a watershed, both in the history of cardiovascular physiology and in the history of the Cambridge School. The myogenic theory and the Cambridge School had grown up together, and they now reached maturity together. This is not to say that the Cambridge School was entirely unknown before. Gaskell's investigation of vasomotor action and Langley's penetrating study of glandular secretion had attracted some international attention, although, especially in Germany, these contributions tended to be associated mainly with the German laboratories in which the two men had studied. It was Gaskell's work on the heart that marked the full-fledged emergence of Cambridge as a significant independent center of physiological research. Apparently, Gaskell's achievement generated a certain curiosity about Foster and the other members of the Cambridge School. After 1883 Foster's papers were cited in most major review articles on the problem of the heartbeat, and by the 1890's T. W. Engelmann was referring conspicuously to the work of "Foster and his pupils."[58] Almost certainly because of Gaskell's references to it, Foster's work was rescued from its previous obscurity.

In directing his students toward the myogenic-neurogenic debate, Foster had chosen a topic of great significance for late nineteenth-century physiology. The issue eventually dominated the entire physiology of the circulation, the more so because of its implications for theories of vasomotor action.[59] Foster's choice, though prescient, also had deep roots in his own past. Long before coming to Cambridge, probably under the influence of William Sharpey, he had come to believe that the dominant neurogenic theory rested on a shaky foundation. He communicated this conviction to his students—to A. S. Lea, to A. G. Dew-Smith, to J. N. Langley, to G. J. Romanes, and to Gaskell himself—and each of them turned for a time to the problem of the heartbeat. Each, moreover, confirmed Foster's conviction, and the Cambridge School gradually developed into a major sanctuary for the generally unpopular myogenic theory of the heartbeat. With Gaskell's treatise of 1883, the Cambridge position reached a new level

[58] Theodor W. Engelmann, "Ueber den myogen Ursprung der Herzthätigkeit und über automatische Erregbarkeit als normale Eigenschaft peripherischer Nervenfäsern," *Archiv für die gesammte Physiologie des Menschen und der Thiere, 65* (1897), 535-578, on 539.

[59] See, e.g., Elie Cyon, "Myogen oder Neurogen?" ibid., *88* (1902), 225-294, on 225.

of completeness and persuasion. Foster felt enough confidence in Gaskell's results to incorporate the myogenic theory into the fifth edition of his textbook (1888-1891).[60] No other revision could have brought him greater personal satisfaction.

To be sure, Gaskell's paper of 1883 did not immediately convince everyone, and the myogenic-neurogenic debate continued for some time, especially in Germany.[61] The difficulties in extending the myogenic theory from the lower vertebrate to the mammalian heart proved more challenging than Gaskell had perhaps expected them to be,[62] and he himself did not contribute to that task. By about 1910, however, the extension had been accomplished, notably through the work of Wilhelm His, Jr., S. Tawara, and Joseph Erlanger on the atrioventricular bundle in mammalian hearts, and of Arthur Keith and Martin Flack on the mammalian cardiac pacemaker.[63] Throughout this period, Gaskell's work served as the major point of departure for studies of heart action. Before long, his myogenic theory and his concept of heart block were incorporated into the pathology, pharmacology, and therapeutics of the heart, and his conclusions have remained at the base of cardiology ever since.

The importance of Gaskell's work has long been recognized, but its intimate connection with Foster's research has been inadequately appreciated. Almost incredibly, Langley himself described the cardiological research of each man in some detail without drawing a connection between them.[64] By 1960, Foster's work had become so utterly forgotten that Carl J. Wiggers could suggest, in a review of "significant advances in cardiac physiology during the nineteenth century," that Gaskell's interest in the problem of the heartbeat must have been inspired by his stay in Ludwig's institute at Leipzig.[65] Fielding Garrison, Walter Langdon-Brown, and Henry Dale perceived a bond between

60 Michael Foster, *A Textbook of Physiology*, 5th ed., largely revised, 4 parts (London, 1888-1891), pp. 285ff.

61 See, e.g., Cyon, "Myogen oder Neurogen?" By this point, Cyon had become the most fervent opponent of the myogenic position. His views receive further attention in Chapter 11 below, pp. 344, 350-351.

62 Cf. W. H. Gaskell, "Observations on the Innervation of the Heart: On the Sequence of the Contractions of Different Portions of the Heart," *Brit. Med. J.*, 1882 (2), 572-573.

63 On the developments after 1883, see inter alia Engelmann, "Ursprung der Herzthätigkeit"; Robert Tigerstedt, *Die Physiologie des Kreislaufes*, 2nd ed., 4 vols. (Berlin, 1921), esp. Bd. II; and J. A. E. Eyster and W. J. Meek, "The Origin and Conduction of the Heartbeat," *Physiological Review, 1* (1921), 3-43.

64 See J. N. Langley, "Foster," esp. pp. 239-240; and Langley, "Walter Holbrook Gaskell," *Proc. Roy. Soc., 88B* (1915), xxvii-xxxvi, esp. xxviii-xxix.

65 Carl J. Wiggers, "Some Significant Advances in Cardiac Physiology during the Nineteenth Century," *Bull. Hist. Med., 34* (1960), 1-15, on 6-7.

Foster's work and Gaskell's with varying degrees of vagueness.[66] Nothing in print even approached a full appreciation of the connection until 1970, when Richard D. French explored "the link between Romanes' jellyfish research and the cardiological research of the Cambridge school of Michael Foster and Walter Gaskell."[67] Besides providing an admirable summary of the major papers of Foster and Gaskell on the problem of the heartbeat, and of Romanes on the locomotor system of the medusae, French briefly but explicitly discussed Foster's influence on his two student-colleagues. In the end, though, French conveyed the impression that Romanes influenced Gaskell's research more profoundly than Foster did. There is room for reasonable disagreement here. The intimate and informal atmosphere of the Cambridge laboratory clearly encouraged a lively exchange of ideas among those working there; and the nearly simultaneous publication of several of the pertinent papers sometimes makes it difficult to determine the precise directions of that exchange.[68] However one comes out on this issue, it affects only minimally my basic theme that the Cambridge School descended from and crystallized around the problem of the heartbeat. Foster, Gaskell, and Romanes all clearly have a place in the early history of the Cambridge School.

Nonetheless, the relative influence of Foster and Romanes on Gaskell's research is worth examining for two reasons: (1) Though George

[66] See Fielding H. Garrison, "Sir Michael Foster and the Cambridge School of Physiologists," *Maryland Medical Journal, 58* (1915), 106-118, on 109; Langdon-Brown, "Gaskell," pp. 4-5; and H. H. Dale, "Sir Michael Foster," *Notes and Records of the Royal Society (London), 19* (1964), 10-32, on 22.

[67] Richard D. French, "Darwin and the Physiologists, or the Medusae and Modern Cardiology," *J. Hist. Biol., 3* (1970), 253-274, quote on 266, and see Acknowledgments above.

[68] Two examples should suffice here. The very similar papers of Foster and Langley on vagus action in the poisoned frog's heart (see Chapter 8 above, "John Newport Langley") appeared within six months of each other, Langley's being in fact the earlier of the two: J. N. Langley, "The Action of Jaborandi on the Heart," *J. Anat. Physiol., 10* (1876), 187-201; and Michael Foster, "Some Effects of Upas-Antiar on the Frog's Heart," ibid., *10* (1876), 586-594. For the month of publication of each, see "Contents," ibid., *10* (1876), iv. Langley's paper preceded by only one month the version of Foster's paper read before the Cambridge Philosophical Society on 15 November 1875. See *Proc. Camb. Phil. Soc., 2* (1876), 398-400.

Similarly, with respect to contractile "block" (see text immediately below), Romanes sent his first description of this phenomenon in medusae to the Royal Society nine months before Foster and Dew-Smith published an account of the same phenomenon in the frog's ventricle. See G. J. Romanes, "Preliminary Observations on the Locomotor System of Medusae," *Phil. Trans., 166* (1876), 269-313; and Foster and Dew-Smith, "Constant Current."

In neither case do I doubt that the basic direction of influence was from Foster toward his younger students, but I concede that this judgment is based on more general evidence (notably Foster's long-standing interest in the problem of rhythmic motion) and not on the precise date of publication of these particular papers.

Romanes was very definitely Foster's student, he worked only sporadically in the Cambridge laboratory and therefore could have participated only indirectly in the founding of a research tradition there. To establish a direct and intimate link between the rise of the Cambridge School and the problem of heartbeat, it seems essential to identify someone who both worked on the problem and remained almost constantly on the scene during the school's embryonic years. On this twofold criterion, no one even approximates Foster's qualifications. (2) More importantly, every challenge to the breadth and depth of Foster's influence on the early cardiological research of the Cambridge School undermines one of my major secondary themes: that Foster's achievement at Cambridge can be rescued from its veil of mystery only if and when it is recognized that he was once a competent research physiologist and influential director of research. That theme underlies my detailed attempt to establish Foster's influence on each vital phase of the Cambridge effort to resolve the problem of the heartbeat. To abandon the argument at the point where that effort gained its most notable success would be to vitiate much of what has gone before.

Most of the evidence leading me to doubt that Gaskell's work on the heart actually "arose from" or was "directly inspired" by Romanes' research has already been presented. Romanes, like Gaskell, studied under Foster, who influenced his research from the outset. The theory of ganglionic action that Romanes set forth in 1880 is, in fact, virtually identical to the view Foster had advanced perhaps as early as 1869 and certainly no later than 1876. Moreover, if Gaskell's work on vasomotor action was inspired from the beginning by Foster, who had long emphasized the analogy between vasodilation and cardiac inhibition, then no separate inspiration from Romanes would have been required to direct Gaskell's attention to the problem of the heartbeat.[69]

Now that Gaskell's cardiological research has been described in detail, it is possible to examine still further the relationship between that work and Romanes' jellyfish investigations. In particular, it is important to assess how far Gaskell may have been indebted to Romanes for the concept of "heart block" and for the experimental techniques by which he resolved the problem of the heartbeat. For it is these presumed debts that have led French, among others, to emphasize Romanes' influence on Gaskell.[70] In arguing against that view, my aim is not to minimize the importance and value of Romanes' work, but to insist once again on Foster's influence and technical competence. This

[69] See Chapter 8 above, "George John Romanes"; and "Gaskell and vasomotor action."

[70] See French, "Medusae," esp. pp. 268-273.

latter point deserves special emphasis because several of Foster's later students have suggested that he lacked skill as an experimenter.[71] That this impression is misleading should already be clear from the account of his basic competence given in preceding chapters. Indeed, he designed the elegant experiment that to Langdon-Brown marked the birth of modern cardiology.[72] But the case becomes more persuasive upon a close comparative analysis of the experimental techniques used by Foster, Romanes, and Gaskell.

In published accounts of his work on the heart, Gaskell refers to Romanes' work only once. The reference appears as a footnote to an important passage in the classic paper of 1883. Prior to this passage, Gaskell had been describing how progressive section of the tortoise's auricular tissue affected the passage of the wave of contraction from the sinus to the ventricle. As the section increased in severity fewer and fewer of the sinus waves were able to reach the ventricle. "Finally," Gaskell continued, "when the tissue is so thin that it often appears impossible to cut further without cutting the ligature itself, then the contraction wave is absolutely unable to pass, then the 'block' is complete and any contractions of the ventricle which may occur are absolutely independent of those of the sinus and sinus-auricle." His footnote to this passage reads as follows:

> In his experiments upon the passage of contraction waves along the muscular tissue of the swimming bell of the Medusae, Romanes . . . has throughout made use of the term "block" to express any artificial hindrance to the passage of the contraction. I therefore make use of the same term in speaking of the results of experiments on the cardiac muscle which are very similar to those which he performed on the muscle of the Medusae.[73]

Romanes' experiments on the medusae obviously impressed Gaskell, probably because of the clarity and decisiveness of their results and certainly because they offered support by analogy to Gaskell's work on the heart. But it is hard to know whether, or how far, his debt to Romanes extended beyond what he himself said it was—namely, the "use of the term 'block' to express any artificial hindrance to the passage of the contraction." Perhaps he did indeed borrow from Romanes only the *term* and not the *concept* of contractile blocking. For, as Langley later pointed out, the concept or phenomenon of blocking had already

[71] See. e.g., Dale, "Foster," p. 13; and J. George Adami, "A Great Teacher (Sir Michael Foster) and His Influence," *Publication no. 7, Medical Faculty, Queen's University (Kingston, Ontario, Canada)*, June 1913, pp. 1-17, on 13.

[72] See this Chapter, p. 272.

[73] Gaskell, "Heart of the Tortoise," pp. 65-66, n. 1.

been "partly worked out in the frog's ventricle by Engelmann."[74] Moreover, in a paper appearing almost simultaneously with the paper in which Romanes introduced his term, Foster and Dew-Smith referred to the disruption of a contractile wave in the frog's ventricle when a tissue bridge there was made sufficiently narrow.[75]

The essential technique common to the work of Romanes and Gaskell is that of progressive section so as to produce a tissue bridge. To some extent, at least, progressive section had probably already become an established technique for determining contractile zones. Romanes himself supplied evidence that the German biologist Theodore Eimer, working simultaneously and independently on the medusae, had used the same technique in his work.[76] As should be obvious from the reference to their work immediately above, Foster and Dew-Smith were at the same time using the technique specifically to produce a narrow tissue bridge in the heart. But what is particularly noteworthy is the striking resemblance between Foster and Dew-Smith's version of the technique and Gaskell's later method of "suspending" the tortoise's heart. As described in their paper of 1876, Foster and Dew-Smith produced a narrow tissue bridge by dividing the frog's ventricle longitudinally, extending the resulting "V-shaped mass" almost into a straight line, and then pinning it in that position—a preparation that must have looked quite similar to Gaskell's final preparation of the tortoise's heart in 1883.[77] To his preparation Gaskell then added two delicate levers connected at one extremity with a kymograph arrangement. At the opposite end, one of the levers was applied to the auricle and the other to the ventricle. This technique, too, closely resembles one Foster had already used. In his paper of 1876 on the cardiological effects of the poison upas antiar, he had recorded the cardiac movements "by means of two light levers, placed directly, one on the ventricle and the other on the auricles." He also made one or two experiments using the standard Continental method of measuring endocardial pressure, but he emphasized that "all the main results were gained by help of the simple lever, which, for general purposes is a most useful instrument deserving to be used much more frequently than it is."[78]

Gaskell's main technical contribution would thus seem to be his skillful combination of techniques already used separately by Foster. But that is by no means a trivial achievement, and Gaskell could justi-

74 Langley, "Gaskell," p. xxix.

75 Foster and Dew-Smith, "Constant Current," p. 736.

76 G. J. Romanes, "Preliminary Observations on the Locomotor System of Medusae," *Phil. Trans., 166* (1876), 269-313, on 305-309.

77 Foster and Dew-Smith, "Constant Current," p. 736. For Gaskell's preparation, see this chapter, Figure 3.

78 Foster, "Upas Antiar," p. 586.

fiably insist upon the originality of his "suspension method," first described in his Croonian Lecture of 1882 on the frog's heart.[79] Writing of Gaskell's method in 1900, Engelmann described it as the one arrangement that had allowed a comprehensive understanding of heart action.[80] And whereas Foster had focused mainly on ventricular events, Gaskell used his method to investigate the problem of the coordinated sequential heartbeat as a whole. Besides this, and besides being the first to exploit fully the unique advantages of the tortoise's heart, Gaskell apparently also introduced the use of a screw clamp to isolate the auricles from the ventricle.[81] But this technique, so important for his work on the frog, was of minor value in his work on the tortoise, where it proved more convenient to hold the heart in place with a simple aortic clasp rather than with a clamp between the auricle and ventricle. In fact, this apparently fortuitous modification in technique may have led him to adopt the method of progressive section previously employed by Romanes and Foster among others.[82]

No part of this argument is meant to suggest that Foster belonged among the leading experimental physiologists of his day. As a research physiologist, he was no match for Ludwig or Bernard, nor even for Engelmann, Heidenhain, or Gaskell himself. Nor is it my intention to imply that Gaskell and Romanes were slavish disciples of Foster, incapable of true originality in thought or technique. Both in fact displayed significant originality, and their published research certainly attracted more attention than did Foster's. As French suggests, perhaps the most fundamental point to recognize about the work of all three is that they shared a common interest in the problem of rhythmic motion and a common commitment to an evolutionary approach. Nevertheless, it is important to emphasize that Foster was a competent research physiologist, to whom Gaskell and Romanes owed specific technical-cum-intellectual debts and from whom their common interests and approach almost certainly derived. Only then can one begin to appreciate just how closely the rise of the Cambridge School was bound up with the problem of the heartbeat and with Foster's effort to resolve it in evolutionary terms. And only then can one insist that Foster's brief career in research played an indispensable part in the hitherto mysterious process by which he created his great school.

[79] See, e.g., Gaskell, "Cardiac Muscle," p. 172.

[80] Theodor W. Engelmann, "Ueber die Wirküngen der Nerven auf das Herz," *Archiv für Anatomie und Physiologie* (*Physiologische Abtheilung*), no vol. number, 1900, pp. 315-361, on 316.

[81] See Gaskell, "Heart of the Frog," pp. 994-996.

[82] See this Chapter, p. 281.

PART FOUR

Denouement and Conclusion

10. The Growth and Consolidation of the Cambridge School, 1883-1903: Foster in His More Familiar Entrepreneurial Role

> The work of a school, like the task of investigating a single problem, involves choices and hence exclusions; and it is only in retrospect, if at all, that the correctness of those decisions can be assessed. But the difference in time-scale and social character between the investigation of an individual problem and the work of a school is the cause of significant differences in their cycle and development. . . . A problem has a natural end, when a conclusion can be drawn. That task is accomplished, and the scientist can then decide whether his next task should be to improve the previous conclusion, or turn to something else. This can also happen in the case of a school, when a grand-ancestor problem is finally solved to the satisfaction of those working on it. But it is more common for the work of a school to lack such a neat conclusion. For, even when it is organized around such a grand problem, its work will tend to be derivative on the insights of its founder. . . . The leadership has a technical, social and emotional commitment to the established methods; and so, unless it is very enlightened, recruits who bring in fundamentally new insights will not be welcomed or absorbed. . . . Only if the work of a school has enough links with that of others for its facts to be capable of some translation into statements about other objects of inquiry will its productions survive in the common stock.
>
> J. R. Ravetz (1971)[1]

IN June 1883, just as Gaskell achieved his resolution of the problem of the heartbeat, Foster became the first professor of physiology at Cambridge. From then until his retirement in 1903, he directed his efforts mainly toward the consolidation, extension, promotion, and more complete organization of the school he had founded. To a certain extent, these efforts were bound up with his more general efforts on behalf of English physiology, and he did not always distinguish between the two. What was good for the Cambridge School was perforce good for English physiology, and vice versa. Not everyone else saw it quite that way, but Foster now had the insitutional base and political power to ensure that his perceptions would often prevail and never lack an attentive audience.

1 Jerome R. Ravetz, *Scientific Knowledge and Its Social Problems* (Oxford, 1971), pp. 231-233.

By 1883, in fact, Foster dominated English physiology. A leading force in the Physiological Society of Great Britain since its formation in 1876, he also joined the executive council of the new Association for the Advancement of Medicine by Research in 1882, when it undertook its quiet but immensely effective campaign to neutralize the Vivisection Act of 1876.[2] He held a chair in what had long been one of the world's most prestigious universities, and he remained a Fellow of its largest, richest, and most scientific college. Editor of the leading journal for physiology in the English-speaking world, he was also the author of its most celebrated physiological textbook. His important friends in important places included T. H. Huxley, the newly elected president of the Royal Society. Last, but far from least, he had succeeded Huxley (and before him, Sharpey) as biological secretary of the Royal Society in 1881.

Even in the absence of any proper study of the role played by secretaries of the Royal Society, we can be confident that this new position made Foster the most powerful man in English biological circles.[3] A paper communicated to the Society under his sponsorship was certain to gain a place in its *Proceedings* or *Transactions*. Any grant proposal having his approval would enjoy an immediate advantage over others. And any Foster-backed candidate for membership or other honors at the Royal Society was likely to succeed. Can it be mere coincidence, for example, that the number of governmental grants awarded to physiologists through the Royal Society increased phenomenally during his secretaryship? Is it fortuitous that the biological sciences in general took an increasingly large share of the grant budget after Foster became senior secretary in 1885?[4] And can it be accidental that the first three

[2] See Richard D. French, *Antivivisection and Medical Science in Victorian Society* (Princeton, 1975), Chapter 7.

[3] For a skeletal version of the statutes governing the Royal Society from its foundation through 1939, see *The Record of the Royal Society of London for the Promotion of Natural Knowledge*, 4th ed. (London, 1940). The formal duties and powers of the secretaries are covered on pp. 295, 308-309, and 325. Almost equally sketchy and formal is the account of secretarial activities in Henry Lyons, *The Royal Society, 1660-1940: A History of Its Administration under Its Charters* (Cambridge, 1944), passim. For one indication that these statutory histories conceal important aspects of the secretarial role, see R. M. MacLeod, "The Royal Society and the Government Grant: Notes on the Administration of Scientific Research, 1849-1914," *The Historical Journal, 14* (1971), 323-358. Without emphasizing the point, MacLeod mentions (p. 340) that the senior secretary of the Royal Society had chief administrative responsibility for the Government Grant to the Society—"the first and until 1890 the major continuous source of direct government finance for original scientific investigation" (p. 324).

[4] According to MacLeod (ibid., p. 353), the "number of grants awarded by Board G (physiology) quadrupled between 1889 and 1914 while grants for Board B (physics) merely doubled." The number of grants to chemists increased much more

Croonian Lecturers chosen by the Society during Foster's secretaryship were former students of his, and that all three—G. J. Romanes (1881), W. H. Gaskell (1882), and H. N. Martin (1883)—favored his position on the problem of the heartbeat?[5] Although the details of the process remain unexamined, there can be no doubt that the Cambridge School of Physiology owed part of its success to Foster's activities as biological secretary of the Royal Society, a position he held simultaneously and coterminously with his chair at Cambridge. When he resigned both offices in 1903, his tenure as biological secretary had become (and still remains) the longest in the long history of the Royal Society.[6]

By that point, a few saw Foster's career as proof of Lord Acton's dictum that power corrupts. One can only imagine the discontented murmurings of authors whose articles were rejected by the *Journal of Physiology*, while Foster's genial and indulgent editorial policy toward others drove the *Journal* toward bankruptcy and eventually, in 1894, forced a total reorganization under the severe but able direction of J. N. Langley.[7] Meanwhile, Foster's famous *Textbook* attracted criti-

---

slowly. Thanks chiefly to the increase in grants for physiology and medicine, the early dominance of physics, astronomy, and chemistry over the Government Grant was reversed during the period Huxley and Foster served as biological secretary. In 1870, when Huxley assumed the post, biology took less than 20% of the Government Grant. By 1880, when he resigned in favor of Foster, the portion had grown to nearly 30%. By 1887-1888, biology was taking more than 50%, and it continued to do so through the rest of Foster's secretaryship (ibid., p. 345). Perhaps, as MacLeod suggests, these results merely reflect a more general pattern of relative disciplinary growth, but the possibility that Huxley and Foster acted as disciplinary advocates surely deserves consideration.

[5] For a list of the Croonian Lecturers and Lecturers during Foster's tenure as biological secretary (1881-1903), see *Record of the Royal Society* (n. 3), pp. 361-362. Besides Romanes, Gaskell, and Martin, at least two others on that list can be identified as students of Foster—H. Marshall Ward (1890) and Charles Scott Sherrington (1897). Yet another, the German-Dutch physiologist Theodor Wilhelm Engelmann (1895), was the leading Continental advocate of the myogenic theory of the heartbeat. For his Croonian Lecture, see T. W. Engelmann, "On the Nature of Muscular Contraction," *Proc. Roy. Soc., 57* (1895), 411-433. In point of fact, neither Engelmann nor any of Foster's students used the Croonian Lecture explicitly to promote the myogenic theory (indeed, as described in Chapter 9, Gaskell took the occasion to announce his temporary defection from Foster's position). Nonetheless, Romanes, Gaskell, and Engelmann were leading investigators of rhythmic motion and leading advocates of myogenicity in their other writings.

[6] For a list of the secretaries of the Royal Society from its foundation through 1938, see *Record of the Royal Society* (n. 3), pp. 342-343. Except for Foster, Peter Mark Roget (21 years), and Sharpey (17 years), no biological secretary ever served longer than twelve years. Two of the secretaries for the physical sciences served even longer than Foster—Joseph Planta (1776-1804) and G. G. Stokes (1854-1885), whose thirty-one years in the office mark the all-time record.

[7] Cf. Walter Morley Fletcher, "John Newport Langley—In Memoriam," *J. Physiol., 61* (1926), 1-27, on 12-13.

cism for its increasing parochialism. F. Gowland Hopkins, himself later to be co-opted by Foster and Cambridge, went so far as to attack Foster's alleged partiality in print. In 1889, in a paper otherwise devoted to the chemical genesis of physiological pigments, Hopkins charged that "our chief physiological text-book is somewhat chary of incorporating new work which has not had its origin on the banks of the Cam!"[8]

The most widely publicized criticisms of Foster concerned his actions in Parliament, where he served as M.P. for the University of London from 1900 to 1906. Although they give no source for the quotation, Arthur Schuster and A. E. Shipley shuddered to recall that "a man of such outstanding merit" as Foster should once have said, "Not till I became a Member of Parliament did I understand what *power* meant."[9] Paradoxically, it was often charged that Foster made too little use of his power in Parliament, and he attracted widespread abuse for his "muddle-headed" decision to desert his original party. Having stood for election as a Liberal Unionist because of that party's stand on Irish home rule, he crossed the aisle to join the Liberal party because of its position on education. His attempt at reelection in 1906 failed by a narrow margin.[10]

But the barbs which Foster probably felt most severely were those directed at his performance as biological secretary for the Royal Society. This is not the place to discuss the internal history of the Society or even of Foster's role in it, but Foster's correspondence with Huxley and Thiselton-Dyer makes frequent reference to his skirmishes there. At the Royal Society, if not in Parliament, Foster apparently exercised his power with some relish, and by 1893 even Thiselton-Dyer, who was otherwise so much in sympathy with Foster and his aims, complained to Huxley that his leading protégé had become part of an obstructive "oligarchy" in the Society.[11] In 1901 the Council of the Society adopted a bylaw whereby its two secretaries could no longer serve in office beyond ten years,[12] a direct slap in the face to Foster, who had already been biological secretary for twice that long. Although he then contemplated immediate retirement, Foster continued as secretary until November 1903, when he was replaced by Archibald

[8] See David Keilin, *The History of Cell Respiration and Cytochrome*, prepared for publication by Joan Keilin (Cambridge, 1966), pp. 106-107.

[9] Arthur Schuster and A. E. Shipley, *Britain's Heritage of Science* (London, 1917), p. 301.

[10] See *The Times* (London), 17 January 1903; and W. H. Gaskell, "Sir Michael Foster," *Proc. Roy. Soc., 80B* (1908), lxxi-lxxxi, on lxxix-lxx.

[11] W. T. Thiselton-Dyer to Huxley, 23 October 1893, Huxley Papers, XXVII. 226.

[12] See Lyons, *Royal Society*, p. 300.

Geikie. Ten months earlier, incensed by some unspecified act on Foster's part, his former student Walter Gardiner wrote Thiselton-Dyer of his anger and disappointment at what Foster had become.[13] Near the end of a remarkable diatribe against Foster, Gardiner wished he could send him a New Year's card bearing this stanza from Tennyson's *In Memoriam*:

> I cannot see the features right
> When on the gloom I strive to paint
> The face I know; the hues are faint
> And mix with hollow masks of night.

Clearly, Foster's efforts at the Royal Society and elsewhere did not win universal admiration or unremitting support. Nonetheless, in an affectionate obituary notice on Foster, Thiselton-Dyer could end by insisting that Foster had not courted the power that fell inevitably upon him, and that while he doubtless enjoyed watching his former students join the Royal Society, "he scrupulously refrained from pressing their claims."[14] Whatever Foster's alleged transgressions, they cannot have been fiscal in nature, for he left Lady Foster a modest inheritance at best.[15] Leaving it for others to examine and assess Foster's performance at the Royal Society and in the various governmental duties entrusted to him, we can only suggest that his staure and power in Victorian scientific circles offered additional scope for his continued efforts to promote the Cambridge School.

Foster's retreat from the laboratory

By the early 1880's, Foster had effectively abandoned laboratory teaching as well as research. For several years, Gaskell, Langley, and Lea had been teaching the advanced course in physiology, with each taking those sections in which his own research interests lay. Lea taught the sections on physiological chemistry, Gaskell those on circulation and the peripheral nerves, and Langley most of the rest, including general

13 W. Gardiner to W. T. Thiselton-Dyer, 21 January 1903, Thiselton-Dyer Papers, Royal Botanic Gardens, Kew. The date of Gardiner's letter suggests that he was among those outraged by Foster's decision to cross the Parliamentary aisle rather than resign as M.P.

14 W. T. Thiselton-Dyer, "Michael Foster—A Recollection," *Cambridge Review, 28* (1907), 439-440.

15 In a letter describing the circumstances surrounding Foster's sudden death, the previously outraged Gardiner wrote: "I understand from Shipley that Lady Foster will continue to live at Shelford and that she is left, unfortunately, rather badly off." Gardiner to W. T. Thiselton-Dyer, 5 February 1907, Thiselton-Dyer Papers, Royal Botanic Gardens, Kew.

histology, glandular secretion, and the central nervous system. This pattern evolved gradually, beginning about 1875, when a separation was first made between elementary and advanced classes in physiology, but by 1881 it had taken "definite shape." After the advanced courses were more completely organized in 1882-1883, the system continued with very little modification throughout the rest of Foster's years at Cambridge.[16]

As Gaskell, Langley, and Lea became increasingly responsible for the advanced teaching, so naturally did they become increasingly concerned with the direction of research in the laboratory. Students who came to Cambridge after 1880 encountered Foster only as the lecturer in the elementary courses who often rushed from the lecture hall to a waiting hansom cab bound for London. The future Nobel laureate Charles Scott Sherrington, who entered Cambridge in 1879, remembered Foster chiefly as an "appalling lecturer," and "never saw him do any demonstration." Sherrington found Gaskell, Langley, and Lea a "more appealing triumvirate," and his own research interests developed under the influence of Langley and Gaskell, especially the latter.[17] Henry Dale, another future Nobel laureate who is responsible for the portrait of Foster rushing to meet his cab, failed to establish any personal contact with Foster while studying at Cambridge from 1894 to 1900.[18] During roughly the same period, Joseph Barcroft and Walter Morley Fletcher were also trained in physiology at Cambridge. Barcroft later gained eminence for his work on respiration and blood gases and for

[16] The evolving pattern can be traced through the *Cambridge University Reporter, 5* (1875), 15; *8* (1878), 13-15; *10* (1880-1881), 28, 339, 485, 552, 671-675; and *12* (1882-1883), 28-29, 336, 594, 698-699. After 1883, the most significant modifications in the pattern involved changes and additions in personnel. Beginning in 1892, Lewis Shore offered an "intermediate course" (ibid., *22* [1893], 368). In 1894 the future anthropologist W. H. R. Rivers joined the staff as lecturer on the physiology of the sense organs and on physiological psychology (ibid., *24* [1895], 699). When A. S. Lea's health broke down in 1895, his course on physiological chemistry went temporarily to Alfred Eichholz (ibid., *24* [1895], 699; *25* [1896], 363). Eichholz, in turn, gave way to F. Gowland Hopkins (ibid., *28* [1898], 33). In 1900-1901 Hugh Kerr Anderson taught an advanced course in each of two terms, while W. B. Hardy, Ivor Tuckett, and Walter Morley Fletcher combined to offer a one-term advanced course (ibid., *30* [1900], 1144). For the official university posts held by these and other members of Foster's staff, see J. R. Tanner, ed., *The Historical Register of the University of Cambridge* (Cambridge, 1917), pp. 115, 119, 124, 128, 140-142.

[17] See Judith P. Swazey, *Reflexes and Motor Integration: Sherrington's Concept of Integrative Action* (Cambridge, Mass., 1969), p. 7; C. S. Sherrington, "Marginalia," *Science, Medicine and History: Essays . . . in Honour of Charles Singer*, ed. E. A. Underwood, 2 vols. (London, 1953), II, 545-553, on 545; and Ragnar Granit, *Charles Scott Sherrington: An Appraisal* (London, 1966), pp. 8-13.

[18] Henry Dale, "Sir Michael Foster," *Notes and Records of the Royal Society (London), 19* (1964), 10-32, on 25, 27.

his writings on the internal environment; Fletcher, for his work on muscle metabolism. On neither man, it seems, did Foster exert a really significant influence, despite some evidence of personal contact between him and Fletcher.[19] In fact, it is quite possible that Foster had no direct and active role in any piece of physiological research during the period of his professorship.

Even the cardiological research carried out at Cambridge after 1883 bears little evidence of Foster's immediate influence. Only one paper, by W. B. Ransom of Trinity College, belongs securely in the Fosterian tradition. This important paper, published in 1885, represents yet another Cambridge contribution to the myogenic theory of the heartbeat.[20] In a brief historical introduction, Ransom acknowledged that the neurogenic theory continued to prevail, though Gaskell's work on the tortoise's heart had "already done much to shake this somewhat insecurely based theory." Ransom suggested that further light might be thrown on "this vexed question" by an "examination from the evolutionary standpoint of first the lower and simpler and then the higher and more complex forms of heart."[21] Foster or Gaskell would hardly have said it otherwise.

Ransom chose to investigate systematically the full range of invertebrate hearts. Nearly all studies in this field uncritically reaffirmed the prevailing neurogenic theory. He knew of only three investigations that had been made from a properly critical point of view—one by Foster, one by Foster and Dew-Smith, and one recent study by the German physiologist, Wilhelm Biedermann. But since all three dealt only with the snail's heart, Ransom felt that his more ambitious study might yield important new results, the more so because he had given particular attention to the cephalopod heart, the most fully developed of invertebrate hearts. The main conclusion he drew from his extensive and persuasive investigation was that "in all the animals examined the cardiac muscle has the power of rhythmical contraction independently of nerve structures."[22]

Ransom's entire approach, as well as his main conclusion, bears the obvious imprint of the Cambridge setting. And yet, the only men acknowledged for their direct assistance are Henri Lacaze-Duthiers and

[19] See Maisie Fletcher, *The Bright Countenance: A Personal Biography of Walter Morley Fletcher* (London, 1957); and Kenneth J. Franklin, *Joseph Barcroft, 1872-1947* (Oxford, 1953). On Foster and Barcroft, see also F. L. Holmes, "Joseph Barcroft and the Fixity of the Internal Environment," *J. Hist. Biol.*, *2* (1969), 89-122, on 93.

[20] W. B. Ransom, "On the Cardiac Rhythm of Invertebrates," *J. Physiol.*, *5* (1885), 261-341.

[21] Ibid., p. 262.

[22] Ibid., p. 336.

Anton Dohrn. Ransom had worked in Lacaze-Duthiers' marine laboratory at Roscoff during July and August 1883, and "at the Cambridge table" of Dohrn's zoological station at Naples during the winter of 1884.[23] Foster may have aroused Ransom's initial interest in the problem of the heartbeat, and he was doubtless pleased by his student's main conclusion, but there is no decisive evidence of his active participation in or supervision of the actual research. In fact, Ransom's work shows considerable independence of spirit, and his many favorable references to Foster's work do not entirely overshadow serious disagreements on several more or less minor points. In view of much other evidence that Foster had already abandoned the laboratory by this time, his immediate influence even on Ransom's research is at least open to question.

The physical growth of the Cambridge School, 1883-1903

By October 1884, Foster's withdrawal from the laboratory had become so complete that Lea apparently felt obliged to write him a letter emphasizing just how crowded conditions in the practical classes had again become.[24] The new building for physiology and comparative anatomy, completed in 1879 and just extended to four stories in 1882, was once again bursting at the seams. Not only had the elementary course in physiology grown to more than 130 students, but the course in elementary biology, which Foster had just turned over to Sydney Vines and Adam Sedgwick, was expected to draw at least 120 students during each of the next two terms.[25] If it required a letter from Lea to alert Foster to the full magnitude of the problem, nothing more was needed to move him to immediate and forceful action. He sent Lea's letter to the Museums and Lecture-Rooms Syndicate along with one of his own, which closed with these words: "I do not wish to say anything which might seem to force matters, but things have really come to such a pass that it is a serious question whether I and my staff should bring discredit on ourselves by pretending to do what we cannot do; whether, that is, it would not be better to give up altogether the attempt to carry on the practical course."[26]

In February 1886, after nothing tangible had been done to mitigate his concern, Foster emphasized the need again in a letter addressed to Professor Newton, chairman of the Special Board for Biology and Geology,

[23] Ibid., p. 263. On the connection between Cambridge and the Naples zoological station, see Chapter 5, n. 37.

[24] See *Cambridge University Reporter* [*CUR*], *14* (1885), 505-507.

[25] Ibid., *14* (1884), 197.

[26] Ibid., *14* (1885), 505-506.

who then sent it on to the Museums and Lecture-Rooms Syndicate.[27] By October 1886, Foster was practically frantic. In another letter to the syndicate, he pointed out that there were now 156 students "attending my lectures and desiring to take part in the practical instruction." Despite every attempt to arrange the teaching in some sensible fashion, he had reached the conclusion "that in justice to my demonstrators, to my subject, and to myself, I must refuse to attempt to teach more than about 80 men with my present accommodation." This left two options: either he must limit his class to 80, or the university must provide him immediately with additional accommodation for about 60 or 70 men. He and J. W. Clark had found that the latter path could be taken on a temporary basis for about £350. This would provide a room to be added to the Museum of Comparative Anatomy, "which would serve my purposes, and which should the University at any future time provide me with adequate permanent accommodation, would prove very useful for work in connection with the Museum of Comparative Anatomy."[28]

After some debate, during which personal hostility toward Foster surfaced,[29] this temporary expedient was adopted.[30] During the next two years, Foster continued to complain, but somewhat less forcefully. Perhaps he now recognized that he was making almost impossible demands upon the university. The Museums and Lecture-Rooms Syndicate was certainly cooperative. It had commissioned and secured architect's plans as early as 1884, but Foster's department grew so fast that the plans became inadequate almost as soon as they were drawn. Finally, in February 1889, the syndicate presented greatly expanded plans for accommodating the departments of human anatomy and physiology. Not including the large lecture room to be shared by both departments, the new plans called for an additional building of three stories, 75 by 31 feet, for physiology alone. If these plans were carried out, the cost of both departments would reach £14,000, with about £5,000 required just for the department of physiology.[31]

This raised the persistent and critical problem of financing. It certainly did not help matters that the land-rich colleges were suffering from the effects of an agricultural depression. In fact, just as the Museums and Lecture-Rooms Syndicate released its expanded plans for new buildings in anatomy and physiology, the University Senate was considering a proposal to reduce the percentage of collegiate income to

27 Ibid., *15* (1886), 504-507.

28 Ibid., *16* (1886), 153-155.

29 Ibid., *16* (1886), 219. As usual, P. W. Latham was the source of this hostility. See Appendix 1 below.

30 Ibid., *16* (1887), 737.

31 Ibid., *18* (1889), 478-480.

Table 2

Number of students in physiology classes at Cambridge, 1883-1903

Academic year	*Michaelmas term*	*Lent term*	*Easter term*
1883-1884	*	*	Elem. 77 Adv. 24 } 101
1884-1885	Elem. 133 Adv. 28 } 161	Elem. 141 Adv. 28 } 169	Elem. 130 Adv. 20 } 150
1885-1886	Elem. 130 Adv. 50 } 180	Elem. 134 Adv. 30 } 164	*
1886-1887	Elem. 156 Adv. *	*	*
1887-1888	*	*	*
1888-1889	*	*	Elem. 190 Adv. 30-40 } c. 225
1889-1899	*	*	*
1899-1900	*	320[a]	320
1900-1901	*	*	*
1901-1902	300	300	300
1902-1903	290	290	290
1903-1904	270	270	270

* No data available

Data from *Cambridge University Reporter, 14* (1885), 762; *15* (1886), 694; *16* (1886), 153; *18* (1889), 598; *29* (1900), 963; *31* (1902) 943; *32* (1903), 925; *33* (1904), 943. Figures are approximate and do not generally include women students.

[a] Hereafter no distinction is made between elementary and advanced students.

be contributed to the University Chest. The outcome of this debate, which dominated the university from 1888 to 1890, was uncertain when the syndicate's report came up for discussion in March 1889. Foster emphasized again the magnitude of his needs. The elementary class had continued to grow and now numbered 190 students, while the advanced class attracted 30-40 students and was prevented from further growth only by want of accommodation. He was particularly distressed by the effect of crowding upon this advanced class, "which was practically driven out of the place, so that the University was suffering in the character of its teaching because it did not afford opportunity for initiating young men into original research."[32]

Foster was told that everything depended on the state of the University Chest. But the collegiate forces were developing new strength, and prospects were not especially bright. In the end, the colleges

[32] Ibid., *18* (1889), 596-600, quote on 598.

largely carried the day, despite eloquent opposition from G. M. Humphry and from the Trinity reformer, Henry Sidgwick.[33] Under these circumstances, Foster might have lost his case for the forseeable future had not Sidgwick himself come swiftly to the rescue. In November 1889 Sidgwick offered to give £1,000 immediately and £500 more within the next two years toward the proposed new buildings in physiology, "provided that the Financial Board . . . feel themselves justified in reporting in favour of commencing these buildings without delay."[34] This generous offer was quickly accepted and the new buildings were finished by May 1891.[35]

In his annual report, published one month later, Foster wrote that "the whole laboratory in fact is now an admirable one."[36] Thanks to Sidgwick's gift, the Cambridge School developed smoothly for nearly a decade. And yet by 1900, when Langley submitted the annual report, signs of overcrowding had once again appeared. In the two most recent terms, he reported, the number of students attending classes in physiology had grown to 320. Simultaneously, research had developed to such an extent that the room originally intended for physiological physics "is now subdivided by curtains and used for research."[37] The problem of crowding was not settled with anything approaching permanence until 1914, when the Draper's Company provided the money for a large and elaborate new laboratory.[38] With the coming of the First World War, nearly all of the young researchers were drawn into the struggle, while Langley kept a quiet vigil, working virtually alone in the spacious new facilities.[39] The Cambridge School of Physiology had by then come a long way from 1870, when Foster drew perhaps twenty students to his first course of lectures.

Research in the Cambridge School, 1883-1903: new problems and directions

During the period of Foster's professorship, the Cambridge School became internationally recognized as a leading center for physiological research. Foreign workers, even Germans, began to visit the labora-

[33] See ibid., *18* (1888-1889), 92-100, 220-227, 530-534, 730-735, 797; ibid., *19* (1889-1890), 684-688.

[34] Ibid., *19* (1889), 170.

[35] Ibid., *20* (1891), 1039.

[36] Ibid., *20* (1891), 1054.

[37] Ibid., *29* (1900), 963.

[38] The new School of Physiology, erected at a cost exceeding £22,000, had its official opening on 9 June 1914. See ibid., *43* (1914), 1202-1205. For a description of the new laboratory facilities, see also Tanner, *Historical Register*, pp. 247-252.

[39] See Harvey Cushing, *The Life of Sir William Osler*, 2 vols. (Oxford, 1925), II, pp. 413-414, esp. n. 1.

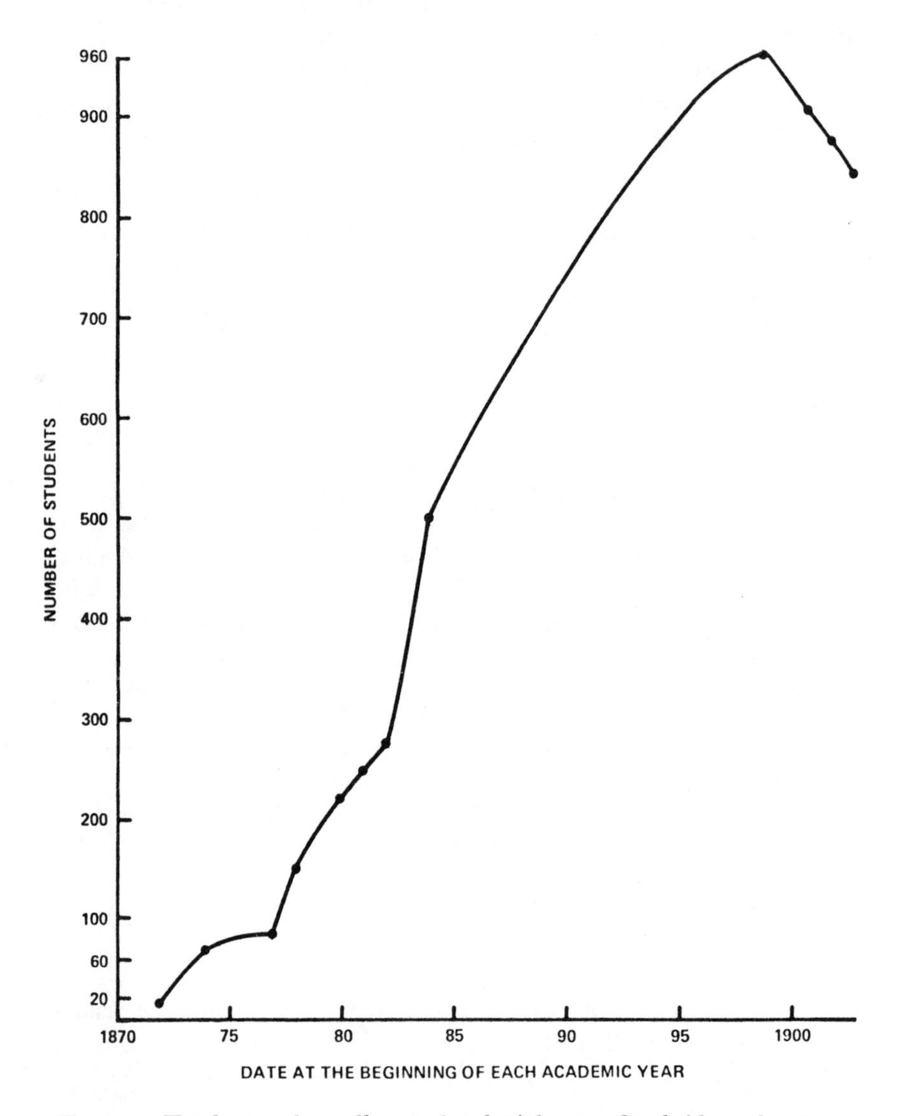

Figure 4. Total annual enrollments in physiology at Cambridge, 1870-1903.[1]

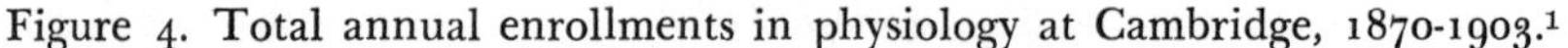

[1] Data from Tables 1 and 2. For each academic year, the total number of enrollments in physiology classes (whether elementary or advanced) has been determined by adding together the enrollments for the Michaelmas, Lent, and Easter terms. For the academic year 1882-1883, the total annual enrollment was clearly somewhat higher (presumably by 15 to 20) than that recorded on the graph, for I was unable to find any enrollment figures for the class in advanced physiology during the Easter term of that academic year. For 1899-1900, it has been assumed

tory. In the academic year 1892-1893, a dozen persons were engaged in original work, "among them Prof. Piotrowski of Lemberg, to whom the hospitality of the Laboratory was given for about six months."[40] P. Botazzi of the University of Florence came for part of 1897-1898.[41] Otto Loewi of Marburg, a future Nobel laureate, visited and carried out some research in 1902.[42] Although he stayed at Cambridge only briefly, he returned to Marburg with a new interest in the sympathetic or autonomic nervous system, an interest that he once described as "imported" from England, and particularly from Gaskell, Langley, and T. R. Elliott (1877-1961), a younger member of the Cambridge School.[43] Indeed, Loewi traced his very interest in basic science partly to Gaskell's Croonian Lecture of 1882 on the isolated frog's heart.[44] With Henry Dale, yet another product of the Cambridge School, Loewi shared the Nobel prize in 1936 for work on the chemical transmission of nervous impulses.

No attempt can be made here to discuss fully the large and increasingly diverse body of research pursued at Cambridge between 1883 and 1903. This would require a separate and extensive study in which neither Foster nor the problem of the heartbeat would find much place. Heart action remained an active field, but with the exception of Ransom's work of 1885 (already described above), this new cardiological research differed in spirit and aim from the work done earlier under Foster's inspiration and direction. Of this new work, the most extensive was the collaborative effort of Charles Smart Roy and John George Adami, culminating in a long paper of 1892 on the physiology and pathology of the mammalian heart.[45] Adami, who entered Cambridge in 1880, later spoke in general and elusive terms of Foster's influence on him,[46] but his research on the heart bears little impress of

---

that the enrollment in the Michaelmas term was the same as it was for each of the other two terms of that academic year (namely, 320). The enormous increase in total enrollments between 1882-1883 and 1884-1885 can be explained in part by the expansion of the class in elementary physiology from a two-term to a three-term course. Foster, who turned the course in elementary physiology over to his deputy J. N. Langley in 1900, resigned in 1903.

---

40 *CUR, 22* (1893), 963. 41 Ibid., *27* (1898), 903. 42 Ibid., *32* (1903), 925.

43 Otto Loewi, "An Autobiographic Sketch," *Perspectives in Biology and Medicine, 4* (1960), 3-25, on 13.

44 Otto Loewi, "Prefatory Chapter: Reflections on the Study of Physiology," *Annual Review of Physiology, 16* (1954), 1-10, on 2. More generally on Loewi, see G. L. Geison, "Otto Loewi," *DSB*, VIII (1973), 451-457.

45 C. S. Roy and J. G. Adami, "Contributions to the Physiology and Pathology of the Mammalian Heart," *Phil. Trans., 183B* (1892), 199-298.

46 See Marie Adami, *J. George Adami: A Memoir* (New York, 1930), pp. 8-10, 42; and J. George Adami, "A Great Teacher (Sir Michael Foster) and His Influ-

that influence. Rather, it was his collaborator Roy, eight years his senior, who probably determined the basic direction of their work.

Roy came to Cambridge in 1881 as first recipient of the George Henry Lewes Studentship.[47] As director of the studentship, Foster was almost certainly responsible for his selection, for, according to a letter written by George Eliot in 1879, Foster considered Roy "by far the most promising" of the applicants. Roy was, however, an Edinburgh graduate "and not a pupil of . . . Foster's."[48] For the previous several years, he had been working on the Continent, at Berlin under Rudolf Virchow and the cardiovascular physiologist Hugo Kronecker, at Strassburg under Goltz, and at Leipzig under pathologist Julius Cohnheim. He had already published two valuable studies in cardiovascular physiology, and had achieved recognition as a master of physiological instrumentation and technique.[49] Roy was, in short, an accomplished physiologist before coming to Cambridge.

Roy's interest in cardiology may have made him particularly attractive to Foster, but neither the style nor the direction of Roy's research seems to have been affected by his transfer to the Cambridge setting. In their long paper of 1892, Roy and Adami said that their observations had begun "some ten years ago in the late Professor Cohnheim's laboratory at Leipsic, and have been continued, with frequent intervals, till the present time, the work being done partly at the Brown Institution, but mainly during the last four years, when we have been working together, in our own laboratory here [at Cambridge]."[50] Their work focused not on the cause of the heartbeat, but rather on the role of the cardiac nerves in the intact, unexcised heart, apparently in the belief that such a heart could better serve the needs of cardiac pathology. Roy's special orientation can perhaps be traced to his training under Cohnheim, a pathologist. In 1884 Roy became first professor of pathology at Cambridge, holding the chair until his early death in 1897.

Others besides Ransom, Roy, and Adami pursued cardiological re-

---

ence," *Publication no. 7, Medical Faculty, Queen's University (Kingston, Ontario, Canada)*, June 1913, pp. 1-17.

[47] On the Lewes Studentship, see Chapter 6 above, "Students and fellowships."

[48] Gordon S. Haight, ed., *The George Eliot Letters*, 7 vols. (New Haven, Conn., 1954-1955), VII, p. 215.

[49] See C. S. Roy, "On the Influences which Modify the Work of the Heart," *J. Physiol.*, *1* (1878) 452-496; and Roy, "The Form of the Pulse-Wave, as Studied in the Carotid of the Rabbit," ibid., *2* (1879), 66-81. More generally on Roy, see C. S. Sherrington, "Charles Smart Roy," *Yearbook of the Royal Society (London)*, 1902, pp. 231-235; Humphry Davy Rolleston, *The Cambridge Medical School: A Biographical History* (Cambridge, 1932), pp. 108-110; and J. J. G. B., "Charles Smart Roy," *Journal of Pathology and Bacteriology*, *5* (1898), 143-146.

[50] Roy and Adami, "Mammalian Heart," p. 200.

search at Cambridge after 1883. The later work of Walter Gaskell deserves examination. For Gaskell's interest in heart action did not end abruptly with his classic treatise of 1883 on the tortoise's heart. Not until 1887 or 1888 did he abandon this interest, and in the meantime he published a couple of papers dealing more or less explicitly with the question of heart action. In 1886, for example, he showed in the tortoise that vagus stimulation produced an electrical variation in the heart opposite in sign to that produced by cardiac contraction, thereby contributing importantly to the rapidly developing field of cardiac electrophysiology, from which the electrocardiogram was soon to emerge.[51]

But as this work of 1886 perhaps already suggests, Gaskell's interests were developing in a way scarcely related to Foster's overriding concern with the origin of the heartbeat. Between 1883 and 1887, Gaskell's work on the heart became intimately bound up with and eventually submerged in a general study of the sympathetic nervous system. Already in the later stages of his work on the problem of rhythm, Gaskell had demonstrated his capacity to work independently of Foster's direction. After 1883 his originality and independence became increasingly apparent. While retaining and making fundamental use of the general evolutionary approach he had learned from Foster, Gaskell pursued the internal logic of his own work until it eventually led him to issues quite divorced from the problem of the heartbeat.

Gaskell, Langley, and the autonomic nervous system

After 1888, in fact, Gaskell devoted all of his efforts to the problem of the origin of the vertebrates. This interest apparently evolved logically out of his work on the involuntary nervous system, and that work, in turn, arose directly out of his earlier investigation of the vasomotor nerves and the problem of the heartbeat. These interconnections deserve to be specified, and not least for a reason that Gaskell himself provides. In tracing the origins of his work on the vertebrates, he suggested that it might be "instructive . . . to see how one investigation leads to another, until at last, *nolens volens*, the worker finds himself in front of a possible solution to a problem far removed from his original investigation, which by the very magnitude and

[51] W. H. Gaskell, "On the Action of Muscarin upon the Heart, and on the Electrical Changes in the Non-beating Cardiac Muscle Brought about by Stimulation of the Inhibitory and Augmentor Nerves," *J. Physiol.*, *8* (1887), 404-415. Cf. Wilhelm Biedermann, *Electro-Physiology*, trans. F. A. Welby, 2 vols. (London, 1896), I, pp. 432-434.

importance of it forces him to devote his whole energy and time to seeing whether his theory is good."[52] What gives the exercise special utility here is that it leads us at least momentarily to Gaskell's important work on the sympathetic or autonomic nervous system. This work is significant not only for its own sake, but also because it can be used to point the way toward a somewhat more focused discussion of the diverse research carried out at Cambridge after 1883. For without in any way denying that great diversity did exist, it seems possible in retrospect to place much of this research within the broad area of the autonomic nervous system.

The autonomic system was in fact the focus of Langley's research from about 1890 until his death in 1925, and it is for his contributions in this area that he is chiefly known. Both he and Gaskell wrote monographs on what the former eventually called the autonomic system and the latter the involuntary system.[53] Their work together marks an epoch in the history of neurophysiology, and it attracted even more attention to the Cambridge School than did Gaskell's work on the heart. It was largely in this new Cambridge setting that Charles Sherrington developed his early ideas on nervous integration and that T. R. Elliott, Henry Dale, and probably even Otto Loewi gained the inspiration for their work on neuromimetic drugs and humoral nervous transmission.

Until all of this work is studied fully in its historical context, our appreciation of the Cambridge School under Foster must remain incomplete. Fundamental insights into the dynamics of this group may emerge from a proper examination of the nature, precise sources, and possible interrelatedness of their work on the involuntary nervous system. So complete an examination will not be attempted here. But a brief examination of the work of Gaskell and Langley may serve to clarify a few of the issues.

To understand the direction of Gaskell's work ater 1883, we must return fleetingly to the history of work on the vagus nerve following the dramatic discovery of vagal inhibition by the Weber brothers in the 1840's. In pursuing this work, physiologists had been bewildered by the wide range of effects produced when the vagus was stimulated in that most popular of experimental animals, the frog. For though the heartbeat was often inhibited, it sometimes responded in precisely the opposite way; that is, both the rate and strength of the cardiac con-

[52] W. H. Gaskell, *Origin of the Vertebrates* (London, 1908), pp. 1-2.

[53] W. H. Gaskell, *The Involuntary Nervous System* (London, 1916); J. N. Langley, *The Autonomic Nervous System*, Part 1 (Cambridge, 1921). The projected second part of Langley's monograph never appeared.

tractions sometimes *increased* upon vagus stimulation in the frog. Arguing by analogy with the mammalian heart, where specific accelerator nerve fibers had been detected, some physiologists offered a simple explanation for the confusing behavior of the frog's heart. Perhaps, they suggested, the frog's vagus is not a simple nerve, composed solely of inhibitory fibers, but a compound nerve containing accelerator fibers as well. If so, vagus stimulation might at different times activate either the inhibitory fibers alone, or the accelerator fibers alone, or both at the same time. The bewildering range of effects would then become intelligible.

Both in his Croonian Lecture of 1882 on the frog's heart and in his classic paper of 1883 on the tortoise's heart, Gaskell had rejected this hypothesis,[54] probably under the influence of Foster, who considered the idea an unnecessary "Deus ex machina."[55] Gaskell argued instead that the vagus is the "trophic" (anabolic) nerve of the heart and produces all of its effects by virtue of this property. On this view, only one kind of fiber (trophic) was involved in vagus stimulation, and the response of the frog's heart in any particular case depended on the degree to which certain molecular events had progressed in the cardiac tissue.[56]

Gaskell must have felt a little unsure of this conclusion, for he continued to explore the possibility that coldblooded animals might possess both inhibitory and accelerator cardiac fibers. By the summer of 1884 he had turned to the crocodile, which "being the most likely of all cold-blooded animals to possess a nervous system closely resembling that of the warm-blooded," was also the most likely to possess both kinds of fibers. And, in fact, he encountered no great difficulty in distinguishing a set of purely accelerator fibers from a set of purely inhibitory fibers in the crocodile's cardiac nerves. Examination of the tortoise's heart brought the same result. So, ignoring his previous objections to the hypothesis, Gaskell decided that both sets of fibers were probably also present in the frog.

Within weeks, Gaskell had found decisive evidence for his new position, which he set forth in an elegant paper of 1884. Tracing the frog's vagus from its origin in the medulla oblongata, he found it to consist for a time of two branches which then joined in a large ganglion outside the cranial cavity to form a single nerve trunk that continued

[54] See esp. W. H. Gaskell, "On the Innervation of the Heart, with Especial Reference to the Heart of the Tortoise," *J. Physiol., 4* (1883), 43-127, on 122-124.

[55] Michael Foster, "Some Effects of Upas-Antiar on the Frog's Heart," *J. Anat. Physiol., 10* (1876), 586-594, on 590-593.

[56] See Chapter 9 above, "Gaskell's defection."

toward the heart. It was this trunk that was ordinarily used to examine the effects of vagus stimulation. Gaskell focused instead on the two preganglionic branches and found that stimulation of one branch resulted always in purely accelerator effects, while stimulation of the other resulted always in purely inhibitory effects. For Gaskell, the riddle of the frog's vagus was now solved. All of the confusion had resulted from the failure to perceive that the so-called vagus, the nerve trunk passing from the large ganglion to the heart, was in fact the "vago-sympathetic." It was a mixture of purely inhibitory fibers and purely accelerator fibers that could be distinguished from one another prior to their merger in the large extracranial ganglion.[57]

The future direction of Gaskell's research was profoundly affected by this discovery. He knew now that in coldblooded animals, as in mammals, there are two anatomically distinct sets of cardiac nerves performing separate, indeed opposing, functions. He determined at once to study "the opposing influences of these two systems of nerves upon all the functions of the heart of the frog, tortoise and crocodile as fully as possible."[58] But to do so he needed first to elucidate and compare "the anatomical distribution of the sympathetic cardiac nerves in as many different species as possible."[59] For this task, he secured the help of his Cambridge colleague, morphologist Hans Gadow. They found that the cardiac nerves were distributed in basically similar ways in all the species they examined. In every case, the nerves seemed to originate in the upper thoracic portion of the sympathetic chain.[60]

From this point on, Gaskell's thought developed rapidly, as structural and functional considerations became intertwined and mutually reinforcing. He moved from his study of the cardiac nerves to a study of the sympathetic nervous system as a whole, and then rather quickly to a new theory of the origin of vertebrates. In its essentials, this intellectual development took place in the brief span of four or five years, between 1883 and 1888. It is as if a flood of ideas converged upon Gaskell almost simultaneously, and it is hard to tell from his published papers the precise way in which one idea led to the next. Without seeking to establish the sequence definitively, we can offer a plausible reconstruction of the overall direction of this thought.

Soon after discovering the existence of cardiac accelerator fibers in

[57] W. H. Gaskell, "On the Augmentor (Accelerator) Nerves of the Heart of Cold-blooded Animals," *J. Physiol.*, *5* (1884), 46-48.

[58] Ibid., p. 48.

[59] W. H. Gaskell and Hans Gadow, "On the Anatomy of the Cardiac Nerves in Certain Cold-blooded Vertebrates," *J. Physiol.*, *5* (1885), 362-372, on 362.

[60] Ibid., p. 370.

coldblooded animals, Gaskell noticed a striking morphological distinction between these accelerator fibers and the inhibitory fibers of the vagus nerve. Using a dissecting lens to follow the course of the cardiac fibers in the tortoise, he observed that while both the accelerator and vagus fibers issued from the spinal cord as medullated fibers, the accelerator fibers lost their medullae on passing through the sympathetic chain, while the vagus fibers retained their medullae throughout their course. By early 1885, he had confirmed this rule in a wide variety of vertebrate and mammalian species.[61]

Already, it seems, Gaskell had integrated these results into a remarkably bold and sweeping hypothesis. For him, as for Foster, the heart was just one example of an involuntary muscle, and he was confident that his results on cardiac innervation could be extended to the smooth muscles of the arterial, alimentary, and glandular systems. Now Gaskell had long believed (again with Foster) that vagal inhibition ultimately produced a beneficial, constructive effect on the cardiac muscle. Immediately, of course, the stimulation of inhibitory fibers slowed the heart or even brought it to standstill. But once that phase had passed, the secondary results of vagus action were to maintain the rhythm in a heart beating at its normal rate, to increase it in a flagging heart, to restore the conduction power, and to repair damaged cardiac tissue.[62] It was these secondary phenomena that Gaskell had in mind when he spoke of the vagus as the trophic or anabolic nerve of the heart. But if inhibitory fibers are ultimately constructive in action, their opposing accelerator fibers must have an antagonistic, destructive action, leading to "increased activity followed by exhaustion, symptoms of katabolic action."[63]

When generalized from the cardiac nerves to the involuntary system as a whole, this concept led to the notion that every involuntary muscle was innervated by two nerves of opposite action, one anabolic and the other katabolic. The cardiac nerves presented "the most perfect types of such opposite nerve actions," but they did not stand alone. Indeed, wrote Gaskell:

> the evidence is becoming daily stronger that *every tissue* is innervated by two sets of nerve fibres of opposite characters so that I look

[61] See W. H. Gaskell, "On the Relations between the Function, Structure, Origin, and Distribution of the Nerve Fibers which compose the Spinal Cord and Cranial Nerves," *Medico-Chirurgical Transactions, 71* (1888), 363-376, on 364-365.

[62] W. H. Gaskell, "On the Structure, Distribution and Function of the Nerves which Innervate the Visceral and Vascular Systems," *J. Physiol., 7* (1886), 1-80, on 50.

[63] Ibid., p. 50.

> forward confidently to the time when the whole nervous system shall be mapped out into two great districts of which the function of the one is katabolic, of the other anabolic, to the peripheral tissues: two great divisions of the nervous system which are occupied with chemical changes of a synthetical and analytical character respectively, which therefore in their action must show the characteristic signs of such opposite chemical processes.[64]

By further analogy with the cardiac nerves, Gaskell expected the anabolic and katabolic nerves to be histologically distinguishable from one another, particularly on the basis of their medullation after passing through the sympathetic chain. It was under the inspiration of these leading themes that Gaskell undertook his full-scale systematic investigation of the involuntary nervous system.

A classic paper of 1886 contains the major results of Gaskell's work on the involuntary system.[65] He found that the visceral or involuntary nerves arose from the central nervous system in three distinct groups. There was a cervico-cranial outflow, a thoracic outflow, and a sacral outflow. In all three groups the visceral fibers left the central nervous system as peculiarly fine, white, medullated fibers. But the fibers issuing from the thoracic region lost their medullae in the sympathetic ganglia and passed to the viscera as nonmedullated fibers. The fibers issuing from the cervico-cranial and from the sacral regions retained their medullae as they passed to the periphery. In action the fibers issuing from the thoracic region were antagonistic to both the cervico-cranial and the sacral outflows. In broad outline, this plan of the involuntary system is still accepted, though significant modifications in detail and terminology were soon made.

From the point of view of physiological thought, however, perhaps the most important result of Gaskell's work was his discovery that the connection between the central nervous system and the chain of sympathetic ganglia was unidirectional, with the peculiarly small white fibers (the "white rami") supplying the sole connection. Earlier in the century it had been thought that a system of gray rami returned from the sympathetic chain to the central nervous system, so that there was an interplay between two essentially independent nervous systems. Bichat had christened these two systems the "organic" (central) and the "vegetative" (sympathetic). Although this conception of the nervous system had come under mounting criticism, no broad generalization took its place until Gaskell clarified the relationship between the sym-

[64] Ibid.

[65] Ibid., pp. 1-80.

pathetic chain and the spinal cord.[66] He showed that the gray rami were in fact merely peripheral nerve fibers that supplied the blood vessels of the spinal cord and its membranes, and that issued not from the sympathetic chain but from the central nervous system, as did the white rami. There was, then, no real separation into "organic" and "vegetative" nervous systems. There is, wrote Gaskell in 1908, "no give and take between two independent nervous systems... as had been taught formerly, but only one nervous system, the cerebro-spinal."[67] So fundamentally did Gaskell alter prevailing conceptions of the involuntary system that Walter Langdon-Brown could insist that "to read an account of this system before Gaskell is like reading an account of the circulation before Harvey."[68]

Within a few years of Gaskell's death in 1914, the tributes to his work on the involuntary system had become frequent and glowing enough to irritate his erstwhile colleague, J. N. Langley. Langley had himself been working on this system since 1889, and felt that Gaskell's contributions were being overestimated at his expense.[69] In point of fact, Langley's work modified and extended Gaskell's results in several important ways. Among other things, he established a vastly improved nomenclature, restricting the by then ambiguous term "sympathetic" to the thoracic outflow and introducing the word "parasympathetic" for the antagonistic cranial and sacral outflows. On the suggestion of Richard Jebb, professor of Greek at Cambridge, he gave the name "autonomic" to the three outflows collectively.[70]

While mapping out this system in unprecedented detail, Langley observed that all autonomic nerves eventually lose their medullae, the sympathetic fibers losing theirs early, in the main sympathetic chain, while the parasympathetic fibers retained theirs until they passed through a peripheral terminal ganglion. And these peripheral ganglia, he argued, are not collected together according to function (as

[66] See e.g., E. F. A. Vulpian, *Leçons sur la physiologie générale et comparée du système nerveux* (Paris, 1866), pp. 14-15, 721-733; Donal Sheehan, "Discovery of the Autonomic Nervous System," *Archives of Neurology and Psychiatry (Chicago)*, *35* (1936), 1081-1115; and Erwin H. Ackerknecht, "The History of the Discovery of the Vegetative (Autonomic) Nervous System," *Med. Hist.*, *18* (1974), 1-8, on 5.

[67] Gaskell, *Origin of Vertebrates*, p. 2.

[68] Walter Langdon-Brown, "W. H. Gaskell and the Cambridge Medical School," *Proceedings of the Royal Society of Medicine (London)*, *33* (1939), Section of the History of Medicine, pp. 1-12, on 6.

[69] J. N. Langley, "The Arrangement of the Autonomic Nervous System," *Lancet*, 1919 (2), 951. See also ibid., pp. 827-832, 873-879, 923-929, 965-970.

[70] See J. N. Langley, "The Sympathetic and Other Related Systems of Nerves," in *Textbook of Physiology*, ed. E. A. Schäfer, 2 vols. (Edinburgh, 1898-1900), II, pp. 616-696, on 659-660.

had been claimed), but rather are associated with definite somatic areas. With consistent sensitivity to the importance of the neurone theory, Langley also showed that every efferent nerve fiber passed through one nerve cell, and one only, on its way to the periphery. As Gaskell himself would have expected, the morphological clarifications contributed by Langley had significant physiological consequences as well. Perhaps most importantly, Langley's imaginative use of drugs as investigative tools eventually gave his work a dimension totally absent from Gaskell's, and led him to his hypothesis that "receptive substances" were involved in the transmission of nervous impulses across synaptic junctions. Although the tangled web of issues and influences has yet to be unraveled, it seems likely that this hypothesis formed part of the background for later theories of the chemical transmission of nervous impulses.[71]

For our purposes, however, it is less important to allocate credit between Gaskell and Langley than to examine how it happened that the two most productive members of Foster's school came to concentrate almost simultaneously on the involuntary or autonomic nervous system. Because they belonged to the same institution and worked in close proximity, this common focus seems too striking to be mere coincidence. It is tempting to suppose that their work on the involuntary system must have been connected and must have had its origin in some common source in the Cambridge setting, perhaps even Foster himself. For though Foster's own research focused almost exclusively on the problem of the heartbeat, he emphasized more than once that the most important branch of physiology was "the physiology of the nervous system—compared with which all the rest of physiology, judged either from a practical or a theoretical point of view, is a mere appendage."[72] Much more significantly, the special tendency of Cambridge neurophysiologists, including Gaskell, Langley, T. R. Elliott, and Henry Dale, to contemplate chemical mechanisms for neurotrans-

[71] See Walter B. Cannon, "The Story of the Development of Our Ideas of Chemical Mediation of Nerve Impulses," in *American Journal of the Medical Sciences*, n.s. *188* (1934), 145-159; H. H. Dale, "T. R. Elliott," in *Biographical Memoirs of Fellows of the Royal Society, 7* (1961), 53-74, esp. 63-64; and John Parascandola and Ronald Jasensky, "Origins of the Receptor Theory of Drug Action," *Bull. Hist. Med., 48* (1974), 199-220, esp. 211-216. According to Dale (pp. 63-64), Langley reacted negatively toward the hypothesis of chemical neurotransmission as originally proposed by T. R. Elliott in 1904. But as Parascandola and Jasensky point out (pp. 214-215), Langley himself suggested in 1906 that neurotransmission might take place through the secretion of a chemical compound at the nerve ends, which compound might then combine with the "receptive substance" in the effector cell.

[72] Michael Foster, "Vivisection," *Macmillan's Magazine, 29* (1874), 367-376, on 373. Cf. his *Textbook of Physiology* (London, 1877), p. ix.

mission seems to echo Foster's claim that the vagus nerve exerts an essentially chemical action on the heart.[73] Nonetheless, there exists no decisive evidence that Foster played a direct role in either Gaskell's or Langley's major research on the involuntary nervous system. Nor indeed is there any evidence of a direct connection between the work of Gaskell and Langley themselves. For while Gaskell's work on this system evolved out of his work on the heart, Langley's had a quite different origin.

Indirectly, of course, Langley had been concerned with the involuntary system from the beginning. His initial work dealt with the effects of jaborandi on the heart and nervous system, and his study of secretion had grown out of this research. In the course of his long and detailed investigation of glandular secretion, he had naturally examined the nervous influences involved in the process. Indeed Gaskell, in his first major paper on the involuntary system, found occasion to make use of Langley's secretory studies. Even though Langley might have denied the argument himself, Gaskell used his results to claim that "the growth of the protoplasm in the gland cell is not brought about by the action of [Heidenhain's] so-called 'trophic' nerves, but rather is under the control of the so-called 'secretory' nerves; that therefore two sets of fibres must be distinguished in these latter nerves the one [katabolic and] secretory, by the action of which the water and granules are discharged from the cell, the other anabolic and concerned in the building up of the protoplasmic network [of the gland]."[74]

But Langley's subsequent concentration on the involuntary system did not grow out of his work on the secretory nerves. Nor was it directly related to Gaskell's work. Rather, it had its origin in the discovery that nicotine could selectively interrupt nerve impulses at the sympathetic ganglia. In a joint paper announcing this discovery in 1889, Langley and William Lee Dickinson emphasized its immense potential as an analytic tool. In particular, it offered a quick and unimpeachable means of distinguishing true sympathetic fibers from others

[73] Note, in particular, Gaskell's effort to establish the vagus nerve as the "trophic" (or anabolic) nerve of the heart, and to compare its effects to those of such chemical substances as lactic acid or muscarin (see esp. Chapter 9, pp. 268-270). Recall also that Langley, at the very outset of his career, had adopted Foster's view that drugs could exert a direct inhibitory action on the cardiac musculature (see Chapter 8 above, "John Newport Langley").

[74] Gaskell, "Structure, Distribution and Function," pp. 51-52. Langley himself consistently maintained that only one kind of nerve fiber was involved in salivary secretion per se, any secondary phenomena being due to local vasomotor action. See, e.g., J. N. Langley, "Note on Trophic Secretory Fibres to the Salivary Glands," *J. Physiol.*, *50* (1916), xxv-xxvi.

by establishing which fibers actually ended in the sympathetic ganglia and which merely passed through.[75] Although Langdon-Brown claimed privately that the discovery was really Dickinson's alone,[76] it is usually ascribed jointly to him and to Langley, and certainly evolved out of experiments conducted at Cambridge on the effects of the narcotic pituri, a sample of which had been sent to Langley from New Zealand by Archibald Liversidge (himself an alumnus of the Cambridge School). With equal certainty, it was Langley who most swiftly, brilliantly, and successfully exploited the new technique. It enabled him to sweep past all competitors in the race to specify the distribution and function of the autonomic nerves.[77] By Langley's own testimony, all of his later work "developed step by step in logical sequence from the result of the observations made by Dickinson and myself on the action of [pituri]."[78]

Just about the time Langley and Dickinson made this discovery, Gaskell turned his attention from the sympathetic system to the problem of the origin of vertebrates. Struck by this coincidence, Ragnar Granit has suggested that Gaskell switched to developmental morphology precisely because of Langley's spectacular success at delineating the involuntary system with the aid of nicotine.[79] But this suggestion seems unlikely on chronological grounds alone. For Gaskell was already formulating his theory of the origin of vertebrates no later than 1887 or 1888, at least a year or two before Langley and Dickinson announced their discovery and put the new technique to work. Even in the absence of this chronological evidence, a persuasive case could be made that Gaskell's switch to vertebrate morphology was due not to competition from Langley but rather to the internal dynamic of his own research.

While still at work on the involuntary system, Gaskell had been struck by the fact that its nerves were confined not only to three distinct regions of the spinal cord, but also to clearly defined segments within each region. Deeply impressed by the similarity between this vertebrate arrangement and the central nervous system of the segmented invertebrates, he gradually elaborated the theory that the vertebrates were descended from an extinct arthropod stock of which the

[75] J. N. Langley and W. Lee Dickinson, "On the Local Paralysis of Peripheral Ganglia, and on the Connexion of Different Classes of Nerve Fibres with Them," *Proc. Roy. Soc.*, *46* (1889), 423-431.

[76] Walter Langdon-Brown to Donal Sheehan, 11 March 1940, copy deposited in obituary files, Yale University Medical Historical Library.

[77] Notably, J. Rose Bradford. Cf. ibid.; and T. R. Elliott, "Sir John Rose Bradford," *DNB, 1931-1940*, pp. 96-98.

[78] Langley, "Arrangement," p. 951.

[79] Granit, *Sherrington*, p. 13.

king crab is the nearest living representative. Although decidedly unusual, Gaskell's theory was not without its contemporary analogues. In particular, his Cambridge colleague Frank Balfour had earlier suggested that the annelids and the vertebrates were descended from a common segmented invertebrate ancestor,[80] and the possibility of a link between Balfour's views and Gaskell's deserves further examination.

Gaskell agreed that earlier attempts to trace the vertebrates to the segmented invertebrates had failed, but only because they had been based on an erroneous assumption—that the transition from invertebrates to vertebrates required an inversion of ventral and dorsal surfaces. This assumption derived in turn from the fact that the vertebrate nervous system is dorsal to the alimentary canal, while the invertebrate arrangement is reversed. To explain that same fact, Gaskell proposed the extraordinary hypothesis that the vertebrates arose through the enclosure of the ancestral arthropod gut by the growing central nervous system, and the formation of a new alimentary canal ventral to it. Gaskell also claimed that this new alimentary canal was formed by invagination from the epiblastic layer. These views directly violated two settled morphological tenets: (1) that the alimentary canal is the one system that persists through evolutionary change; and (2) that in all cases the alimentary canal arises from the hypoblastic germinal layer, never from the epiblastic. Against the second tenet Gaskell argued that morphologists applied the germ layer theory in a circular manner, often deducing the layer from which a structure arose merely from its ultimate morphological destination. Against the first, he argued that it was folly to insist upon the importance of the alimentary canal in evolution when the central nervous system, and especially the brain, was so obviously the engine of evolutionary advance. "The race," he wrote, "is not to the swift, nor to the strong, but to the wise."[81]

After 1888, Gaskell devoted his research efforts almost exclusively to this ill-fated hypothesis, developing it in a series of papers through 1906 and then gathering the evidence together in a full-length book, *The Origin of Vertebrates* (1908). From the beginning, his ideas met a chilly reception from most morphologists, with one opponent accusing him of "diabolical ingenuity."[82] Nor can it be said that the di-

[80] Cf. E. S. Russell, *Form and Function* (London, 1916), p. 282; and T. E. Alexander, "Francis Maitland Balfour's Contributions to Embryology," unpublished Ph.D. dissertation (University of California, Los Angeles, 1969), p. 116.

[81] Gaskell, *Origin of Vertebrates*, p. 19.

[82] See the lively discussion (pp. 15-50) following Gaskell's paper, "Origin of Vertebrates," *Proceedings of the Linnaean Society (London)*, session 122 (1910), pp. 9-15. For an anonymous critique of Gaskell's book, see *Nature, 80* (1909), 301-

rection of research since has vindicated Gaskell's bold attempt to trace the vertebrates to an arthropod ancestor. Most morphologists today consider the vertebrates of common origin with the echinoderms, and such influence as Gaskell's book enjoyed probably stemmed mainly from its suggestive material on the endocrine system. If Langley's final years were clouded by irritating allegations that his work on the autonomic nervous system merely filled in a few details in the grand Gaskellian scheme, Gaskell's were clouded by the knowledge that his beloved theory of the origin of the vertebrates had failed to win a sympathetic audience, even at Cambridge.[83]

Conclusion

This brief examination of the later research of Gaskell and Langley, together with my prior sketch of the new cardiological research going on beside them, adds a few broad strokes toward a tentative group portrait of the Cambridge School between 1883 and 1903. It presents a picture strikingly different from the school during its formative years. In that period Foster was the dominant influence over both the general organization and scientific research of the school. With one really notable exception (Langley's work on secretion), the research effort was directed mainly toward establishing the myogenic theory of the heartbeat. The whole style and orientation of the school were virtually indistinguishable from Foster's own: the fundamental problem was that of rhythmic motion, and the common approach was evolutionary.

By 1883 Gaskell, Langley, and Lea had assumed primary direction over the day-to-day activities of the laboratory, while Professor Foster retreated to his more familiar role as its institutional promoter. For perhaps two reasons, Foster's retreat from the laboratory helped to ensure the continued growth and success of his school. In the first place, it gave him the opportunity to exploit the institutional base and political power he had gradually built up with general admiration and peer approval during the late 1870's and early 1880's. Although his later efforts did not always enjoy an equally favorable reception outside Cambridge, or even within, they almost certainly

---

303. Gaskell's response is in ibid., pp. 428-429. Even more critical of Gaskell's work was Bashford Dean in *Science*, n.s. *29* (1909), 816-818. For the most favorable evaluation of Gaskell's theory I have seen, see the comments by his Cambridge colleague Hans Gadow in J. N. Langley, "Walter Holbrook Gaskell," *Proc. Roy. Soc., 88B* (1915), xcii-xxxvi, on xxxii-xxv.

[83] See G. L. Geison, "Walter Holbrook Gaskell," *DSB*, v (1972), 279-284, on 283.

contributed to the increasing dominance of his school of physiology. Together with the other advantages they enjoyed at Cambridge, and quite apart from their generally high and sometimes exceptional talent, Foster's students could now rely on their mentor for his patronage, and he responded with an almost paternal interest in their careers.

Inevitably, there was some dispute as to whether Foster exercised his power responsibly or not. Thiselton-Dyer, himself not at Cambridge, ended by insisting that on the whole he did, and that he "scrupulously refrained from pressing [his students'] claims." On that issue, perhaps the most appropriate contemporary testimony—neither fawning nor disloyal—comes from John George Adami, one of Foster's moderately endowed students who built a distinguished career as a professor of pathology and bacteriology at McGill University in Montreal:

> I have left to the last what was the basis of Foster's success as the founder of a school, namely, his keen interest not merely in the work accomplished by his pupils, but in those pupils themselves. Once he recognized, as he thought, the right spirit in a man, he was that man's steady friend, willing to help him forward by his counsel and by his influence. Indeed, those outside Cambridge, who had not come under his influence, made it a sore point that he was apt to regard his "geese as swans" and by his influence to gain posts for them over better men. This is a matter that the future must determine, whether his judgment was right or no, but assuredly he was a most loyal friend to the men who came under him. We regarded him with an almost filial affection, and in any difficulty we tramped or bicycled out to his house at Little Shelford, or sought him at the laboratory, sure that he would hear and discuss our matter and give us wise impartial advice.[84]

In the end, the immense success of Foster's school speaks to the collective wisdom of his judgment without endorsing it in every individual case.

But the Cambridge School probably also benefited in a second, less obvious way from Foster's declining role in the laboratory. Having earlier argued that Foster's career in research was crucial to the rise and early success of the school, it may seem paradoxical to argue now that his retreat from the laboratory can help to explain its continued success. Yet the paradox is only apparent. The early focus of the Cambridge School on the problem of the heartbeat gave it cohesion, identity, and esprit de corps during its years as a small emergent force in

[84] Adami, "Great Teacher," p. 15.

English physiology. By great good fortune, Gaskell's resolution of that problem also brought it international attention and acclaim. But as Ravetz suggests, a research school faces the danger of "coasting on the insights of its founder, with the possibility of degeneration if the school does not dissolve when its time is up":

> In such situations, the result is that over the generations the work of the school becomes less original, less significant, and eventually obsolete. Its detour has become a cul-de-sac. If in spite of this it retains political power, it can then become a serious distorting influence on the progress of the field.[85]

By Ravetz's criteria, Foster thus made an "enlightened" move when he withdrew from the laboratory and turned it over to others with somewhat different research styles and an interest in "descendant" research problems. One of the difficulties besetting physiology in late nineteenth-century Germany was precisely the extent to which its chairs were dominated by men who did their most important work before 1870 but who continued to impose their research programs on younger aspirants to the field.[86] As early as 1873, Foster was complaining about the increasing "monotony" of Ludwig's efforts at Leipzig.[87] If that now seems a surprisingly prescient judgment, as does Foster's decision to focus on the problem of the heartbeat, it would be going too far to claim that Foster acted deliberately and consciously when he made his enlightened withdrawal from the laboratory. Almost certainly, the process by which Cambridge retained and enhanced its eminence as a center for physiological research was as unplanned, informal, and unconscious as the process by which it had become a research center in the first place.

The benefits, nonetheless, are there for all to see. In spite of, if not because of Foster's retreat, Gaskell and Langley contributed prolifically and fundamentally to physiological research. Indeed, between them they were responsible for fully 40 percent of the papers produced by the Cambridge School between 1878 and 1900.[88] Gaskell's work on the heart, Langley's on secretion, and their separate but complementary work on the autonomic nervous system brought Cambridge to the leading edge of physiological research. Even their very

[85] Ravetz, *Scientific Knowledge*, p. 232.

[86] Cf. A. Zloczower, "Career Opportunities and the Growth of Scientific Discovery in 19th Century Germany with special Reference to the Development of Physiology," unpublished M.A. thesis (Hebrew University, Jerusalem, 1960).

[87] Foster to E. A. Schäfer, 21 July [1873], Sharpey-Schafer Collection, Wellcome Institute of the History of Medicine, London.

[88] See Appendices II and III.

different styles perhaps reinforced each other and stimulated a useful divergence in the objects of investigation. Certainly one senses an emerging fragmentation in the Cambridge School during the last two decades of the nineteenth century and beyond. As it grew in size, its early corporate identity was lost or greatly diluted. Each member of the school was now pursuing the internal dynamic of his own work, and an increasing diversity of aims and approaches evolved.

Gaskell and Langley, the two dominant figures in the laboratory, seem to have communicated surprisingly little with each other, either socially or intellectually. They were entirely different personalities, and they represented fundamentally different styles of research. Gaskell, more nearly in the Fosterian tradition, was open and genial in disposition, while his evolutionary, almost intuitive, and speculative approach to physiology led him ultimately to his ill-fated theory of the origin of the vertebrates. By contrast, Langely was aloof, cautious, almost severe, and firmly committed to simple but effective techniques. For one of his less avid admirers, Langley's prowess as a figure-skater perfectly captured the spirit of his personality and his work—"brilliant technique on an icy background."[89] As a teacher, he was too exacting to carry most novices with him, and even a few of his advanced students and colleagues apparently resented his reserve and alleged egotism. Between Gaskell's protégés and Langley's, there is some evidence of strained interpersonal relationships, and they tended to divide themselves into Gaskellian and Langlian camps.[90] But in his remaining role as official director and genial gray eminence, Foster preserved a modus vivendi between the two, and perhaps the tension had its creative as well as its destructive side. As he was no longer so firmly committed to particular research styles, Foster could steer his school through these troubled waters. And especially since he did not match either Gaskell or Langley in his capacity for research, his retreat from the laboratory must be included among the explanations for the increasing preeminence of the school he had founded by working there.

[89] Langdon-Brown to Sheehan (n. 76). For a rather more favorable portrait of Langley's personality, see Fletcher, "Langley" (n. 7).

[90] Langdon-Brown's letter to Sheehan (n. 76) provides perhaps the most striking evidence of the fragmentation of the Cambridge School toward the end of Foster's career, but the differences between Gaskell and Langley as personalities and investigators emerge clearly from their published works and from almost all of the obituary notices on them. See also J. F. Fulton to Donal Sheehan, 27 March 1940, copy deposited in obituary files, Yale University Medical Historical Library.

11. Concluding Reflections

> We know so little about late Victorian science. At present it lies in uneasy limbo—just beyond the range of any living memory, but not far enough away to have yet been subject to rigorous historical enquiry. . . . We know even less about the growth of the characteristic social institutions of modern science. The teaching laboratory, the center of excellence, and the invisible college are the embarrassed subjects of historical ignorance—and we do not even begin to understand the way ideas and institutions interact to create the very intellectual texture of science itself.
>
> Arnold Thackray (1970)[1]

TOWARD the end of his valuable and suggestive attempt to account for the success of Liebig's school of chemistry at Giessen, J. B. Morrell concedes that his model "suffers from the limitation that it tends to be an idealizing rationalization of [that] success."[2] My account of Foster's school of physiology suffers from the same tendency, and perhaps more severely. For whereas Morrell identifies the factors in Liebig's achievement partly by comparing his school with Thomas Thomson's relatively unsuccessful school of chemistry at the University of Glasgow, I have made no systematic attempt to measure Foster's accomplishment against that of competitors at home or abroad.

Nonetheless, it would be both surprising and disappointing if the story of the Cambridge School had no significance beyond itself, and the preceding chapters in fact contain some hints as to those wider implications. In this concluding chapter, those hints will become more explicit as the factors in Foster's achievement at Cambridge are brought into sharper focus. Tentative steps will also be taken toward a comparison of late Victorian physiology with its German counterpart, and toward a fuller explication of the relationships between internal scientific issues and external social factors.

Oddly enough, it will prove convenient to begin with these larger themes. For the argument has already been made, with regard to English physiology as a whole, that both its early Victorian stagnancy and its late Victorian ascendancy were simultaneously conceptual and institutional phenomena.[3] That argument begins with an effort to

[1] Arnold Thackray, "Commentary," *Historical Studies in the Physical Sciences, 2* (1970), 145-149, on 145-146.

[2] J. B. Morrell, "The Chemist Breeders: The Research Schools of Liebig and Thomas Thomson," *Ambix, 19* (1972), 1-46, on 45.

[3] See Chapters 2 and 6 above.

show that the early Victorian stagnancy had its main conceptual base in the traditional anatomical bias of English physiology, and its associated institutional base chiefly in the peculiar features of English medical education. In particular, England alone among European countries had evolved a pattern of medical licensure that was almost entirely independent of the universities.[4] Except for the special privileges bestowed upon the mere trickle of medical graduates from Oxford and Cambridge, licensure rested essentially in the hands of three private medical corporations, the Royal College of Physicians, the Royal College of Surgeons, and the Worshipful Society of Apothecaries. The requirements of these medical corporations tended to be quite narrowly "pragmatic," with surgical anatomy being particularly stressed by the most important of them, the Royal College of Surgeons. Moreover, since neither the College of Surgeons nor the Society of Apothecaries required university education of its licentiates, the overwhelming majority of medical practitioners received a highly utilitarian training in proprietary medical schools, usually in one of the dozen affiliated with the large hospitals of London. At a time when laboratory science in general, and experimental physiology in particular, could claim scant relevance to medical practice, the teaching in these hospital schools naturally assumed a didactic and anatomical emphasis. The peculiar strength of natural theology and antivivisection sentiment in early Victorian England both reflected and reinforced this anatomical bias. And thus, especially between 1840 and 1870, as physiology became a rigorous experimental science in Continental laboratories, English physiology became isolated from the mainstream.

During the next three decades, for reasons that remain incompletely probed, medical anatomy lost much of its powerful grip over physiology in England. Perhaps the single most important factor was an 1870 change in the statutes of the Royal College of Surgeons, whereby "practical" (i.e.. laboratory) physiology became required of all candidates for membership in the College. When imitated the following year by the degree-granting body of government examiners known as the University of London, this measure made laboratory physiology obligatory for virtually all English medical students. That either the University of London or the ordinarily pragmatic College of Surgeons should have proposed this new requirement seems to demand some special explanation, for experimental physiology remained of dubious utility to medical practice. Until the origins of the new statute receive closer attention, one can only surmise that its passage and implemen-

[4] Among contemporaries, this point was emphasized with special force by G. M. Humphry, "President's Address," *Brit. Med. J.*, 1880 (2), 241-244, on 242.

tation depended on the combined advocacy of T. H. Huxley (examiner in anatomy and physiology at both the University of London and the College of Surgeons in the late 1860's), Michael Foster (Huxley's successor as examiner at the University of London), and especially Foster's good friend and ally, George Murray Humphry, professor of human anatomy at Cambridge and a member of the executive council of the Royal College of Surgeons at the time it adopted the statutory change.

In any case, the new requirement had profound and immediate consequences. Throughout Great Britain, institutions for medical teaching introduced courses in practical physiology and created laboratories to accommodate them. The course at University College London—developed by Foster in the late 1860's when practical physiology had been an elective subject—immediately became the leading model.[5] Within two decades, as the comparative examination success of students taught by specialists with experimental training became clear, even the London hospital schools had begun to remove physiological teaching from the hands of the clinical staff and to appoint full-time physiologists in their stead. An immense (though as yet uncharted) growth in subprofessorial positions transformed English physiology from a handmaiden of medical anatomy and a leisurely avocation into an independent discipline and a viable career possibility. At the professorial level, the emerging independence of the discipline received formal recognition through the creation of four new chairs of physiology in the English universities between 1873 and 1883—the Brackenbury Chair at Owen's College, Manchester (1873), the Jodrell Chair at University College London (1874), the Waynflete Chair at Oxford (1882), and the Cambridge chair to which Foster was elected in June 1883. With these institutional developments, and with the decline of natural theology following Darwin's *Origin of Species* (1859),[6] the old anatomical bias of English physiology gave way to an increasing reliance on experiment.

Indeed, the rise of an organized antivivisection movement in the 1870's is itself a conspicuous reflection of the extent to which animal experiments were becoming de rigueur in English physiology. And though the antivivisection forces did persuade Parliament to appoint a Royal Commission in 1875 and to enact legislation in 1876, they also galvanized the small band of British physiologists into organized

[5] See esp. *Lancet*, 1870 (2), 578. On Foster's course at University College, see Chapter 3, "University College."

[6] On the connection between natural theology and the anatomical bias, see Chapter 2, "The anatomical bias of English physiology."

action of their own. It was no accident that the British Physiological Society, the first of its kind in the world, held its founding meeting in the same year (1876) that Great Britain became the first government to legislate the practice of animal experimentation. Never a very serious obstacle for English physiologists (if slightly more so for their Scottish colleagues), the Vivisection Act became little more than a minor bureaucratic annoyance after 1882, and the number of animal experiments increased exponentially throughout the late Victorian period.[7]

Toward a "national style" for late Victorian physiology

And yet, even as English physiology joined the experimental mainstream flowing from France and especially from Germany, it retained a distinctive character. Still too young to bear a complete separation from its anatomical mother, "the clever child took with her the microscope and the finer study of structure, leaving nothing but the cadaver for anatomy."[8] English physiologists, that is to say, transferred to histology what remained of their traditional affiliation with anatomy, while their more assertive Continental counterparts often left the histological sibling behind to fend for itself or to return to its home in the anatomical hinterlands.[9] As recently as 1971, in his historical essay on the Cambridge School of Physiology during the late nineteenth century, the English histologist D. H. M. Woollam seemed to reveal a continuing nostalgia for those halcyon days when histology and physiology marched arm in arm together.[10]

[7] Richard D. French, *Antivivisection and Medical Science in Victorian Society* (Princeton, 1975). In his list of other countries that subsequently enacted vivisection legislation (p. 234), French makes no mention of Germany. According to P. Hoffmann, however, antivivisection laws did emerge during the Third Reich, presumably with the support of the vegetarian Hitler. Writing in 1953, Hoffmann went on to say that because these laws were passed during the Nazi era, "nobody thinks of enforcing them." See P. Hoffmann, "Germany," in *Perspectives in Physiology: An International Symposium, 1953* ed. Ilza Veith (Washington, D.C., 1954), pp. 81-93 on 93.

[8] Harvey Cushing, *The Life of Sir William Osler*, 2 vols. (Oxford, 1925), I, p. 92.

[9] See ibid.; Lloyd G. Stevenson, "Anatomical Reasoning in Physiological Thought," in *The Historical Development of Physiological Thought*, ed. Chandler McC. Brooks and Paul F. Cranefield (New York, 1959), pp. 27-38, esp. 33, 37; Kenneth J. Franklin, "Physiology and Histology," in *A Century of Science*, ed. Herbert Kingle (London, 1951), pp. 222-238; and J. G. McKendrick, "The Scientific and Social Relations of Anatomy and Physiology," *Brit. Med. J.*, 1876 (2), 379-381, on 379. See also n. 19 below.

[10] See D. H. M. Woollam, "The Cambridge School of Physiology, 1850-1900," in *Cambridge and Its Contribution to Medicine*, ed. Arthur Rook (London, 1971), pp. 139-154, esp. 139-140, 153.

Another relatively peculiar feature of late Victorian physiology, also with roots in the early Victorian period, was its comparative and broadly biological outlook. This emphasis can be traced to William Sharpey and William Benjamin Carpenter, whose lectures and text-books dominated mid-century British physiology. Sharpey and Carpenter, prominently among others, were convinced that physiology belonged part and parcel to the discipline of biology. Later and more specific formulations of this view, especially by T. H. Huxley, implored biologists (and therefore physiologists) to recognize that all living organisms—plants as well as animals—had a common structural basis in the cell, a common functional basis in the structureless protoplasm within each cell, and a common history in the evolutionary sense of that term. Through Huxley's immensely influential course on elementary biology at South Kensington, first offered in the summer of 1871, this comparative perspective found concrete expression in a laboratory setting.[11]

Like the alleged distinction between the "vitalistic materialism" of French physiology and the "mechanistic materialism" of its German counterpart, the national style of late Victorian physiology could easily be exaggerated, especially by focusing on programmatic slogans at the expense of the actual practice of laboratory research.[12] In fact, even the most cursory inspection of Continental research will soon disclose striking examples of histological perspective and a broad comparative approach. Thus Carl Ludwig (1816-1895), the most famous living physiologist after Claude Bernard's death in 1878, certainly did not forget the importance or needs of histology when planning his new institute at Leipzig.[13] Completed in 1870, this laboratory quickly became a leading center for the extension of his histological-cum-physio-

[11] See Chapter 5. On Sharpey, see Chapter 3, "William Sharpey." On Carpenter, see W. B. Carpenter, *Nature and Man: Essays Scientific and Philosophical*, with an introductory memoir by J. Estlin Carpenter (London, 1888).

Earlier in the century, shorn of its Darwinian aspects and supports, the broadly "biological" approach to physiology had prominent Continental representatives—from whom, indeed, Sharpey and Carpenter drew their initial inspiration. Its leading German spokesman, Johannes Müller, retained the comparative orientation of the *Naturphilosophen* even after declaring his independence from the rest of their program. Foster, who owed his most immediate debt to Sharpey, also emphasized his early admiration for the breadth of vision represented by Müller and Carpenter. See Michael Foster, "Verworn on General Physiology," *Nature, 51* (1895), 529-530, on 529.

[12] See Chapter 2, "National styles in Continental physiology."

[13] See H. P. Bowditch, "The Physiological Laboratory at Leipzig," *Nature, 3* (1870), 142-143; Charles Wurtz, *Les hautes études pratiques dans les universités allemandes*, rapport présenté à son exc. le Ministre de l'Instruction Publique (Paris, 1870), pp. 59, 67; and John Scott Burdon Sanderson, "Ludwig and Modern Physiology," *Proceedings of the Royal Institution, 15* (1896), 11-26, on 18.

logical investigations of secretory processes and of the relationship between the so-called "sympathetic" nervous system and cardiovascular function.[14] Claude Bernard himself, though perhaps less adept with the microscope than Ludwig, also established a section for histology in his laboratory at the Collège de France.[15] Moreover, through his toxicological and histogenic concerns, Bernard became the most famous representative of the biological approach to physiology, especially insofar as that approach centered on the phenomena common to plants and animals.[16] His disciple and successor, Paul Bert (1833-1886), further promoted Bernard's conception of "general physiology," while the comparative biological approach could also claim leading German exponents in E. F. W. Pflüger (1829-1910), Wilhelm Kühne (1837-1900), Theodor Wilhelm Engelmann (1843-1909), and Max Verworn (1863-1923).[17]

As these Continental examples suggest, the histological and broadly biological style of late Victorian physiology was probably less pronounced and certainly less isolating than the anatomical bias of the early Victorian period. Lest one begin to doubt its existence altogether, however, it is well to emphasize that the putative conceptual phenomenon had institutional correlates—on the one hand, in the peculiarly English arrangement whereby histology became absorbed into departments of physiology, and on the other, in the powerful local impact of Huxley's laboratory course in elementary biology, which

14 Cf. Heinz Schroer, *Carl Ludwig, Begründer der messenden Experimentalphysiologie, 1816-1895* (Stuttgart, 1967). In addition to these two major research areas, Ludwig's institute also devoted considerable attention to the problem of blood gases, where histological techniques played no appreciable role.

15 Joseph Schiller, *Claude Bernard et les problèmes scientifiques de son temps* (Paris, 1967), p. 38. On Bernard as the prodigiously talented anatomical operator who was less comfortable with strictly histological technique, see Michael Foster, *Claude Bernard* (London, 1899), pp. 93-94, 230-238.

16 See Claude Bernard, *Leçons sur les propriétés des tissus vivants* (Paris, 1866); Bernard, *Rapport sur les progrès et la marche de la physiologie générale en France* (Paris, 1867); and Bernard, *Leçons sur les phénomènes de la vie communs aux animaux et aux végétaux*, ed. A. Dastre, 2 vols. (Paris, 1878-1879). Schiller, *Claude Bernard*, Chapter 2, links Bernard's "general physiology" to his histologic concerns, as does Michael Gross, "Function and Structure in Nineteenth Century Physiology," unpublished Ph.D. dissertation (Princeton, 1974), Chapter 4. By contrast, it is argued in M. D. Grmek, *Raisonnement expérimental et recherches toxicologiques chez Claude Bernard* (Genève, 1973), p. 389, that Bernard's toxicological work led to his concept of general physiology.

17 See Nikolaus Mani, "Paul Bert," *DSB*, II (1970), 59-63; K. E. Rothschuh, "Eduard Friedrich Wilhelm Pflüger," ibid., X (1974), 578-581; Rothschuh, "Wilhelm Friedrich Kühne," ibid., VII (1973), 519-521; Rothschuh, "Theodor Wilhelm Engelmann," ibid., IV (1971), 371-373; and Rothschuh, "Max Verworn," ibid., XIV (1976), 2-3.

apparently had no Continental counterparts as late as 1928.[18] Moreover, it is entirely possible that the special histological and biological orientation of late Victorian physiology forms part of the explanation for its remarkably rapid rise to a competitive and in some ways preeminent place in the physiological world. Thus, without yielding fully to Harvey Cushing's judgment that histology represented the golden path to England's physiological future, one certainly can perceive an unusually pronounced histological dimension in the important work of J. N. Langley on glandular secretion, of Walter Gaskell on the problem of the heartbeat, of both Langley and Gaskell on the involuntary nervous system, and of Charles Scott Sherrington on the "integrative action of the nervous system."[19]

But it is the biological tone of late Victorian physiology that deserves special attention here. More specifically, it was the extent to which English physiologists espoused and utilized Darwinian evolutionary theory that set them apart most clearly from their European colleagues. Continental physiologists of this period generally ignored or disdained Darwin's work, and some were openly hostile to it. Even the otherwise "biological" Claude Bernard revealed little or no sympathy for Darwinian thought.[20] Like Louis Pasteur, whose celebrated investigations of fermentation, spontaneous generation, and disease certainly touched on physiological issues, Bernard mistrusted a theory so insulated from ordinary experimental verification or refutation.[21] For his part, Henri Milne-Edwards (1800-1885) could scarcely admit into physiology the Darwinian doctrines he so decisively opposed in zoology, his main subject.[22] Not dissimilar was the reaction of the

[18] Cf. Ludwig von Bertalanffy, *Modern Theories of Development: An Introduction to Theoretical Biology*, translated and adapted by J. H. Woodger (New York, 1962), p. v.

[19] The importance of histological talent and concerns in the work of Langley and Gaskell should be clear from Chapters 9 and 10 above. Sherrington's classic work represents a particularly striking example of the persistent interpenetration of histology and neurophysiology. As John Fulton once put it, "throughout his life [Sherrington] always sought to turn anatomical facts into physiological language." John F. Fulton, "Reflections on the Historical Backgrounds of Neurophysiology," in Brooks and Cranefield, *Historical Development*, pp. 67-79, quote on 68. See also Judith P. Swazey, *Reflexes and Motor Integration: Sherrington's Concept of Integrative Action* (Cambridge, Mass., 1969); Ragnar Granit, *Charles Scott Sherrington: An Appraisal* (London, 1966); and E. G. T. Liddell, *The Discovery of Reflexes* (Oxford, 1960).

[20] See Schiller, *Claude Bernard*, pp. 139-154.

[21] See ibid.; Robert E. Stebbins, "French Reactions to Darwin, 1859-1882," unpublished Ph.D. thesis (University of Minnesota, 1965), pp. 16-18, 114-124; and Reino Virtanen, *Claude Bernard and His Place in the History of Ideas* (Lincoln, Nebraska, 1960), esp. p. 44.

[22] See Henri Milne-Edwards, *Introduction à la zoologie générale* (Paris, 1851); and his *Leçons sur la physiologie et l'anatomie comparée de l'homme et des animaux*, 14 vols. (Paris, 1857-1881), XIV, pp. 269-335.

leading German physiologists, even though the "mechanistic," anti-teleological implications of natural selection might seem superficially appealing to the spokesmen for German "biophysics."[23] In fact, German physiologists generally perceived no connection between evolutionary thought and their specialized experimental pursuits, and they were suspicious of a theory that recalled the more extravagant speculations of the now despised *Naturphilosophen*.[24] Perhaps by 1870 Carl Ludwig recognized the prematurity of his campaign to reduce physiology to analytical mechanics, but he showed no greater sympathy for evolutionary theory than he had for other "biological" principles in his *Lehrbuch* of the mid-1850's.[25] He shares with Pasteur the curious distinction of having mentioned Darwin's name in print precisely once.[26]

English physiologists, by contrast, frequently cited and consistently admired Darwin's work. Indeed, if English physiology did have a "national style," and if that style were to be captured in a single word, no better choice could be made than "evolutionary." Francis Darwin insisted that his famous father's own concerns, "physiological and evolutionary, were indeed so interwoven that they cannot be sharply separated."[27] The leading figures in the transformation of late Victorian physiology clearly agreed. At its very first formal meeting, in the late spring of 1876, the British Physiological Society sought some appropriate way of expressing its sense of indebtedness to Darwin. In a letter of 1 June 1876, young George Romanes described the occasion for him:

> I am sure the Physiological Society will be very pleased that you like being an hon. member, for it was on your account that honorary membership was instituted. At the committee meeting which was called to frame the constitution of the Society, the chairman (Dr.

[23] Cf. Owsei Temkin, "The Idea of Descent in Post-Romantic German Biology: 1848-1858," in *Forerunners of Darwin*, ed. Temkin et al. (Baltimore, 1959), pp. 323-355, on 353.

[24] Pierce C. Mullen, "The Preconditions and Reception of Darwinian Biology in Germany, 1800-1870," unpublished Ph.D. thesis (University of California, Berkeley, 1964), esp. pp. 68-69, 215-216, 242-250. More generally, the German response to Darwinian theory seems to reveal quite marked disciplinary differences. See William M. Montgomery, "Germany," in *The Comparative Reception of Darwinism*, ed. Thomas F. Glick (Austin, Texas, 1974), pp. 81-116.

[25] For an English translation of Ludwig's programmatic manifesto of the 1850's, see Morton H. Frank and Joyce J. Weiss, "The 'Introduction' to Carl Ludwig's *Textbook of Human Physiology*," *Med. Hist.*, *10* (1966), 76-86. On Ludwig's consistent aversion to "biological principles," see Burdon Sanderson, "Ludwig" (n. 13), pp. 19-20.

[26] See G. L. Geison, "Louis Pasteur," *DSB*, x (1974), 350-416, on 382, n. 100; and Schroer, *Ludwig*, p. 281.

[27] Francis Darwin, "Darwin's Work on the Movement of Plants," in *Darwin and Modern Science*, ed. A. C. Seward (Cambridge, 1909), pp. 385-400, quote on 385.

> Foster) ejaculated with reference to you—"Let us pile on him all the honour we possibly can." . . . when it came to considering what form the expression of it was to take, it was found that a nascent society could do nothing further than make honorary members.[28]

Foster's famous *Textbook of Physiology* (1877) probably borrowed from Darwin its "dominant principle" of the physiological division of labor,[29] and quickly became a major vehicle and source of the evolutionary outlook of English physiology.

One can also find support for the existence and fertility of the evolutionary style in evidence discussed by Richard D. French. Reviewing the work of Charles Darwin himself and Burdon Sanderson on insectivorous plants, of George Romanes on the locomotor physiology of the medusae, and of Foster and Gaskell on the heart, French argues that these three apparently disparate lines of research were generated under the impulse of an "evolutionary climate of ideas," and that they were more immediately linked by their common concern with the evolutionary relationship between structure and function in nervous elements.[30] With respect to the work of Foster, Romanes, and Gaskell, my discussion in Chapters 7-9 confirms and reinforces that conclusion.

More speculative support for the English evolutionary style may be drawn from a consideration of perhaps the four most exciting conceptual developments in physiology during the early years of the twentieth century—namely, hormones and chemical coordination, the integrative action of the nervous system, homeostasis, and vitamins. Because they were developed in a period that slightly postdates the chronological focus of this book, they have thus far been ignored. But it is at least plausible to conjecture that these four concepts are related by virtue of their significance for evolutionary adaptation, that evolutionary principles may therefore have played a role in their genesis and diffusion, and that the evolutionary bias of late Victorian physiology may help to explain why English physiologists contributed more fundamentally to all four concepts than their French or German counterparts.[31]

28 Ethel Romanes, *The Life and Letters of George John Romanes* (Cambridge, 1896), p. 51.

29 See Chapter 7, "The significance of the Royal Institution lectures."

30 Richard D. French, "Darwin and the Physiologists, or the Medusa and Modern Cardiology," *J. Hist. Biol.*, *3* (1970), 253-274.

31 I am thinking, in particular, of the contributions of E. A. Schäfer, W. M. Bayliss, E. H. Starling, H. H. Dale, and (the American) Walter B. Cannon to the study of hormones and "chemical coordination"; of C. S. Sherrington's magisterial work on nervous integration; of the contributions of J. S. Haldane, Joseph Barcroft, and Cannon to the elucidation of homeostatic mechanisms; and of F. G. Hopkin's

In the case of hormones, the evidence is more than preliminary and strikingly pertinent to the themes of this book. E. H. Starling, then at University College London, introduced the term "hormone" and the concept of "chemical coordination" in a series of lectures in 1905. In doing so, he criticized the tendency of (Continental) physiologists "to ascribe every nexus between distant organs to the intervention of the nervous system," and emphasized the extent to which lower, nerveless organisms already displayed a capacity for coordinated activity through "chemical adaptations."[32] Like Foster, that is to say, Starling wondered why differentiated nervous mechanisms should be required in higher organisms to serve a function that already existed in the relatively undifferentiated tissues of lower forms. In the phylogenetically higher forms, chemical coordination took place partly through those "internal secretions" that Starling now christened "hormones." Significantly, German-speaking physiologists had shown little interest in these "internal secretions" and were later to admit their embarrassment that "the fundamental questions regarding the chemical control of physiological processes had [so] long been overlooked."[33] As one might expect from their preoccupation with the phylogenetically recent nervous system, German physiologists were unprepared for this novel concept of "the actual seniority of hormonal, as compared with nervous, control of the organism."[34]

Throughout Part Three of this book, a very similar distinction is drawn between English and German approaches to the problem of the heartbeat. The following section argues more explicitly that the problem of the heartbeat does indeed reveal national styles in the physiology of the late nineteenth century and offers some tentative suggestions as to the possible relation of these national styles to the wider sociopolitical environment.

Nobel prize-winning work on vitamins. See, e.g., Chandler McC. Brooks, "Discovery of the Function of Chemical Mediators in the Transmission of Excitation and Inhibition to Effector Tissues," in Brooks and Cranefield, *Historical Development*, pp. 169-181, esp. 172-173; Brooks and Harold A. Levy, "Humorally-Transported Integrators of Body Function and the Development of Endocrinology," in ibid., pp. 183-238; John F. Fulton and Leonard G. Wilson, eds., *Selected Readings in the History of Physiology* (Springfield, Illinois, 1966), esp. Chapters VIII-XII; and W. P. D. Wightman, *The Emergence of General Physiology* (Belfast, 1956, 20 pp. pamphlet), esp. pp. 18-19.

[32] E. H. Starling, "The Chemical Correlation of the Functions of the Body," *Lancet*, 1905 (2), 339-341, 423-425, 501-503, 579-583, on 339.

[33] See Merriley Borell, "Origins of the Hormone Concept: Internal Secretions and Physiological Research, 1889-1905," unpublished Ph.D. dissertation (Yale, 1976), p. xii.

[34] Humphry Davy Rolleston, *The Endocrine Organs in Health and Disease* (London, 1936), p. 34.

Speculative interlude: national styles and the problem of the heartbeat

The Cambridge campaign on behalf of the myogenic theory reached its climax in Gaskell's classic treatise of 1883. In a sense, Gaskell was able to achieve his resolution of the problem, and thus to make a cogent case for Foster's most cherished belief, only because he had earlier deserted him. For Gaskell's resolution, though unthinkable in the absence of his general Fosterian approach, had been reached only after he reconceived the problem of the sequential vertebrate heartbeat as a whole and then managed to deprive the cardiac ganglia even of the subordinate coordinating role that Foster had felt obliged to grant them.

On the surface, Gaskell's achievement rested on new histological and physiological evidence, revealed by new experimental techniques. But the novelty of that evidence and those techniques depended crucially on his switch in experimental animal from the frog to the tortoise, and that switch in turn arose out of the conviction that special insights might come from the comparative evolution of function in different species. Moreover, Gaskell developed his new histological and experimental evidence in a distinctly evolutionary context. Even in its most specific features, that context reveals his debt to Foster, for Gaskell's leading principle was the same "physiological division of labour" that dominated Foster's *Textbook* of 1877 and lay implicit even in his Royal Institution lectures of 1869 on the involuntary movements of animals. Along with Foster, and through his influence, Gaskell was impressed by the spontaneous contractile capacity displayed by the undifferentiated protoplasm of amoebae and cilia, and by the relatively undifferentiated tissue of the heart and arterial walls as compared to that of ordinary skeletal muscles. Like Foster before him, Gaskell wondered why differentiated nervous mechanisms should be required for the rhythmic beat of the vertebrate heart when a simple ciliated cell already displayed automatic rhythmicity in their absence. Indeed, differentiation ought to be *inimical* to rhythmicity, for differentiation proceeds at the expense of prior and more basic functions (including rhythmicity) for the sake of subsequent and more specialized ones (including speed of impulse conduction). To allow the energy required for differentiation to be spent on a function that existed in its absence would show, in Foster's words, "a wasteful want of economy."[35]

[35] Michael Foster and A. G. Dew-Smith, "The Effects of the Constant Current on the Heart," *J. Anat. Physiol., 10* (1876), 735-771, on 770.

Curiously, Foster did not allow his evolutionary perspective to carry him quite as far as it might have. For even though he insisted on the coordinated activity of cilia, and even though he located the sequential character of the snail's heartbeat solely in its primitive contractile tissue, he assigned the faculty of coordination in the frog's heart to its differentiated ganglia. By doing so, as Gaskell showed in his treatise of 1883, Foster momentarily forgot his usual respect for the animal "economy." After Gaskell demonstrated that the ganglia were no more essential to the coordinated sequence of the vertebrate heart than to its rhythmicity, it could be seen that the energy expended in the production of the ganglia represented an investment of an entirely different order. And the investment, if costly, returned handsome dividends. For the real and only role of the ganglia, Gaskell argued, lay in their association with the inhibitory processes unknown in simpler organisms, but of great utility to the more specialized requirements of the vertebrate and mammalian heart.[36] Indeed, then as now, some saw the relative strength of "inhibitions" in humans as a measure of their advance from lower "animality," and perhaps the historian may forgive (if not excuse) Claude Bernard for once suggesting, in a popular lecture on the heart, that the more highly developed inhibitory power of the nervous system in men explained their physical and moral superiority over women.[37]

To insist that an evolutionary perspective underlay and contributed importantly to the myogenic position of the Cambridge School is not to say that Darwinian concepts were applied consistently or rigorously. Most notably, the mechanism of natural selection was rarely if ever mentioned at all. To Foster and his students, it probably seemed irrelevant to the issues in dispute. In this case as in others, a tendency to

[36] W. H. Gaskell, "The Contraction of Cardiac Muscle," in *Textbook of Physiology*, ed. E. A. Schäfer, 2 vols. (Edinburgh, 1898-1900), II, 169-227, on 197.

[37] Claude Bernard, "Étude sur la physiologie du coeur," *Revue des deux mondes, 56* (1865), 236-252, on 252. In its general thrust, this remarkable lecture sought to unite the physiological conception of heart action with the imagery used by poets and other artists—thereby to deny any antithesis between art and science or between reason and sentiment. Not so incidentally, Bernard also insisted that the physiological investigation of such topics as the relationship between the heart and brain need not and should not lead to "materialism." Quite clearly, this evening lecture at the Sorbonne can be situated in the same sociopolitical context as Pasteur's Sorbonne lecture of 1864 on spontaneous generation. See John Farley and G. L. Geison, "Science, Politics and Spontaneous Generation in Nineteenth-Century France: The Pasteur-Pouchet Debate," *Bull. Hist. Med., 48* (1974), 161-198, esp. 188-190. See also J. M. D. and E. Harris Olmsted, *Claude Bernard and the Experimental Method in Medicine* (New York, 1961), pp. 126-128. For a suggestive comment on Bernard's position in the materialist-spiritualist debate, see Claude Bernard, *Cahier de notes*, 1850-1860, ed. M. D. Grmek (Paris, 1965), pp. 243-247, n. 153.

emphasize natural selection as the novel feature of Darwin's work may distort the nature of his influence on the actual practice of biologists not directly concerned with the process of speciation per se. Even among those who did focus on speciation, controversy long raged over the full adequacy of natural selection as an explanation of evolution.[38] Whatever their views on this controversy, evolutionary morphologists, taxonomists, and physiologists doubtless often found natural selection of little pertinence to the specific research problems they faced. But because Darwin had argued so persuasively for the reality of the evolutionary process itself, such biologists often did make confident use of his more general evolutionary concepts. Among physiologists, Michael Foster and his Cambridge students serve as particularly striking examples.

Because the myogenic theory of the heartbeat depended so much on biological and evolutionary concepts, we should not be surprised to discover that different national attitudes toward myogenicity corresponded to national differences in the response of physiologists to evolutionary theory itself. In its modern version, the myogenic theory had its roots chiefly (if not quite exclusively) in Britain. Indeed, William Sharpey (Foster's own mentor) and the Edinburgh histologist, John Hughes Bennett, had retained myogenic sympathies even in the face of the Continental shift to neurogenicity.[39] Between 1859 and the 1880's, with the notable exception of T. W. Engelmann, it was Foster and his students who sought to establish the myogenic theory, while outside Britain even passive skepticism toward the neurogenic position was rare. After 1883, when Gaskell's classic treatise appeared, English physiologists and cardiologists rapidly accepted and extended his basic conclusions. In the task of extending his work from the simple vertebrate to the mammalian heart, a task that proved more difficult than

[38] See, e.g., John E. Lesch, "The Role of Isolation in Evolution: George John Romanes and John T. Gulick," *Isis, 66* (1975), 483-503.

[39] On Sharpey's views, see Chapter 7, n. 50. Bennett's myogenic bias seems implicit in his emphatic opposition to the view that "muscular contractility is . . . dependent on the nervous system." Insisting that contractility is "a vital power inherent in the tissues which possess it," Bennett traced his position to Haller and argued that this view was "supported by all that is now known of the subject." Even without reference to Claude Bernard's important new curare experiments (see below), Bennett seemed supremely confident of the available evidence as to the capacity of muscle or even individual cells (e.g., cilia) to contract when isolated from the influence of the nervous system. See John Hughes Bennett, "Physiology," *Encyclopedia Britannica*, 8th ed. (London, 1853-1860), XVII (1859), pp. 648-703, esp. 652-653. On the other hand, because Bennett did not explicitly confront and refute the new evidence concerning intracardiac ganglia, one cannot be absolutely certain that his position on the problem of the heartbeat was any more decisive than Claude Bernard's.

Gaskell had expected, his compatriots participated actively. Particularly important was the work of J. W. Pickering on the physiology of the embryonic heart, of A. F. S. Kent on the atrioventricular bundle of the mammalian heart, and of Arthur Keith and Martin Flack on the cardiac pacemaker. On the more strictly clinical side of cardiology, the myogenically based contributions of Sir Thomas Lewis, Sir James Mackenzie, and Sir Arthur Cushny gained widespread attention and support.[40]

Outside of Britain, it was only in North America that the myogenic theory was quickly adopted and developed. H. Newell Martin (Foster's own first student) and his younger co-workers at Johns Hopkins contributed to it through their efforts to sustain the mammalian heart in isolation from its nervous system. The mechanical devices they developed along the way are sometimes cited as prototypes of the heart-lung machines that have made it possible to do open-heart and heart-transplant surgery, and W. J. Meek went so far (in 1928) as to call Martin's method of isolating the mammalian heart "possibly the greatest single contribution ever made from an American physiological laboratory."[41] Another major American contributor to the myogenic theory was Joseph Erlanger, who established the dependence of the sequential heartbeat on the bundle of artrioventricular muscles in the mammalian heart.[42] The general myogenic orientation of American physiologists probably catalyzed the debate that erupted at the 1905 meeting of the American Physiological Society, when A. J. Carlson of Chicago advanced the neurogenic theory in the case of the *Limulus* crab heart.[43]

In Europe, by contrast, the myogenic theory had never managed to elicit widespread support. At the outset, around 1860, the English myogenicists seemed to have a partial ally in zoologist Henri Milne-

[40] On the developments after 1883, see inter alia Robert Tigerstedt, *Die Physiologie des Kreislaufes*, 2nd ed., 4 vols. (Berlin, 1921), esp. vol. II: J. A. E. Eyster and W. J. Meek, "The Origin and Conduction of the Heartbeat," *Physiological Review, 1* (1921), 3-43; and A. P. Fishman, ed., *Circulation of the Blood: Men and Ideas* (New York, 1964), passim.

[41] Walter J. Meek, "The Beginnings of American Physiology," *Annals of Medical History, 10* (1928), 111-125, on 124. Further on Martin and the isolated mammalian heart, see C. S. Breathnach, "Henry Newell Martin (1848-1893): A Pioneer Physiologist," *Med. Hist., 13* (1969), 271-279.

[42] See, e.g., Joseph Erlanger, "The Localization of Impulse Initiation and Conduction in the Heart," *Archives of Internal Medicine, 11* (1913), 334-364. More generally on Erlanger, see Joseph Erlanger, "A Physiologist Reminisces [1964]," in *The Excitement and Fascination of Science: A Collection of Autobiographical and Philosophical Essays* (Palo Alto, 1965), pp. 93-106; and A. M. Monnier, "Joseph Erlanger," *DSB*, IV (1971), 397-399.

[43] See Carl J. Wiggers, "Physiology from 1900-1920: Incidents, Accidents and Advances [1951]," in *Excitement and Fascination of Science*, pp. 547-566, on 550-551.

Edwards, promulgator of the doctrine of the physiological division of labor. But his position on the problem of the heartbeat became increasingly equivocal and irrelevant as time went on.[44]

Other French-speaking physiologists offered even less support. Charles-Edouard Brown-Séquard, whose contributions to the study of vasomotor phenomena won him sometimes exaggerated acclaim, advanced the somewhat peculiar but basically neurogenic theory that the rhythmic heartbeat was determined by the action of blood-borne carbon dioxide on the nervous system.[45] Obviously of special interest are the views of Claude Bernard, who died five years before Gaskell produced his myogenic resolution of the problem. On the surface, it might seem that Bernard had already overthrown the neurogenic theory through his famous discovery of the mid-1850's that curare destroys the action of the motor nerves without abolishing muscular irritability in general and without interrupting the hearbeat in particular. This discovery, made independently by the Swiss histologist and physiologist Albert von Koelliker, went a long way toward establishing Haller's long-debated claim that muscular "irritability" could exist independently of nervous "sensibility." Perhaps because it was perceived as a threat to neurogenicity, Bernard's work met a surprisingly obstinate opposition from some German physiologists, notably Conrad Eckhard of Giessen.[46] In fact, however, Bernard himself never

[44] In 1859, with special emphasis on the curare experiments of Bernard and Koelliker, Milne-Edwards suggested that cardiac contractility might be independent of the nervous system, belonging rather to the cardiac musculature itself. He also expressed doubt about the role of the intracardiac ganglia, though he presumed they were responsible for coordinating the sequential character of the heartbeat. See Milne-Edwards, *Leçons sur la physiologie et l'anatomie comparée* (n. 22), IV (1859), pp. 142-143, 164. By 1879, however, Milne-Edwards had apparently adopted the view that ganglia were directly responsible for rhythmic motion, in the heart as elsewhere, and in all organisms that had a nervous system of whatever sort. Interestingly enough, he cited Romanes' early work on the medusae in partial support of this view. See ibid., XIII (1878-1879), pp. 101, 189-199. Cf. Chapter 8, "George John Romanes." As usual, Milne-Edwards hedged his position, and his suggestion that the ganglia release periodic discharges (ibid., p. 101) is followed immediately by the disclaimer that "we know almost nothing on this subject." Nonetheless, the overall pattern of his remarks on the nervous control of involuntary movements (including the heartbeat and vasomotor action) tend now toward a somewhat unusual version of neurogenicity, not wholly unlike that of Brown-Séquard. Cf. ibid., XIII (1878-1879), pp. 259-283, passim.

[45] *J. M. D. Olmsted, Charles-Edouard Brown-Séquard: A Nineteenth Century Neurologist and Endocrinologist* (Baltimore, 1946), pp. 60-61. As one has come to expect, Brown-Séquard's views of vasomotor action were consistent with his conception of the cause of the rhythmic heartbeat. See Vulpian (n. 48 below), passim.

[46] For an exhaustive and penetrating analysis of Bernard's work on curare, see Grmek, *Raisonnement expérimental,* Chapter 3. The opposition of Eckhard and other German physiologists is discussed on pp. 306-307, 342.

took a really decisive position on the cause of the rhythmic heartbeat. And since he sometimes insisted that curare had no effect on the nerves and ganglia of the sympathetic or involuntary system, the neurogenically inclined could easily maintain that the continued beating of the heart in curarized animals depended on these nervous structures.[47] Indeed, if Bernard's curare experiments seemed superficially myogenic in their thrust, he also contributed indirectly to the neurogenic theory through his putative demonstration of a reflex nervous mechanism in the lining membrane (or endocardium) of the heart and through his "ganglionic" theory of vasomotor action.[48] Especially because of the fre-

47 In his private notebook of 1850-1860, Bernard expressed uncertainty as to whether the heartbeat should be ascribed to "a *primum movens*, or to direct excitation of blood, air or CO_2 . . ." See Bernard, *Cahier de notes* (n. 37), p. 176. Grmek's impressive account of Bernard's curare experiments reflects Bernard's own ambivalence by never really confronting the problem of the heartbeat head-on. Had Bernard taken a decisively myogenic position, Foster would surely have enlisted his authority on behalf of his own lonely campaign for myogenicity. In fact, however, Foster pursued that campaign without reference to Bernard or his curare experiments.

The most developed and explicit statement known to me of Bernard's position on the problem of the heartbeat is to be found in his popular Sorbonne lecture of 1865 (see n. 37). After general introductory remarks, Bernard presented the standard embryogenic argument for the myogenic theory—namely, that since the heart begins to beat in the embryo before the nervous system is anatomically distinguishable or physiologically active, one must conclude that the contractions of the heart do not require the intervention of the nervous system. In making this point, Bernard did not go beyond such earlier statements of the argument as, say, that of Sharpey. He then cited the results of his curare experiments to argue that the heartbeat should not be ascribed to [central?] nervous influences even in the fully developed heart. His discussion of intracardiac nervous structures did not directly or necessarily threaten the myogenic position, for he suggested that these nervous elements act only to *inhibit* (rather than to produce) the contractions of the heart. Nonetheless, Bernard's subsequent emphasis on "reflex connections" between the heart and the central nervous system (an emphasis required by his main goal of uniting the physiological and poetic conceptions of heart action) had the effect of muting or obscuring the force of his seemingly myogenic stance up to that point. Indeed, midway through his lecture, Bernard introduced a generalization that sounds, on the face of it, purely neurogenic in its implications: "All the involuntary movements of the heart . . . have no other source than the reaction of sensibility on the pneumogastric motor nerves of this organ." Bernard, "Étude" (n. 37), p. 246. See also n. 48 immediately below.

48 For one example of the way in which Bernard's endocardial reflex was used on behalf of neurogenicity, see William Rutherford, "Lectures on Experimental Physiology," *Lancet*, 1871 (2), 841. In his casual acceptance of the neurogenic theory, Alfred Vulpian depended heavily on the presumed analogy between cardiac inhibition and vasodilation, and on Bernard's ganglionic theory of the latter. See Vulpian, *Leçons sur l'appareil vasomoteur (physiologie et pathologie)*, ed. H. C. Carville, 2 vols. (Paris, 1875), esp. 1, pp. 178ff, 321-322. In the end, whatever Bernard's own position on the problem of the heartbeat, the most important point is that the neurogenic theory retained its hegemony even in the wake of his curare experiments.

quently "biological" direction of Bernard's thought, his reaction to Gaskell's work would be interesting to have. In its absence, the French generally took little part in the myogenic-neurogenic debate.[49]

Almost by definition, Germany (or the German-speaking world) was the leading center of the dominant neurogenic view, for it led the field of physiology in general. Even so, the strength of neurogenic sentiment there seems peculiarly marked and persistent. After long ignoring the work of Foster and his students, most German physiologists responded to Gaskell's treatise of 1883 along a continuum ranging from indifference through doubt to outright opposition, and some of them continued to respond this way long after the rest of the world seemed satisfied. Ludwig's school at Leipzig was a major cradle of neurogenic sympathies. In fact, the two most vociferous opponents of the myogenic theory even into the twentieth century were Hugo Kronecker, an assistant in Ludwig's laboratory in the late 1860's and 1870's; and Elie Cyon, a Russian-born physiologist who also worked under Ludwig, with whom in 1866 he co-discovered the "depressor" nerve, the vasodilator branch of the vagus nerve.[50]

To be sure, a very few German workers did contribute to the genesis and extension of the myogenic theory. In particular, Theodor Wilhelm Engelmann entered the campaign on behalf of myogenicity per-

[49] Note, for example, the rarity of French sources in Engelmann's thoroughly documented review of 1897. Theodor W. Engelmann, "Ueber den myogen Ursprung der Herzthätigkeit und über automatische Erregbarkeit als normale Eigenschaft peripherischer Nervenfäsern," *Archiv für die gesammte Physiologie des Menschen und der Thiere, 65* (1897), 535-578. See also the sources in n. 40 above.

[50] See Wilhelm His, Jr., "The Story of the Atrioventricular Bundle with Remarks concerning Embryonic Heart Activity [1933]," trans. T. H. Bast and W. D. Gardner, *J. Hist. Med., 4* (1949), 319-333, on 331. On Kronecker, see K. E. Rothschuh, "Hugo Kronecker," *DSB*, VII (1973), 504-505. On the bizarre and neglected career of Cyon, see Jonas R. Kagan, "Elie Cyon, physiologiste et diplomate (1843-1912)," *Revue d'histoire de la médicine hebraïque, 18* (1965), 149-154; Léon Delhoume, "Claude Bernard et Elie de Cyon," *Concours médical, 79* (1957), 1439-1442, 1551-1553; and H. S. Koshtoyants, *Essays on the History of Physiology in Russia*, ed. D. B. Lindsley (Washington, D.C., 1964), passim.

Carl J. Wiggers, "Some Significant Advances in Cardiac Physiology during the Nineteenth Century," *Bull. Hist. Med., 34* (1960), 1-15, on 12, besides also identifying Kronecker and Cyon as leading opponents of myogenicity, mentions two other physiologists who actively defended the neurogenic theory well into the twentieth century—Oskar Langendorff and Johannes Dogiel. Of this pair, too, one (Langendorff) was German, the other (Dogiel) Russian-born; and both had worked with Ludwig at Leipzig. On Langendorff, see Hugo Kronecker, "Oskar Langendorff (1830-1909)," *Travaux de l'Association de l'Institut Marey (Paris)*, 2 (1910), 326-327. Dogiel, whom we have already encountered briefly as an opponent of Foster's work on the snail's heart (Chapter 8, "Francis Darwin"), is a more obscure figure. For the merest sketch of his career (which fails even to give his date of death), see *Biographisches Lexikon der hervorragenden Aerzte aller Zeiten und Völker*, ed. August Hirsch et al., 3rd ed. (München, 1962), II, pp. 287-288.

haps a decade later than Foster, and he remained an active participant in it long after Foster had ended his brief career in physiological research. To judge solely from the impact of their published work, Engelmann also contributed far more significantly to the triumph of the myogenic theory, though his early efforts (like Foster's) were directed mainly toward establishing the myogenicity of the isolated ventricular beat and lacked the cogency of Gaskell's more thoroughly myogenic resolution of the problem of the rhythmic and sequential vertebrate heartbeat as a whole.[51] Another German worker, Wilhelm His, Jr., contributed fundamentally to the unexpectedly difficult task of extending Gaskell's results from the lower vertebrate to the mammalian heart. In fact, His is generally regarded as co-discoverer at least (with A. F. S. Kent) of the atrioventricular bundle in mammalian hearts. Moreover, in making this contribution (c. 1890), His relied on evidence and principles that immediately invite comparison with the views of Foster and Gaskell. Having found that rhythmic pulsations could be detected in the cardiac tissue of mammalian embryos before their hearts had developed any distinct nervous structures whatever, His found it impossible to imagine that the nervous structures appearing subsequently could be essential to rhythmicity or the other basic properties of the mammalian heartbeat. Thus, in a retrospective account of his research, he could write as follows:

> That a morphologically yet unfinished tissue possesses already all of the later characteristics [of the heartbeat] is in itself a biological problem of great interest and it is no less interesting to follow [the changes accompanying] the course of its differentiation.
>
> If one assumes with the proponents of neurogenic automaticity, that these characteristics [i.e., rhythmicity, coordinated sequence, etc.] at a later time are conveyed by the nervous elements, then it would be a gripping biological problem how one tissue could take over the function of a second.[52]

But what is most interesting and important about the occasional German contributors to the myogenic theory was the extent to which they set themselves apart from the general direction of German physiology. Engelmann did so literally by moving to Utrecht, where he produced his most important work on the problem. Even earlier, he had made himself a conspicuous exception among German-trained physiologists by embracing Darwinism. In Utrecht, where he became first

[51] See J. N. Langley, "Walter Holbrook Gaskell," *Proc. Roy. Soc., 88B* (1915), xxvii-xxxvi, on xxviii.

[52] His, "Story of the Atrioventricular Bundle," p. 331.

professor of histology and general biology in 1877, Engelmann often wrote for a zoological audience, believing that his work on lower organisms would be of little interest to (German) physiologists concerned so exclusively with mammals and frogs.[53] Similarly, His, an anatomist and embryologist by training, spoke contemptuously of the approach of German physiologists to the problem of the heartbeat:

> ... the problem of learning the characteristics of the beating of nerveless hearts did not lure the German physiologists. This conforms to their slight inclination toward comparative physiology with which only Engelmann, Kühne, Biedermann and later Verworn concerned themselves. This one-sidedness was desired because, as Schmiedeberg said once to me, "We have such difficulties in recognizing vital phenomena that we are satisfied to know them only in two or three kinds of animals."[54]

Another noteworthy dissenter from the neurogenic trend of German biology was the Prussian cellular pathologist, Rudolf Virchow (1821-1902). Even further outside the German physiological mainstream than Engelmann or His, Virchow did not give the problem of the heartbeat any central place in his own research. Rather, his myogenic sympathies were grounded in his concept of "the autonomy of the cell," which led him to suppose that cardiac cells should be able to function independently of the nervous system.[55]

Although we now accept the myogenic theory, we should not suppose that the work of the German neurogenicists was necessarily or fundamentally "wrong." No small part of the interest of the myogenic-neurogenic debate lies precisely in the fact that the positions taken had less to do with purely experimental evidence than with convictions about what sort of evidence and approach should be used in pursuit of physiological problems. There was no single "crucial experiment" capable of resolving the differences between the myogenicists and neurogenicists. To some extent, they asked different questions of the same material. What most impressed the (chiefly English) myogenicists was the appearance of rhythmicity in the absence of differentiated nervous elements. What most impressed the (chiefly German) neurogenicists was the obvious influence of nervous action on the beat of the fully developed heart. Some neurogenicists, at any rate, could accept

[53] See Helmut Kingreen, *Theodor Wilhelm Engelmann (1843-1909): Ein bedeutender Physiologe an der Schwelle zum 20. Jahrhundert,* nr. 6 in Münstersche Beiträge zur Geschichte und Theorie der Medizin, ed. K. E. Rothschuh et al. (Münster, 1972), esp. pp. 16-17, 39.

[54] His, "Story of the Atrioventricular Bundle," p. 331.

[55] See this Chapter, p. 351.

the occasional appearance of nerveless rhythmicity without perceiving the relevance of this to the normal intact vertebrate heart, where nervous elements were so obviously present. And most myogenicists agreed that nervous influences played an important role in the intact vertebrate heart even as they denied that this had anything to do with the real issue in dispute. For them the crucial question was whether, why, and how the function of rhythmicity could pass from muscular to nervous tissue.

But reluctance to pass judgment as to which side was ultimately "right" or "wrong" in the myogenic-neurogenic debate should not obscure my deep conviction that Gaskell's work was vastly more *fertile* than that of his German adversaries. And because his approach depended so fundamentally on evolutionary concepts, perhaps the peculiarly evolutionary tone of late Victorian physiology really can help to explain its rapid rise in international status. Before developing this idea further, it should be admitted that the English ascendancy can be explained at least partly without any reference to its specific conceptual style. The contemporary retreat of French physiology into relative insignificance, for example, can probably be associated with the more general decline of French science now attracting so much historical attention.[56] Meanwhile, though Germany's famous "army" of state-supported research workers remained productive in quantitative terms, one begins to wonder how many of these Ph.D. theses and their sequelae advanced (or at least pursued) precision at the expense of new directions and fresh insights. In physiology, at any rate, the young officer corps grew restless under the aging generals who continued to monopolize the major chairs. Eventually, many of the most promising young investigators sought brighter career prospects in such expanding neighboring fields as pathology, pharmacology, bacteriology, hygiene, and experimental psychology.[57] The soldiers who remained behind were perhaps more likely and more willing to coast on the insights of their aging mentors, and thus to carry their degenerating research schools into narrower and narrower cul-de-sacs.[58] In late Victorian England, by contrast, research schools of physiology were still in the

[56] See, e.g., Jerome R. Ravetz, *Scientific Knowledge and Its Social Problems* (Oxford, 1971), p. 226, n. 20; and Robert Fox, "Scientific Enterprise and the Patronage of Research in France 1800-70," *Minerva, 11* (1973), 442-473.

[57] See Abraham Zloczower, "Career Opportunities and the Growth of Scientific Discovery in Nineteenth-Century Germany, with special reference to Physiology," unpublished M.A. thesis (Hebrew University, Jerusalem, 1960); and Hans-Heinz Eulner, *Die Entwicklung der medizinischen Spezialfächer an den Universitatën des Deutschen Sprachgebietes* (Stuttgart, 1970).

[58] Cf. Ravetz, *Scientific Knowledge*, pp. 231-233.

process of being established and career opportunities were expanding rapidly. This institutional setting, it could be argued, virtually guaranteed the observed rise of English physiology.

In seeking to put some conceptual flesh on this institutional skeleton, some will think I have erred in choosing evolutionary thought as the main pablum. Others might emphasize instead the hoary vitalist-mechanist debate. And their choice would doubtless have some justification, even in the particular case of the problem of the heartbeat. Certainly the notion that muscular contraction could take place in the absence of nervous action, and more specifically that "innate" properties of the cardiac tissue were responsible for its rhythmic beat, carried vitalistic connotations for some nineteenth-century physiologists.[59] To the extent, then, that German physiologists had become prisoners of their "mechanistic" rhetoric or their elaborate physical instrumentation, they could be expected to mistrust the myogenic theory—a claim that perhaps gains in credence from the fact that so many of the leading neurogenicists studied with Ludwig or other "biophysicists." But my general belief that the vitalist-mechanist distinction has been overdrawn and overemphasized, together with my even firmer conviction that these large quasi-philosophical categories have not yet been linked persuasively and concretely with evidence about the actual practice of research, has led me to focus instead on the possible role of evolutionary concepts.

Moreover, evolutionary thought seems to me a more promising direction for the pursuit of the most ambitious and difficult task facing the historian of science—namely, to establish tangible and cogent bonds between the internal content of mature science and its social environment. For we already have considerable knowledge of the comparative reception of Darwinian thought in various social settings,[60] and we even have a putative social explanation for its English origins and popularity under laissez-faire capitalism. This alleged social explanation—somewhat undermined by the fact that socialists (including Marx and Engels themselves) found support in "Darwinism" for their own views—can and has given rise to all sorts of nonsense. Nonetheless, and without indulging the patently absurd notion that Darwin's achievement can be explained in any meaningful way by referring it to the economic

[59] See, e.g., A. E. Best, "Reflections on Joseph Lister's Edinburgh Experiments on Vaso-motor Control," *Med. Hist.*, *14* (1970), 10-30, esp. 22-29. It is also suggestive that the seventeenth-century "iatromechanist" Giovanni Borelli was among the rare advocates of neurogenicity before the mid-nineteenth century. Kenneth J. Franklin, *A Short History of Physiology*, 2nd ed. (London, 1949), p. 101.

[60] See esp. Thomas F. Glick, ed., *The Comparative Reception of Darwinism* (Austin, Texas, 1974).

system of Victorian England, one may insist that he expressed his views in an idiom with peculiarly English resonances. As Charles Gillispie has suggested, the language of Darwin's *Origin of Species* is such that "none but a Victorian Englishman" could have written it:

> So ordinary is the language that it almost seems as if we could be in the midst of reading a lay sermon on self-help in nature. All the proverbs on profit and loss are there, from pulpit and from counting-house—On many a mickle making a muckle . . . On the race being to the swift . . . On progress through competition . . . On handsome is as handsome does . . . On saving nine . . . On reflecting that in the midst of life we are in death . . . On the compensation that all is, nevertheless, for the best.
>
> . . . . It is, as a German critic said in a remark meant to be scathing, classical political economy applied to biology. Or as Darwin said himself, "This is the doctrine of Malthus applied to the whole animal and vegetable kingdom."[61]

English biologists, immersed in the same sociocultural milieu, probably appreciated the nuances of Darwin's thought more fully than their counterparts abroad and were probably better prepared to use those concepts in pursuit of particular research problems. At the very least, his frequently metaphorical language would have been vastly more familiar to them, and the comparative reception of Darwinian thought doubtless reflects that circumstance.

By a perhaps bold but plausible extension, I consider it only natural that the myogenic theory should have been so peculiarly English in its genesis and diffusion. J. T. Merz suggested long ago that "physiology and economics joined hands" in the Victorian period through the concepts of the "autonomy of the cell" and of the "physiological division of labour."[62] We have already seen how important the latter doctrine was to Foster's campaign for myogenicity. So, very probably, was the concept of cellular autonomy, which Virchow had already used in support of the independent activity of nerveless cardiac cells. In his *Textbook* of 1877, Foster combined an implicit appeal to the doctrine of cellular autonomy with explicit use of the concept of the physiological division of labor in such a way that they seemed mere corollaries of each other. For only if one granted the independent vitality and manifold physiological capacities of the simple unicellular

[61] Charles Coulston Gillispie, *The Edge of Objectivity: An Essay in the History of Scientific Ideas* (Princeton, 1960), pp. 303-304.

[62] John Theodore Merz, *A History of European Thought in the Nineteenth Century*, 4 vols. (Edinburgh, 1904-1912), II, Chapter X, esp. pp. 395-396, 415.

amoeba was it possible to follow or accept Foster's further claim that the physiological activities of more complex organisms arose from "these fundamental qualities of protoplasm peculiarly associated together [through the] dominant principle of the physiological division of labour."[63] In forging this association between the two doctrines, both of foreign origin, Foster probably relied mainly on the works of Charles Darwin, who had used the physiological division of labor in the *Origin of Species* and who later, in 1868, developed his ill-fated hereditary hypothesis of "pangenesis" partly on the basis of the concept of cellular autonomy.[64]

Like Darwin, Foster sometimes couched scientific concepts in the ideologically sensitive language of political economy. Cultural anthropologists may find it significant, for example, that he should speak of mammals as "living up to their physiological income," while cold-blooded animals save for the future; or that he should refer to wise and unwise investments in the "animal economy," emphasizing the "wasteful want of economy" it would show if the energy required for differentiation were to be spent for the sake of a function (like rhythmicity) that existed fully in its absence.[65] Perhaps more importantly, even in one of his technical research papers with A. G. Dew-Smith, Foster did not entirely conceal his irritation at the notion that cardiac cells, having been granted autonomy in the adult snail and in the vertebrate embryo, should be required to yield to the control of a "nervous master" at some later stage in phylogeny or ontogeny.[66]

Given the ideological sensitivity of such language and the undeniable ideological overtones of the evolutionary debate, it becomes less surprising that the "merely" scientific controversy between the myogenicists and the neurogenicists took on—for at least a few of its participants—the sort of passion ordinarily associated with political or religious issues. Thus, that rare German advocate of myogenicity, Wilhelm His, Jr., recalled the days "when the disputed question, neurogenic or myogenic, was fought over with a dogmatic zeal." When His presented his arguments for myogenicity to Hugo Kronecker, the latter, "even though he was a personal friend, never wanted to let any of my argu-

[63] Michael Foster, *A Textbook of Physiology* (London, 1877), p. 3.

[64] On Darwin's use of the physiological division of labor, see Camille Limoges, *La sélection naturelle: étude sur la prèmiere constitution d'un concept (1837-1859)* (Paris, 1970), pp. 135-136. On cellular autonomy and pangenesis, see Charles Darwin, *The Variation of Animals and Plants under Domestication*, 2 vols. (London, 1868), II, pp. 368-371; and G. L. Geison, "Darwin and Heredity: The Evolution of his Hypothesis of Pangenesis," *J. Hist. Med.*, *24* (1969), 375-411, esp. 391-392.

[65] Chapter 7 above, "The Royal Institution lectures"; and Foster and Dew-Smith, "The Effects of the Constant Current," p. 770.

[66] See Foster and Dew-Smith, ibid., pp. 770-771.

ments bear any weight." Elie Cyon, says His, was "even more severe."[67] In fact, Cyon's fervent hatred of the myogenic theory was appropriately linked to an equally passionate hatred of Darwinism. In Cyon's own published words, Gaskell's ideas on the heart invited "anarchy," while Darwinism was indicative of moral and psychological "decadence."[68] And so Cyon—who, according to His, led a "double existence" in Paris as scientist and as political agent for the Czarist Russian government—"could not refrain, even in his article on world opinion, from speaking out against the evil myogenic adherents."[69]

By this point, one begins to be struck by the special appeal of the neurogenic theory in Bismarckian Germany and Czarist Russia, where "Darwinism" was perceived as a threat to the social order, and where autocratic regimes may well have found the notion of central nervous domination congenial to their self-assigned role as manipulator and conditioner of the body politic. In this context, it is significant that Virchow, a decidedly atypical "Prussian," insisted on the "autonomy of cardiac contraction" in an 1855 essay that reflects his vision of the organism as a democratic cell-state of autonomous individuals with equal rights if not equal endowments.[70] Moreover, as Donald Fleming points out, Virchow was "the only great scientist in history to play a major role in politics, as chief of the liberal opposition to Bismarck."[71] His "deep commitment to democracy" can be detected even in his celebrated research in cellular pathology and more specifically in his glimpse of "an emblem of democracy in the human body—a society of individual cells, profiting from the association but each retaining its basic character as an independent focus of vitality."[72]

Because Virchow's political concerns sometimes overshadowed his commitment to scientific "professionalism," he was unusually self-conscious and open about the ideological interests he hoped to serve through his scientific work. Scientists more concerned to establish their

[67] His, "Story of the Atrioventricular Bundle," p. 331.

[68] See Elie Cyon, "Myogen oder Neurogen?" *Archiv für die gesammte Physiologie des Menschen und der Thiere, 88* (1902), 225-294, on 260-261, 285-290; and Cyon, *Gott und Wissenschaft* (Berlin, 1912).

[69] His, "Story of the Atrioventricular Bundle," p. 331. On Cyon's turbulent political views and activities, see esp. Kagan, "Cyon."

[70] Rudolf Virchow, "Cellular Pathology [1855]," in *Disease, Life, and Man: Selected Essays by Rudolf Virchow*, trans. Lelland J. Rather (N.Y., 1962), esp. pp. 106-110.

[71] Fleming's introduction to Jacques Loeb, *The Mechanistic Conception of Life*, ed. Donald Fleming (Cambridge, Mass., 1964), p. x.

[72] Ibid., p. xi. Further on Virchow and his use of metaphor, see Owsei Temkin, "Metaphors of Human Biology," in *Science and Civilization*, ed. R. C. Stauffer (Madison, Wis., 1949), pp. 169-194; and Erwin H. Ackerknecht, *Rudolf Virchow: Doctor, Statesman, Anthropologist* (Madison, Wis., 1953), esp. pp. 44-45.

professional credentials, including Foster and his students, present a much more difficult problem for the historian. Even if and when they did serve ideological interests through their research, they often did so unselfconsciously and in any case naturally avoided any explicit reference to such "extrinsic" concerns. Nonetheless, it is easy to imagine that a Victorian Englishman, especially one of Foster's social background and political sympathies, would be attracted to the metaphorical language of Bismarck's political adversary. Like Virchow, Foster doubtless took ideological comfort from the notion that cardiac cells could function "autonomously," in "democratic" independence of any "nervous master."

At certain points, to be sure, Foster's own writings do seem at least to acknowledge, and sometimes to emphasize, the crucial physiological role of the nervous system. In the 1870's, even as he struggled to establish the myogenic theory of the heartbeat, Foster could insist that the nervous system was the most important branch of physiology, "compared with which all the rest of physiology judged either from a practical or a theoretical point of view is a mere appendage."[73] Late in life, he drew an explicit analogy between living organisms and the "body politic" that also seems to compromise the principle of cellular autonomy:

> The two main features of the organisation of an animal body are, on the one hand, division of labour, the differentiation of the body into members, each of which is set apart to do a special work in the best possible way; and, on the other hand, integration, through which the several members are, by ties such as the vascular and nervous systems, bound together into one body in such a way that each member is helped or checked by its fellows and so guided to work for the common good of all. In a body politic the members are individuals or groups of individuals, and the ties which integrate them are in part unwritten customs, in part formally enacted laws.[74]

Here Foster seems undisturbed by the notion that the nervous system might "guide" the other parts of the body "to work for the common good of all." Indeed, on one interpretation, this sort of analogy might even seem to suggest "the inevitability (if not the desirability) of centralization of control, and of the subjugation of the parts (that is to

[73] Michael Foster, "Vivisection," *Macmillan's Magazine, 29* (1874), 367-376, on p. 373. Cf. Foster, *Textbook*, p. ix.

[74] Michael Foster, "The Organisation of Science," in *Atti Dell'*XI *Congresso Medico Internazionale*, Roma, 29 Marzo-5 Aprile 1894, vol. I (Torino, 1895), pp. 246-256. Cf. *Nature, 49* (1894), 563-564, which gives an abridged, slightly different version of Foster's address.

say, individuals) to the interests of the whole as perceived by the central [nervous] organ."[75]

Yet Foster's analogy, like others of its kind, is open to several alternative interpretations, and one of them offers a way of reconciling his firm commitment to myogenicity with his recognition of the physiological importance of the nervous system in higher organisms. This proposed reconciliation takes as its point of departure Foster's conception of the mode of action of the vagus nerve. As pointed out above, in connection with his discovery of nerveless inhibition, Foster suggested in the 1870's that the vagus nerve exerted its inhibitory effect on the heart not directly, but indirectly through the mediation of some chemical influence on the cardiac muscle.[76] Langley adopted a similar view of vagus action, as did Gaskell, who extended it to the "katabolic" accelerator nerves of the heart.[77] If Foster or other members of the Cambridge School privately held a similar conception of nervous action in general, it may have been easier for him to acknowledge the importance of the nervous system, even as he denied it any role in the origin of the heartbeat. For on such a "chemical" view of nerve action, the nervous system becomes not so much a dominant "master" as a benign "integrative" mechanism required to share one part of its power with the vascular system (to which Foster's analogy specifically refers) and another part with the mediating chemical influences through which its activity is made known.

This proposed reconciliation is admittedly speculative and is not explicitly advanced by Foster himself. Yet it does have the virtue of inviting us to consider the possibility that the myogenic theory and Starling's concept of chemical coordination were mutually reinforcing in ways that went beyond their common grounding in evolutionary principles. For Foster's "chemical" conception of vagus action, however undeveloped, might later be seen as one basis for extending the doctrine of chemical coordination to the nervous system itself. And this, in turn, would further emphasize the "actual seniority of hormonal, as compared with nervous, control of the organism"—a phrase we owe, not so incidentally, to one of the graduates of the Cambridge School during Foster's career there, Humphry Davy Rolleston.[78] It is

[75] Stanislav Andreski, ed., *Herbert Spencer: Structure, Function and Evolution* (London, 1971), p. 28. Actually, Andreski (borrowing, he says, from T. H. Huxley) makes this point in reference to Spencer's famous organismic analogy, but it seems on first sight equally applicable to Foster's analogy as well.

[76] See Chapter 7 above, "Foster's later research on the problem of the heartbeat."

[77] See Chapter 10, nn. 71, 73; and G. L. Geison, "John Newport Langley," *DSB*, VIII (1973), 14-19.

[78] See Rolleston (n. 34); and Appendix III below.

certainly suggestive, to say the least, that Cambridge became such a fertile soil for speculation about the chemical transmission of nervous impulses. Among those exposed (albeit briefly) to this stimulating environment was future Nobel laureate Otto Loewi, who established the concept experimentally in the early 1920's—at first, curiously enough, in the case of the very cardiac nerves to which Foster had drawn attention a half century before.[79]

To determine whether the myogenic theory of the heartbeat or chemical theories of nervous transmission attracted any serious opposition after the 1920's, much more research would be required. But the most likely place to look for such opposition would be Russia, where the Czarist regime had been replaced by a very different brand of autocracy, and where Pavlov's "neurocism" was elevated to official ideology. As the Russian physiologist K. M. Bykov noted in 1953, "from [Pavlov's] study of the physiology of digestion, the principle of neurocism was developed, i.e. 'a line of physiological thinking that tends to emphasize the influence of the nervous system upon the various activities of the organism.' "[80] Perhaps, indeed, some evidence actually does exist for Hoff and Guillermin's passing and undocumented assertion that Gaskell's work on the heart and Starling's on hormones helped preserve England from an "all-pervading nervism" (presumably Germanic in its origins) that ultimately survived "only in Russia under the protection of a regime that saw the promise of a basic change in human nature in . . . a happy combination of the conditioned reflex and the inheritance of acquired characters."[81]

Whatever future research may do to establish or refute these broader speculations, there surely is a prima facie case that an evolutionary style can help to explain not only the general rise of late Victorian physiology, but also the ascendancy of the Cambridge School in particular. Certainly the evolutionary approach played a crucial role in all of Gaskell's influential work, and certainly it helps to explain the

[79] See G. L. Geison, "Otto Loewi," *DSB*, VIII (1973), 451-457.

[80] K. M. Bykov, "A Russian View," in *Perspectives in Physiology*, pp. 25-28, quote on 26. Among others, Brooks and Levy, "Humorally-Transported Integrators," pp. 186-188, suggest that Pavlov's "preoccupation with the concept of reflex response" led him to deny or minimize the role of hormones in digestion and other physiological processes. Intriguingly, Pavlov followed the example of those two Russian neurogenicists, Elie Cyon and Johannes Dogiel, by pursuing cardiovascular work of an apparently neurogenic character under Ludwig at Leipzig. See Wiggers, "Significant Advances in Cardiac Physiology," p. 6.

[81] Hebbel H. Hoff and Roger Guillermin, "Claude Bernard and the Vasomotor System," in *Claude Bernard and Experimental Medicine*, ed. F. Grande and M. B. Visscher (Cambridge, Mass., 1967), pp. 75-104, on 79.

breadth of Foster's influence on the general biomedical enterprise at Cambridge. In short, partly because he most clearly perceived the relevance and potential fertility of biological and evolutionary thought for physiological research, Foster led Victorian physiology out of its stagnancy and produced his notable achievement at Cambridge. It is to the other elements in his achievement that I must now turn by way of summarizing the central themes of this book. Because the groundwork has already been more thoroughly laid, that final task can be accomplished in much less space than this preliminary discourse on national styles in late nineteenth-century physiology.

Foster and the rise of the Cambridge School: the factors in his success

To begin at the most general level, Foster's achievement at Cambridge clearly depended on the Victorian social reforms that brought the middle classes and religious dissenters more fully into the mainstream of English political and cultural life. The Royal Commissions of 1850 and 1872 on the universities of Oxford and Cambridge were part of this movement, and the phenomenal rise of science at late Victorian Cambridge could scarcely have occurred apart from the changes that this governmental intervention supported, encouraged, or imposed. A mere generation earlier, for a man of Foster's background, so distinguished a career in science would have been highly unlikely, and literally impossible at "unreformed" Cambridge.

What is perhaps more surprising is the extent to which Foster's achievement depended on the more traditional features and advantages of Cambridge. For despite the profound changes that swept over the two ancient English universities during the late nineteenth century, they retained important vestiges of their past, and some of them proved distinctly beneficial to Foster's efforts. Wealth was one. To be sure, the university per se was relatively poor, but Cambridge had rich colleges and remained attractive to men of private wealth. Foster's own two favorite students, Frank Balfour and Walter Gaskell, had sufficient family resources (Balfour phenomenally so) to undertake a career in research without concern for its future financial return. So did several others, including A. G. Dew-Smith, from whom Foster received the funds to launch the *Journal of Physiology* even after Dew-Smith had abandoned a career in research. Although never the recipient of a private gift quite so lavish as that which created the Cavendish Laboratory of Physics, Foster did get £1,500 from Henry Sidgwick (Frank Balfour's brother-in-law and Knightsbridge Professor of Moral Philosophy at Cambridge) at a point when the further expansion of his laboratory appeared otherwise doomed by the effects of the agricultural depression of the 1880's. It helped, too, to have effective control over the George Henry Lewes Studentship of Physiology created by novelist George Eliot in 1879.

More important by virtue of their number and regularity were the scholarships and fellowships offered by the individual colleges at Cambridge. These prizes, which had no counterparts at University College London, attracted men of talent and allowed some of them to remain after graduation as research workers and assistants in Foster's laboratory. The natural science fellowships of Trinity College, in particular—awarded after 1873 mainly on the basis of research promise rather than Foster's bête noire, formal written examinations—were vital to the rise of the Cambridge School of Physiology. Admittedly, the college fellowships would have been nearly useless to Foster had they not been extended to students of science and opened up to general competition as part of the Victorian reforms, but their survival in modified form represented one of the most important institutional advantages the Cambridge School enjoyed over its leading rival at University College.

Oddly enough, Victorian Oxford (despite comparable fellowships and collegiate wealth) never posed a really serious threat to the supremacy of Cambridge in physiology, not even after John Scott Burdon Sanderson went to Oxford from University College as first Waynflete Professor of Physiology in 1883. The preeminence of Cambridge, I have argued, must find its explanation less in economic or statutory factors than in the subtle but powerful and persistent operation of local traditions. In fact, Victorian Oxford and Cambridge had very similar statutes, resulting mainly from the Royal Commissions that twice investigated both universities simultaneously. At both, moreover, the Victorian period brought active efforts to create, expand, and encourage full-time, life-long academic careers. In at least one major respect, however, the two universities diverged sharply. Late Victorian Cambridge vigorously promoted and sustained science; Oxford did not. To no small degree, this divergence merely extended the longstanding but informal distinction between mildly latitudinarian, mildly reformist, mathematical Cambridge and high-church, high-Tory, classical Oxford. This informal distinction found a sort of microcosmic reflection in the differences between Trinity College, Cambridge, and Christ Church, Oxford, the largest, wealthiest, and most influential foundations at their respective universities during the Victorian period.

By virtue of his appointment and fellowship at Trinity College, Foster was at once a conspicuous symbol and major beneficiary of the new and more scientific Cambridge being shaped under its leadership. Indeed Trinity College, which also played a vital role in the rise of physics at late Victorian Cambridge, contributed so crucially to Foster's enterprise that his early "Cambridge School" might almost have been called the "Trinity School of Physiology." Even in the securing of what was formally university support, Foster relied heavily on his allies at Trinity College—notably Coutts Trotter, Henry Sidgwick, and J. Willis Clark, the last of whom in fact supervised and administered the university science buildings.

Outside of Trinity College, Foster had two immensely influential friends and patrons—George Murray Humphry, professor of human anatomy at Cambridge, and Thomas Henry Huxley, professor of natural history at the Royal School of Mines and perhaps the most powerful man in Victorian scientific circles. Quite probably, Humphry played an indirect role in bringing Foster to Cambridge in the first place, and we have clear documentary evidence of Huxley's decisive contribution to that end. In Humphry, Foster had a vital ally within the Cambridge medical faculty, while Huxley smoothed his path to the pertinent governmental bodies and to the Royal Society in particular. Both worked with him to make laboratory physiology a required subject for English medical students. And both guaranteed his research school direct access to important channels of publication.

With allies such as these, and with the advantages he automatically enjoyed at Cambridge and Trinity College, Foster's achievement might begin to seem preordained. John Scott Burdon Sanderson, the next most important individual force in late Victorian physiology, never quite secured comparable support or resources and presents an interesting contrast. At University College (1870-1883), he was hampered by relative institutional poverty, and more particularly by meager endowments for research. At Oxford (1883-1895) he faced a disdain for science that made it difficult for him to attract the ablest research talent, especially since the Oxford colleges chose to devote a much smaller percentage of their scholarships, fellowships, and other resources to the scientific enterprise. A more immediate obstacle, not unrelated to the first, lay in the strength of antivivisection sentiment at Oxford. Inspired by the eloquence of John Ruskin, and presumably sustained by the Oxonian emphasis on *Literae Humaniores*, the antivivisection forces very nearly won their battle against the initial funding of a physiological laboratory at Oxford, and later harassed Burdon Sanderson's efforts to obtain the money needed to operate the laboratory from year to year.[82] Meanwhile, at Cambridge, Foster heard scarcely an antivivisectionist murmur.[83]

But it would be folly to ignore the extent to which Foster himself created or nourished his "institutional" advantages, including some of those which seemed to come to him automatically. Doubtless for good reason, personal charm finds little place in modern historical explanation. Yet Foster's success clearly depended importantly on this elusive quality. If he "made" important friends, he also kept them, and neither Huxley nor Humphry would have succumbed to any fawning or blatantly self-serving attempt to win their support. Foster was very decidedly ambitious, but not overtly so. Even Foster's critics, who were clearly increasing in number by the end of his career, did not accuse him of exercising his power for anything but the sake of science as he perceived it. Nor could anyone have denied his clarity of aim, even (or especially) when they disagreed with him. Compared with Burdon Sanderson, in particular, Foster was more favorably endowed with the personal qualities required to create and lead a successful research school.

Whatever he may have said about Foster in private, Burdon Sanderson could not have returned Foster's barb (to Thiselton-Dyer) that he seemed to have "maggots in his head" when it came to formulating or executing a strategy to meet the antivivisectionist challenge.[84] More generally, while Burdon Sanderson moved deliberately and was almost legendary for his muddled absent-mindedness, Foster was more likely to be criticized as too swift in action and too bold in plan. In the crucial early 1870's, as Foster campaigned with customary vigor for the rapid expansion of laboratories and the creation of full-time careers in research, Burdon Sanderson defended the existing provisions for physiological research at Edinburgh and University College, limited his plea for more workers in physiology to those who might spend a few years at the task, and, in what proved to be a life-long indecision, divided his time between physiology at University College and pa-

[82] See Lady Ghetal Burdon Sanderson, *Sir John Burdon Sanderson: A Memoir*, ed. J. S. and E. S. Haldane (Oxford, 1911); and J. B. Atlay, *Sir Henry Wentworth Acland: A Memoir* (London, 1903), pp. 420-429.

[83] To my knowledge, the vivisection question only once entered university discussion during Foster's years at Cambridge, and then only as a passing comment from an individual who opposed G. M. Humphry's effort of 1875 to get Foster appointed to the Jacksonian Chair in Natural Philosophy. See *Cambridge University Reporter, 4* (1875), 267-268.

[84] Foster to Thiselton-Dyer, 23 July 1876, Thiselton-Dyer Papers, Royal Botanic Gardens, Kew. Cf. French, *Antivivisection*, pp. 137-138.

thology at the newly created Brown Institution, of which he was first professor-superintendent.[85]

Surprisingly, since he was a witty and much-recruited after-dinner speaker, Foster did not enjoy the sort of reputation as a lecturer that one might have expected of a man remembered mainly as a "great teacher." In his early Cambridge years, before he became absorbed in his London duties, his lectures were probably not so "appalling" as those C. S. Sherrington heard in the 1880's.[86] But Foster did seem distinctly less comfortable before a class than after dinner. Less aloof and awe-inspiring than the aristocratic Burdon Sanderson, his genial disposition served him better in an informal atmosphere. And a small research laboratory provided precisely the sort of informal setting in which Foster's muted but undeniable charismatic powers could best emerge.[87]

It was, in fact, through his steadfast care and feeding of the research ethos that Foster made his single most important individual contribution to the rise of the Cambridge School. At the most superficial level, this effort took the form of promoting the work of his students by citing them in his *Textbook*, and by creating two new channels of publication for them—*Studies from the Physiological Laboratory in the University of Cambridge* (1873) and the *Journal of Physiology* (1878), which was in certain respects the offspring of the former. But these ploys would have gone for naught had Foster not also inspired his students to undertake a permanent career in scientific research—at a point when they could not have expected much fiscal return or social reinforcement—and had he been less successful in creating a sense of

[85] See John Scott Burdon Sanderson, "Physiological Laboratories in Great Britain," *Nature, 3* (1871), 189; Burdon-Sanderson, "Address to the Section of Anatomy and Physiology," *Brit. Assoc. Rep., 42* (1872), Transactions, pp. 145-150; and, on Burdon Sanderson and the Brown Institution, cf. "The Brown Institution," *Nature, 5* (1871), 138-140. Partly because of Burdon Sanderson's activities at the Brown Institution, much of the teaching burden at University College fell on the shoulders of his assistant E. A. Schäfer, who apparently complained of his status as a result. See Burdon Sanderson to Schäfer, undated letter (B.8.1) and letter of 16 April 1877 (B.8.60), Sharpey-Schafer Collection, Wellcome Institute of the History of Medicine, London. Interestingly enough, Burdon Sanderson left money in his will for the promotion of pathology at Oxford, but not for physiology.

[86] On Sherrington's opinion of Foster as a teacher, see C. S. Sherrington, "Marginalia," in *Science, Medicine and History: Essays in Honour of Charles Singer* (London, 1953), pp. 545-553, on 545; and Swazey, *Reflexes*, p. 7. For evidence that Foster's lectures had probably once been better, see Henry Dale, "Sir Michael Foster," *Notes and Records of the Royal Society (London), 19* (1964), 10-32, on 25-26; and especially A. E. Shipley, "Introduction," in Alan E. Munby, *Laboratories: Their Plannings and Fittings* (London, 1921), p. xiv.

[87] Cf. Dale, ibid., p. 20; and, for the same point in a different context, Morrell, "Chemist Breeders," p. 37.

esprit de corps among them. Moreover, he staked no proprietary claims over the research carried out in his laboratory and allowed his students to follow their own evolving interests wherever they might lead. Like Carl Ludwig at Leipzig (and the chemist Justus von Liebig before him), Foster adopted the at once generous and strikingly successful practice of encouraging his often very young students to publish under their own names, even when he played an active role in their research.[88] While Burdon Sanderson's students seemed to have some difficulty emerging from his shadow to find their own new research problems,[89] Foster's students did so with such success that they eventually forgot he had ever had a direct hand in their research.

But if Foster imitated several of the leading institutional features of Ludwig's famous school, he firmly rejected its conceptual style. Disappointed by what he called the increasing "monotony" of the Leipzig school, he expressed his dissatisfaction with the more general direction of German physiology when he wrote, in the preface to his textbook, that the laboratory at "Hinab" seemed to be directed mainly toward refuting the work of the laboratory at "Hinauf."[90] Meanwhile, Burdon Sanderson pursued valuable research on the common properties of contractile waves in muscle and the insectivorous plants. He, like Foster, displayed a lively interest in Darwinian evolutionary theory and a corresponding sensitivity to the importance of adaptation in physiological phenomena. But he had a much greater admiration for Ludwig's allegedly physicochemical research style than did Foster, and he seemed less prepared to grant evolutionary theory a truly *explanatory* role in physiology, suggesting that its chief value to the physiologist was as a

[88] On Liebig's use of this practice, see Morrell, ibid., p. 30. For accounts of Ludwig's particularly generous attitude toward his student collaborators, see Burdon Sanderson, "Ludwig" (n. 13), pp. 17-18, 21; Langley, "Gaskell," p. xxviii; and Schroer, *Ludwig*, passim. That Foster followed the same practice should be clear from Chapter 8.

[89] Of Burdon Sanderson's students, Francis Gotch, F. J. M. Page, and G. J. Burch published jointly with him on electrophysiological topics, often using the capillary electrometer, and later pursued similar problems with continually refined versions of the same basic technique. Gotch, the most eminent of the three, succeeded Burdon Sanderson in the Waynflete Chair at Oxford in 1895, but his own research did not make enough of an impact to earn him an entry in the *Dictionary of Scientific Biography*. Only three of Burdon Sanderson's physiological students produced major contributions on their own: G. J. Romanes, on whom Foster doubtless exerted the more telling influence; William Maddock Bayliss, who owed much of his eminence to his joint investigations with his brother-in-law E. H. Starling; and A. D. Waller, whose important contribution to the evolution of the electrocardiogram occurred after he was brought to St. Mary's Hospital, London, on Foster's recommendation. On Bayliss, see Charles L. Evans, "William Maddock Bayliss," *DSB*, II (1970), 535-538. And on Waller at St. Mary's Hospital, see Chapter 6, n. 14.

[90] M. Foster, *Textbook*, p. vii.

guide in the choice of problems rather than in their solution.[91] More generally, his tendency to theoretical caution and programmatic ambiguity probably made it difficult for students to perceive any banner around which he would have them rally.

The story of the Cambridge School illustrates forcefully that the fate of a research school need not always depend on the research eminence of its director. For Foster had never ranked among the most creative physiologists of his day, even in England itself. Burdon Sanderson, for one, certainly had a greater capacity for original research and a firmer command of the more sophisticated aspects of experimental technique. But if Foster lacked major research talent himself, he perceived it with remarkable speed and accuracy in others, and he never forgot what his more talented students would need in the way of spiritual and financial support. Meanwhile, Burdon Sanderson pursued his own fine research partly at the expense of his responsibilities to his students. Indeed, precisely on these grounds, most of the leading English physiologists (including Foster) publicly opposed Burdon Sanderson's appointment to the Waynflete Chair at Oxford, and his subsequent election has a decidedly odd aroma about it.[92]

Ultimately, however, the factors discussed above do not wholly account for Foster's achievement at Cambridge. A really satisfying solution to the mystery is found only when his brief but competent career in research is raised from its former obscurity. Besides underlying his judgment of talent in others and his profound respect for the research enterprise, Foster's superficially meager contributions to physiological knowledge played a direct and decisive role in the creation of a re-

[91] Although French, "Darwin and the Physiologists," manages to relate some of Burdon Sanderson's work to a more general "climate of evolutionary ideas" in English physiology, Burdon Sanderson's own published research papers scrupulously avoid the use of any explicitly evolutionary principles. Indeed, if they convey any general theoretical principle at all, it is that of the essential unity of electrophysiological processes in plants and animals (a unity ultimately to be sought at the molecular level). In his more general lectures and addresses, Burdon Sanderson purveyed a somewhat elusive (and perhaps even self-contradictory) conception of physiological method. Despite his interest in Darwinian evolutionary theory, he usually drew a sharp distinction between morphology (where Darwinian evolution is the fundamental doctrine) and physiology (where physical laws are the ultimate aim). He also clearly mistrusted "vitalism" even in its new and more moderate late nineteenth-century form. For him, peculiarly biological principles or entities—whether adaptation, evolutionary theory, or that "Deus ex machina," protoplasm—could be used as guides to problems but not for purposes of *explanation*. For his perhaps most aggressively "physicalist" statement, see his 1896 address on Ludwig and modern physiology (n. 13). For his nearest approach to a "biological" stance, see John Scott Burdon Sanderson, [Presidential Address], *Brit. Assoc. Rep.*, *63* (1893), 3-31.

[92] See *Brit. Med. J.*, 1882 (2), 956, 971, 1021, 1118, 1160-1161.

search tradition at Cambridge. Quite simply, the Cambridge School of Physiology arose from and matured while pursuing Foster's abiding interest in the problem of the heartbeat. A lonely sanctuary for the myogenic theory during fifteen years of struggle for both, the Cambridge School took the first large step along its future golden path when it became identified with the triumph of that theory through Walter Gaskell's cogent and sweeping resolution of the issues.

At the same time, Gaskell's work vindicated the broadly biological and evolutionary approach to physiology that Foster had done so much to promote. Although Foster gradually shifted his own programmatic stance away from his once aggressively biological position,[93] he did so

[93] Indeed, those who read only Foster's 1885 contribution to the *Encyclopedia Britannica* [9th ed., *19* (1885), 8-23] and his 1895 review of Verworn's *Allgemein Physiologie* [in *Nature, 51* (1895), 529] could reasonably conclude that he sought above all to promote a quasi-mechanistic "molecular" physiology. The review of Verworn's book is particularly interesting because Foster there explicitly renounced the "wild dreams" of his "rash youth," when he had hoped to create "a new physiology by beginning with the study of the amoeba, and working upwards. . . ." Chastising Verworn for exaggerating the value of "cellular physiology" and for denigrating "all the knowledge which . . . has been gained by the application of exact physical and chemical methods to the study of the phenomena of life," Foster now admitted that the "physiological world is wise in spending its strength on the study of the higher animals." In the end, he conceded, the study of physiological phenomena (including muscular contraction!) yields more valuable insights in the more highly differentiated organisms precisely because basic processes are more sharply defined in them.

On the other hand, those who focus on the works of Foster's "rash youth" (extending into his forties) will readily perceive the more comparative and "biological" approach on which I have repeatedly insisted. To say nothing further here of his famous *Textbook* (1877), a lecture of 1878 on Claude Bernard reveals a Foster who insisted that physiological phenomena, "even if at bottom molecular," must be approached as if they were sui generis. In the same lecture, Foster said that he explicitly "indoctrinated" his Cambridge students with "the view that the bodies of the complex animal and complex plant are mere instances of the differentiation along two diverging lines of a common protoplasm," and he invited those students to "remember how, in our lectures here, we insist on the blood being regarded as the great 'internal medium' on which, and in which, and by which the tissues live"—the teaching at Cambridge being "largely built upon this [Bernardian] conception." See Michael Foster, "Claude Bernard . . . ," *Brit. Med. J.*, 1878 (1), 519-521, 559-560.

As late as 1881, Foster remained sufficiently biological in his programmatic stance to speak with admiration of Stephen Hales for pushing "beyond the Cartesian ideas of mechanism worked by a central force," and for allegedly embracing "those conceptions of the animal body as an exquisitely adapted self-regulating machine, which we prize and justly prize, as our leading views of to-day." See Foster [Presidential Address to the Section of Physiology], ibid., 1881 (2), 587. Indeed, I would go so far as to suggest that even the "molecular" Foster of the mid-1880's and 1890's had not entirely abandoned his youthful biases. By then, his explicit position on physiological method scarcely differed from that of Burdon Sanderson, but their respective biases perhaps still had echoes in their choice of heroes. Foster chose Bernard; Burdon Sanderson chose Ludwig.

only after he had himself abandoned research and only after he had already launched the Cambridge School on its fertile original path. From 1880 or so, as the school became firmly established and internationally recognized, there was in any case less need for unity of purpose and novelty of approach. Foster's absorption in his other duties permitted a useful diversification of interests and styles between his own spiritual descendant, W. H. Gaskell, and the rather more "Ludwigian" John Newport Langley, his eventual successor in the Cambridge chair. But if Foster's retreat from the laboratory can thus become a partial explanation for the continued eminence of his research school, that school arose in the first place and achieved its first notable success only because Foster spent several years working side by side with his students on the problem of the heartbeat.

In the end, then, the success of Foster's Cambridge School of Physiology involved a complex web of institutional, personal, and intellectual factors. If that seems a merely obvious conclusion, it has been reached only by giving extensive attention to some of the less familiar aspects of Foster's career in their full institutional and scientific context. His hitherto spare image as a "great teacher," organizer, and talent scout is now, I hope, more thoroughly fleshed out. More importantly, the exaggerated conception of him as one who "worked *for* rather than *at* physiology" and as a "discoverer of men rather than facts" should no longer persist in historical myth. As Abraham Flexner noted nearly fifty years ago, Foster produced his achievement at Cambridge in subtle ways. But those ways do not quite, as Flexner thought, "defy expression."

Appendix I

Foster as "Inefficient Teacher": The Debate over Clinical Teaching at Cambridge

In 1878 Foster advocated the founding of what he called "a complete medical school" at Cambridge, saying more specifically that he "would strain every effort to establish the custom of the whole curriculum up to the winning of the official license to practise being carried on at Cambridge and Cambridge alone."[1] On the surface, this position would seem to align him directly against G. M. Humphry's insistence on a limited, preclinical medical program at Cambridge.[2] In fact, however, Foster's remark is embedded in the context of a thinly veiled attack on the existing clinical teaching at Cambridge (especially that of Downing Professor Peter Wallwork Latham), and he openly criticized the efforts of Latham and Regius Professor George Paget to introduce the medical student to clinical studies immediately after completion of the Mathematical Tripos. Foster proposed instead that medical students enter Cambridge at an earlier age than had become customary and that they prepare for clinical work by taking the Natural Sciences Tripos (not so incidentally including his own subject of physiology). If Foster agreed with Latham and Paget that clinical medicine *could* be taught at Cambridge, he disagreed profoundly with them as to the proper character and curricular position of clinical teaching. At bottom he shared Humphry's view that the preclinical natural sciences should enjoy the preeminent place in the Cambridge medical curriculum.

In the long and sometimes passionate debate over clinical teaching at Cambridge, there can be little doubt that Foster and Humphry stood essentially together against Peter Wallwork Latham in particular. They gained a major ally in 1892, when Clifford Allbutt succeeded George Paget as Regius Professor of Physics. Allbutt's appointment marked the first time the Regius Professorship had gone to someone not already resident in Cambridge and not already affiliated with Addenbrooke's Hospital. As such, noted the *Lancet*, "it must have come on many as a surprise, for even 'the calm and serene' air of Cambridge may have been slightly ruffled by the news, where more than one physician may have had hopes in connection with the vacancy."[3] In fact, Allbutt's selection was probably "due largely to the advocacy of

[1] M. Foster, *On Medical Education at Cambridge* (Cambridge, 1878, brochure of 36 pp.), quotes on pp. 35, 31.

[2] See Chapter 6, "Foster's ambassador at large."

[3] Lord Cohen of Birkenhead, "The Rt. Hon. Sir Thomas Clifford Allbutt, F.R.S. (1836-1925)," in *Cambridge and Its Contribution to Medicine*, ed. Arthur Rook (London, 1971), pp. 173-192, on 173.

Sir Michael Foster,"[4] and from the outset Allbutt joined Foster as a target of Latham's personal hostility. Also from the outset Allbutt opposed the founding of a "complete medical school" at Cambridge on the grounds of insufficient clinical material, urging instead a firm grounding in the preclinical sciences. Whether out of personal pique or because of his own position on clinical teaching at Cambridge, Latham is said to have told his colleagues at Addenbrooke's Hospital that Allbutt would gain access to the hospital wards "over his dead body."[5] Until 1900, for some reason, Allbutt definitely was denied affiliation with the hospital, a privilege hitherto routine for the Regius Professor. In the event, he was not forced to step over any prostrate bodies, for Latham had resigned from the hospital staff the year before Allbutt gained access.

This background helps to set the stage for a remarkable episode near the end of Foster's life. Soon after his retirement in 1903, a group of friends and colleagues (including Allbutt) petitioned the university to grant him a pension in recognition of his services to physiology and to Cambridge. But no professorial pension fund existed at the time, and the Council of the Senate rejected the petition. The petitioners then submitted a memorial that would have resulted in Foster's being appointed an extraordinary professor with a stipend to be determined by the General Board of Studies. This memorial also failed, thanks to the violent opposition of Foster's old nemesis, Latham, whose own earlier retirement now afforded him the leisure to produce one of his rare literary efforts—a fly-sheet denigrating Foster's contributions to Cambridge medical education on the grounds that fewer than half of the registered medical students at Cambridge ever received the M.B. and that the number of new registrants had declined from 138 to 104 during the past decade.[6] In point of fact, Latham's critique was dubious even on its own terms, for in 1903 the number of Cambridge M.B.'s reached an unprecedented level of ninety, a level not to be reached again until the 1920's.[7] The main point to be emphasized here, however, is that Latham's real opposition to Foster lay less in the latter's alleged "inefficiency" as a teacher than in longstanding personal hostilities arising out of their differing positions in the debate over clinical teaching at Cambridge.

[4] Ibid., p. 173. Cf. Sir Humphry Davy Rolleston, *The Right Honourable Sir Thomas Clifford Allbutt, K.C.B.: A Memoir* (London, 1929), p. 109.

[5] Lord Cohen of Birkenhead, "Allbutt" p. 173. Further on Latham, see also Sir Humphry Davy Rolleston, *The Cambridge Medical School: A Biographical History* (Cambridge, 1932), pp. 208-210.

[6] Rolleston, *Allbutt*, p. 158.

[7] Ruth G. Hodgkinson, "Medical Education in Cambridge in the Nineteenth Century," in *Cambridge and Its Contribution*, ed. Rook, pp. 79-106, on 105.

Appendix II

Institutional Loci of Research Published in the *Journal of Physiology*, 1878-1900 (vols. 1-25)[1]

Between 1878 and 1900, the *Journal of Physiology* published c.700 articles by c.300 individuals working at 62 identifiable institutions.[2] Of these 700 articles, at least 409 (c. 58%) were based on research carried out primarily in English institutions. The data in the following graph and table pertain only to leading institutions located in England itself.[3] Using both the institutional affiliation supplied by the author and internal evidence (e.g., expressions of indebtedness), every effort has been made to identify the institutions in which the research was actually conducted, but the data should certainly be regarded as approximate rather than precise or irrefragable. Especially when presented graphically, the data suggest a variety of hypotheses to be tested (most obviously, the possible effect on productivity of shifts in leading personnel), but the tentative and limited nature of the data precludes any systematic attempt along those lines here. In particular, it is crucial to

1 I would like to thank my erstwhile research assistant, Sergio Sotolongo, for his diligent assistance in the collection of this data.

2 Somewhat arbitrarily, I have excluded the published abstracts of papers presented before annual meetings of the Physiological Society of Great Britain. These abstracts (which began to appear in the fourth volume of the *Journal of Physiology* and of which nearly 300 had appeared by 1900) seemed to belong to a different category from ordinary articles, even in Foster's own mind, for they are paginated in Roman numerals rather than the Arabic numbers used for the latter. Inclusion of these abstracts would expand the number of contributors by about 40 (i.e., about 40 individuals published abstracts without ever publishing an article). It is also worth mentioning here that no distinction is made in the data between contributions based on research in physiological laboratories *per se* and those based on work in other sorts of settings (e.g., anatomy departments or pathological laboratories, the latter of which became an increasingly significant site of research).

3 The less productive English institutions (with total contributions, 1878-1900, indicated in parentheses) include the following: Owens College (10), Liverpool (8), Royal College of Physicians or Royal College of Surgeons, London (7), the Army Veterinary School (4), and the Brown Institution (2). In addition, about 20 clearly English contributions were based on research conducted in private or otherwise unidentified laboratories. Because foreign physiologists had so many local alternatives to the *Journal of Physiology*, I doubt that frequency of publication in the latter can serve even crudely as an indicator of the relative productivity of foreign institutions. Only in the case of Scottish or Irish institutions does there seem to be any basis for hoping to learn something from such data. Perhaps most interesting here is the fact that the physiological laboratory of the Royal College of Physicians and Surgeons in Edinburgh is represented more frequently (15 times) than any other Scottish or Irish institution, including University College, Dundee (12), or the universities at Belfast (9), Aberdeen (8), or Edinburgh itself (7).

recognize that English physiologists could and did publish in journals other than the *Journal of Physiology*, and any conclusions that seem to be suggested by the data below could well be rendered invalid by an examination of other journals.

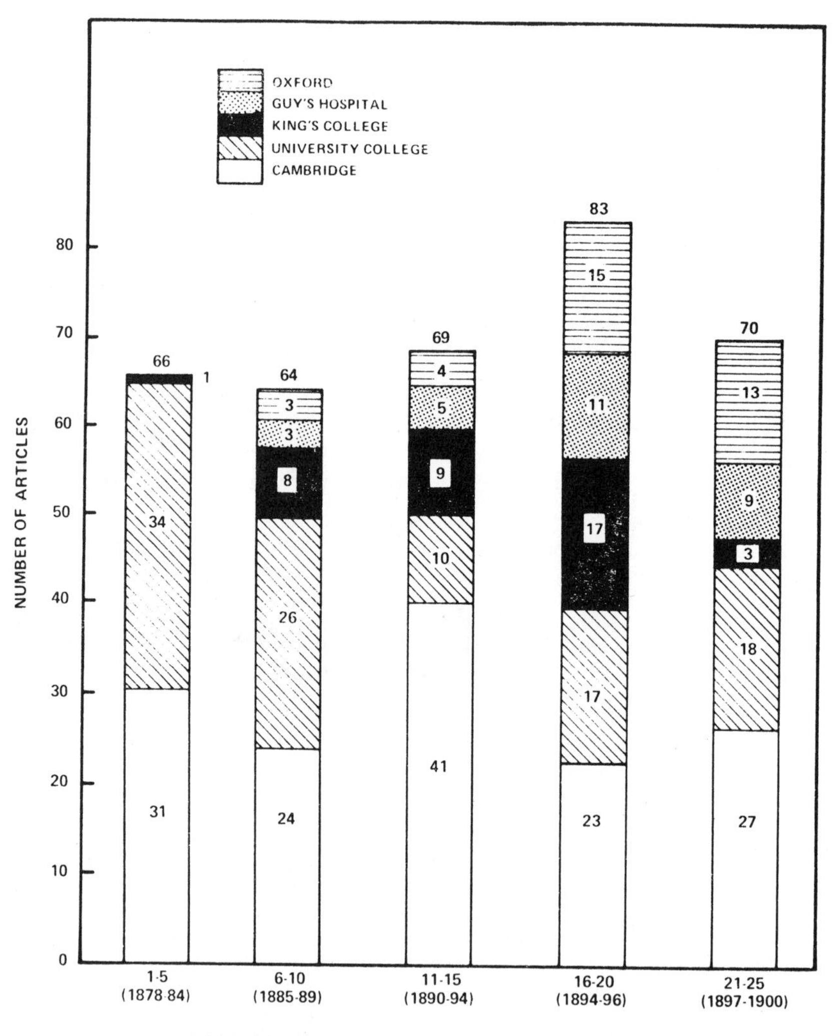

J. Physiol. vol. number	Cambr.	U. Coll. London	King's Coll. London	Oxford	London Hospital Schools: Guy's	St. Thomas	Others[a]
	Number of articles per volume						
1 (1878)	8	4	0	0	0	0	0
2	4	6	0	0	0	1	0
3	7	7	0	0	0	0	0
4	5	9	0	0	0	0	4
5 (1884)	7	7	1	0	0	0	1
6	4	6	5	0	0	0	0
7	4	10	1	0	0	0	0
8	8	7	0	0	1	0	0
9	3	0	1	2	0	0	1
10 (1889)	5	3	1	1	2	0	1
11	14	4	1	1	1	1	0
12	9	3	5	1	2	0	0
13	9	1	0	0	2	1	0
14	4	2	2	1	0	2	0
15 (1894)	5	0	1	1	0	0	0
16	4	5	3	3	4	0	1
17	4	4	5	2	2	2	0
18	3	7	4	5	1	0	0
19	5	0	1	1	3	0	1
20 (1896)	7	1	4	4	1	0	0
21	1	4	0	0	1	4	2
22	7	4	3	3	2	0	0
23	5	2	0	2	2	0	3
24	3	2	0	3	3	0	1
25 (1900)	11	7	0	5	0	1	0
Total	146	105	38	35	27	12	15
					54 (London Hospital Schools)		
% of *English* contributions (n = 409)[b]	36%	26%	9%	9%	7%	3%	4%
					14% (London Hospital Schools)		= 94%[c]
% of *all* articles (n = 700)	21%	15%	5%	5%	4%	2%	2%
					8% (London Hospital Schools)		= 54%

[a] The other hospital schools, with total contributions (1878-1900), are as follows: St. Bartholomew's (6), Charing Cross (5), London Hospital (3), and London Fever Hospital (1).

[b] This figure excludes c. 20 English contributions based on research conducted in private or otherwise unidentified laboratories.

[c] The other 6% includes the contributions from the less productive English institutions listed in footnote 3 of this appendix.

APPENDIX III

Cambridge Graduates and Faculty Who Published Articles in the *Journal of Physiology*, 1878-1900[1]

1. *John George Adami* (1862-1926), 1 article
Christ's, 1880; John Lucas Walker Student in Pathology, 1890; Demonstrator in Pathology, 1888-90; Fellow of Jesus, 1891; Strathcona Professor of Pathology and Bacteriology at McGill University, 1892-1919.
2. *Hugh Kerr Anderson* (1865-1928), 12
Caius, 1884; Fellow, 1897-1912; Demonstrator in Physiology, 1894, 1896; Lecturer in Physiology, 1903-12; Master of Caius, 1912-28.
3. *Joseph Barcroft* (1872-1947), 2
King's, 1893; Fellow, 1899; Demonstrator in Physiology, 1904; Reader in Physiology, 1919-25; Professor of Physiology, 1925-37.
4. *Sydney Arthur Monckton Copeman* (b. 1862), 3
Corpus Christi, 1879; Lecturer in Physiology and Demonstrator of Pathology, St. Thomas's Hospital, 1883-91; Medical Inspector under the Local Government Board, 1891-1919.
5. *Henry Hallet Dale* (1875-1968), 1
Trinity, 1894; Coutts Trotter Student, 1898; Director, Wellcome Physiological Research Laboratories, 1904-14; Secretary of the Royal Society, 1925-35; Nobel Prize for Medicine, 1936.

[1] Based on a comparison of the list of authors in the index to volumes I-XXV of the *Journal* with the entries in *Alumni Cantabrigiensis*, compiled by J. A. Venn, Part II, 1752-1900, 6 vols. (Cambridge, 1940-1954). The main purpose of this appendix is to offer one crude indicator of the central role Cambridge played as a training ground or nursery for experimental biologists. Fully 22 of the 55 men on the list became professors, usually in physiology or pathology. It should not be supposed, however, that every man on the list was initiated into physiological research at Cambridge, or that the research leading to his publications in the *Journal* was necessarily conducted there. In some instances, that is patently not the case. Among other things, the *Alumni Cantabrigiensis* contains entries for men who received their first degree or training elsewhere and only later migrated to Cambridge as students or to assume university positions. For men who received their first degree (or medical qualification) at another institution before coming to Cambridge, I have begun my abbreviated entries by listing the name of that other institution. For those who did get their first degree at Cambridge, the entries begin with the Cambridge college they entered and the date of their admission. The remainder of each entry seeks to reproduce information most useful to historians of science. Individuals who became educators receive more complete entries than those who became medical practitioners or civil servants. Even in the case of later eminent scientists, I have relied on the (sometimes obscure or incomplete) *Alumni Cantabrigiensis* for the sake of consistency. Whenever a position is given without institutional identification (e.g., Demonstrator in Physiology, 1883-85), it is a Cambridge University position.

6. *Ralph Francis D'Arcy* (1864-1940), 1
Caius, 1884; Lecturer in Mathematics and Physics.
7. *William Lee Dickinson* (1863-1904), 4
Caius, 1881; Physician to various hospitals in London.
8. *Edward Travers Dixon* (1862-1935), 1
Trinity, 1889. Military career.
9. *Walter Ernest Dixon* (d. 1931), 2
St. Thomas's Hospital, D.P.H.; Assistant to the Downing Professor of Medicine, 1899; Lecturer in Pharmacology, 1909; Professor of Materia Medica, King's College, London.
10. *Francis Henry Edgeworth* (1864-1943), 1
Caius, 1883; Lecturer on Physiology and Professor of Medicine, Bristol University.
11. *John Sydney Edkins* (1863-1940), 3
Caius, 1882; Demonstrator in Physiology, Owens College, Manchester; Lecturer in Chemical Physiology and Demonstrator, St. Bart's Hospital; Professor of Physiology, Bedford College, London University, 1925-30.
12. *Alfred Eichholz* (1869-1933), 2
Emmanuel, 1888; Fellow, 1893 (the first Jew to be elected Fellow of a Cambridge college); Additional Demonstrator of Physiology, 1896; Chief Medical Inspector of the Board of Education, 1908-30.
13. *Frank Cecil Eve* (dates unknown), 1
Emmanuel, 1890; Demonstrator in Physiology, Leeds University, 1896; House Physician, Royal Infirmary, Hull.
14. *Walter Morley Fletcher* (1873-1933), 2
Trinity, 1891; Fellow, 1897; Coutts Trotter Student, 1896; Senior Demonstrator in Physiology, 1903; Secretary of the Medical Research Council, 1914-33.
15. *Hans Gadow* (1855-1928), 1
Ph.D., Jena, 1878; Natural History Dept., British Museum, 1880-82; Strickland Curator, 1884-1928; Lecturer on Advanced Morphology of Vertebrates, 1884-1920; Reader in Morphology of Vertebrates, 1920.
16. *George Campbell Garratt* (1869-1940), 1
Trinity, 1887; House Physician, Royal Hospital, London; medical career.
17. *Walter Holbrook Gaskell* (1847-1914), 11
Trinity, 1864; Fellow of Trinity Hall, 1889-1914; Lecturer in Physiology, 1883-1914.
18. *Joseph Reynolds Green* (1848-1914), 4
B.Sc., London, 1880; Trinity, 1881; Demonstrator in Physiology,

1886-7; Professor of Botany, Pharmaceutical Society of Great Britain, 1887-1907. Fellow and Lecturer of Downing, 1902.

19. *Albert Sidney Frankau Grünbaum* (1869-1921), 2
Caius, 1887; Director of clinical laboratory, Addenbrooke's Hospital, Cambridge; Senior Demonstrator in Physiology, University College, Liverpool, 1898; Director of Cancer Research, Liverpool, 1903; Professor of Pathology and Bacteriology, Leeds University, 1904-17.

20. *Otto Fritz Frankau Grünbaum* (1873-1938), 4
Trinity, 1892; Physician and Lecturer in Therapeutics, London Hospital.

21. *William Bate Hardy* (1864-1934), 9
Caius, 1884; Fellow, 1892-1934; Junior Demonstrator in Physiology, 1894; Senior Demonstrator, 1896; Lecturer in Physiology, 1912; Secretary of Royal Society, 1915.

22. *Henry Head* (1861-1940), 2
Trinity, 1880; Physician, London Hospital; Editor of *Brain*, 1905-23; distinguished career in neurology.

23. *Frederick Gowland Hopkins* (1861-1947), 7
Guy's Hospital, M.R.C.P., L.R.C.P., 1894; Demonstrator in Physiology and Chemistry, Guy's Hospital; Lecturer in Chemical Physiology, 1898; Fellow of Emmanuel, 1907-10; Fellow of Trinity, 1910-47; Reader, 1902-14; Professor of Biochemistry, 1914-43; Nobel Prize, 1929.

24. *Percival Horton Horton-Smith (later Hartley)* (b. 1867), 2
St. John's, 1886; Fellow, 1891-97; Physician at St. Bart's.

25. *William Hunter* (1861-1937), 1
Edinburgh M.D., 1886; John Lucas Walker Student, 1887; Pathologist, then Physician and Dean of the Medical School, Charing Cross Hospital.

26. *Ernest Lloyd Jones* (1863-1942), 3
Non-Collegiate, 1882; Scholar and House Physician, St. Bart's; Demonstrator in Pathology and Medicine, 1890-93; Demonstrator in Medicine, 1912-14; Practiced medicine in Cambridge, and Hon. Consulting Physician to Addenbrooke's Hospital, Cambridge, for 40 years.

27. *Alfredo Antunes Kanthack* (1863-1898), 3
Educated at Liverpool Univ. College, St. Bart's, and University of London. John Lucas Walker Student, 1891; Medical Tutor, University College, Liverpool, 1892-93; Lecturer in Pathology and Bacteriology, St. Bart's, 1894-97; Professor of Physiology and Bacteriology, Bedford College, London University, 1895-96; Deputy Professor of Pathology, 1896-7; Professor of Pathology, 1897-98.

28. *John Newport Langley* (1852-1925), 48
St. John's, 1871; Fellow of Trinity, 1877; Lecturer in Physiology, 1883-1903; Professor of Physiology, 1903-25.

29. *Walter Sydney Lazarus-Barlow* (d. 1950), 3
Downing, 1884; House Physician, Consumption Hospital, Brompton; Lecturer in Pathology, Westminster Hospital; Demonstrator in Pathology, 1895-97; Director of the Cancer Research Laboratories and Professor of Experimental Pathology, Middlesex Hospital Medical School.

30. *Arthur Sheridan Lea* (1853-1915), 4
Trinity, 1872; Fellow of Caius, 1885; Demonstrator in Physiology, 1883; Lecturer in Physiology, 1883-96.

31. *Frank Spiller Locke* (1866-1949), 4
St. John's, 1886; Sometime Reader in Physics, King's College, London.

32. *John Smyth Macdonald* (1867-1941), 2
Emmanuel, 1886; Holt Fellow in Physiology, University College, Liverpool, 1891; House Physician, Royal Infirmary, Liverpool, 1897; Lecturer in Physiology, Dundee University College, 1897-99; Research Scholar, B.M.A., 1899-1901; Assistant Lecturer in Physiology, Liverpool University, 1899-1902; Professor of Physiology at University College, Sheffield, 1903-14; Holt Professor of Physiology, Liverpool University, 1914-32.

33. *Charles Robertshaw Marshall* (1869-1947), 2
Owens College, Manchester; Research Fellow in Pharmacology, Owens College, 1892-95; Assistant to the Downing Professor of Medicine, 1894-99; Professor of Materia Medica, St. Andrews, 1899-1919; Regius Professor of Materia Medica, Aberdeen, 1919-1930.

34. *Henry Newell Martin* (1848-1896), 4
University College, London; Christ's, 1870; Demonstrator to Michael Foster, Trinity College Praelector in Physiology, 1870-76; Professor of Biology, Johns Hopkins, 1876-93.

35. *William Nicolls* (b. 1856), 1
Trinity College, Dublin; Cambridge B.A. 1886, M.D. 1896; practiced medicine in Cambridge.

36. *William North* (b. 1854), 1
Sidney, 1874; Sometime Assistant to Professor J. S. Burdon Sanderson; Research Scholar, Grocers' Company; Joint Lecturer on Physiology, Westminster Hospital Medical School; Translator, Ministry of Food and Imperial Bureau of Entomology.

37. *Arthur George Phear* (b. 1867), 1
Trinity, 1884; House Physician, various hospitals in London, and medical practitioner in London.
38. *William Bramwell Ransom* (1861-1909), 2
Trinity, 1879; Fellow, 1885; House Physician, University College Hospital; Physician to the General Hospital, Nottingham.
39. *Edward Weymouth Reid* (1862-1948), 13
Non-Collegiate, 1879; Assistant Demonstrator in Human Anatomy, 1882-3; Assistant electrician, St. Bart's, 1885; Demonstrator in Physiology, St. Bart's, 1885; Assistant Lecturer in Physiology, St. Bart's, 1887-1889; Professor of Physiology, University College, Dundee, 1889.
40. *William Halse Rivers Rivers* (1864-1922), 1
St. Bart's M.R.C.S., 1886; House Physician, St. Bart's 1889-90; physician at other hospitals; Lecturer in Physiological Psychology, 1897-1907; Lecturer in Physiology of the Senses, 1907-16; Fellow of St. John's, 1902-22; President, Royal Anthropological Institute.
41. *Humphry Davy Rolleston* (1862-1944), 2
St. John's, 1883; Fellow, 1889-95 and 1925-32; Junior Demonstrator in Physiology, 1886; Demonstrator in Pathology, 1887-88; House Physician, St. Bart's, 1892; Senior Physician at St. George's; President, Royal College of Physicians, 1922-26; Regius Professor of Physic, 1925-32.
42. *Charles Smart Roy* (1854-1897), 8
Edinburgh M.D., 1878; George Henry Lewes Student, 1881; Professor-Superintendent, Brown Institution, London, 1882-84; Professor of Pathology, 1884-97.
43. *Daniel West Samways* (1857-1931), 1
St. John's, 1878; Fellow, 1885-91; House Physician at Guy's Hospital; practiced medicine in London.
44. *Charles Scott Sherrington* (1857-1952), 17
Non-Collegiate, 1879; Fellow of Caius, 1887-93; George Henry Lewes Student, 1884; taught physiology at St. Thomas's Hospital; Brown Professor of Pathology, London University, 1891-95; Professor of Physiology, Liverpool, 1895-1913; Waynflete Professor of Physiology, Oxford, 1913-35; President, Royal Society, 1920-25; Nobel Prize, 1932.
45. *Lewis Erle Shore* (1863-1944), 3
St. John's, 1882; Fellow, 1890-1944; Demonstrator in Physiology, 1887-1896; Lecturer in Physiology, 1896-1930.
46. *James Lorrain Smith* (1862-1931), 7
Edinburgh M.D., 1892; John Lucas Walker Student in Pathology,

1892; Demonstrator in Pathology, 1894; Lecturer in Pathology, Queen's College, Belfast, 1895-1901; Professor of Pathology, Queen's College, Belfast, 1901-1904; Professor of Pathology, University of Manchester, 1904-1912; Professor of Pathology, Edinburgh, 1912-13.

47. *Howard Henry Tooth* (1856-1925), 2
St. John's, 1873; House Physician and Assistant Demonstrator in Physiology, St. Bart's; medical practice in London.

48. *Coutts Trotter* (1837-1887), 1
Trinity, 1855; Fellow, 1861; Trinity Lecturer in Physics, 1869-84; Vice-Master, Trinity, 1885-87.

49. *Ivor Lloyd Tuckett* (1873-1942), 3
Trinity, 1890; Fellow, 1895; Assistant to the Downing Professor of Medicine, 1899; Additional Demonstrator in Physiology, 1899 and 1905; Senior Demonstrator in Physiology, 1905-07; House Physician at University College Hospital.

50. *Sydney Howard Vines* (1849-1934), 1
Guy's Hospital and University of London; Christ's, 1872; Fellow, 1876-88; Reader in Botany, 1884-88; Sherardian Professor of Botany, Oxford, 1888-1919.

51. *James Ward* (1843-1925), 1
Trinity, 1873; Fellow, 1875-1925; Teacher in the Theory of Education, 1880-88; Professor of Mental Philosophy and Logic, 1897-1925.

52. *William Horscraft Waters* (1855-1887), 1
Christ's, 1875; Demonstrator in Physiology, 1879-1882; Demonstrator and Assistant Lecturer in Physiology, Owens College, Manchester, 1882-87.

53. *Frank Fairchild Wesbrook* (1868-1918), 1
Manitoba M.D., 1890; John Lucas Walker Student in Pathology, 1893-95; Professor of Pathology, University of Manitoba; Professor of Public Health, University of Minnesota; President of the University of British Columbia.

54. *Richard Norris Wolfenden* (b. 1854), 4
Christ's, 1873; Lecturer in Physiology, Charing Cross Hospital School; physician to several London hospitals and practitioner in London.

55. *Almroth Edward Wright* (1861-1947), 1
Trinity College, Dublin M.B., 1883; Demonstrator in Pathology, 1887-89; Professor of Physiology, Sydney University, 1889-92; Professor of Physiology at the Army Medical School, Netley, 1892-1902, Director of the Institute of Pathology and Research, St. Mary's Hospital, 1908; Professor of Experimental Pathology, University of London.

APPENDIX IV

Cambridge University Positions in Physiology, 1870-1910[1]

[*Trinity College Praelectorship*
1870-1883 Michael Foster]

1. Professorship in Physiology, f. 1883
1883 Foster
1903 John Newport Langley

2-4. Lectureships in Physiology, f. 1883

2. 1883 Langley
1903 Hugh Kerr Anderson

3. 1883 Walter Holbrook Gaskell

4. 1883 Arthur Sheridan Lea
1896 Lewis Erle Shore

5-8. Demonstratorships in Physiology

5. Senior, f. 1883
1883 Lea
1886 Joseph Reynolds Green
1887 Shore
1896 William Bate Hardy
1903 Walter Morley Fletcher
1905 Ivor Lloyd Tuckett
1907 Joseph Barcroft

6. Junior, f. 1883
1883 D'Arcy Wentworth Thompson
1886 Humphry Davy Rolleston
1887 Hugh Edward Wingfield
1889 Hardy
1896 Anderson
1904 Barcroft
1907 Sydney William Cole

7. Additional, f. 1890
1894 Anderson
1896 Alfred Eichholz
1899 Tuckett
1904 Cole
1905 Tuckett, succeeded by Cole
1907 Keith Lucas

8. Second Additional, f. 1907
1907 Victor James Woolley
1909 Vernon Henry Mottram

[1] Adapted from J. R. Tanner, ed., *The Historical Register of the University of Cambridge* (1917). The bracket around the Trinity praelectorship serves to indicate that this position (unlike the others on this chart) was not an official university post. The positions are numbered and arranged in such a way as to indicate the number of separate positions available at a given time. By giving each man's full name only once on the chart, I hope to make it easier to follow the shifts in personnel and to identify the number of different men who circulated through the available positions.

9-10. Lectureship in Physiological and Experimental Psychology, f. 1897
1897-1907 William Halse Rivers Rivers

9. Lectureship in Physiology of the Senses, f. 1907
1907 Rivers

10. Lectureship in Experimental Psychology, f. 1907
1907 Charles Samuel Myers

11. Lectureship in Chemical Physiology, f. 1898
1898-1902 Frederick Gowland Hopkins

Readership in Chemical Physiology, f. 1902
1902 Hopkins

Appendix V

Cambridge University Positions in Zoology, Comparative Anatomy, and Morphology, 1870-1910[1]

1. Professorship in Zoology and Comparative Anatomy, f. 1866

1866 Alfred Newton
1907 Adam Sedgwick
1909 John Stanley Gardiner

2. Demonstratorship in Comparative Anatomy, f. 1873

1873 Thomas William Bridge
1876 Arthur Milnes Marshall
1877 Bridge
1880 Alfred Cort Haddon
1881 Joseph Jackson Lister
1882 William Hay Caldwell
1883 Sidney Frederic Harmer
1884 Walter Frank Raphael Weldon
1884 Harmer
1886 Arthur Everett Shipley
1887 Montague Arthur Fenton
1888 Lister
1902 Reginald Crundall Punnett
1904 Lister
1908 Frank Armitage Potts

3. Professorship in Animal Morphology, f. 1882

1882 Francis Maitland Balfour (chair terminated upon Balfour's death in that same year)

Lectureship in Animal Morphology, f. 1883

1883-1890 Sedgwick (terminated in 1890)

Readership in Animal Morphology, f. 1890

1890-1907 Sedgwick (terminated in 1907)

Readership in Zoology, f. 1907

1907 William Bateson
1908 Shipley

[1] Adapted from J. R. Tanner, ed., *Historical Register of the University of Cambridge* (1917). As in Appendix IV above (and VI-VIII below) an effort is made to indicate the number of separate positions available at a given time and the number of different men who circulated through those positions.

4. *Lectureship in Advanced Morphology on Vertebrates, f. 1884*
1884 Hans Friedrich Gadow

5. *Lectureship in Advanced Morphology of Invertebrates, f. 1884*
1884 Weldon
1891 Sydney John Hickson
1894-1908 Shipley (terminated in 1908)

Lectureship in Zoology, f. 1908
1909 Gardiner
1910 Lancelot Alexander Borradaile

6. *Demonstratorship in Animal Morphology, f. 1884*
1886 Walter Heape
1886 Harmer
1892 Lister
1893 Ernest William MacBride
1897 John Graham Kerr
1902 Gardiner
1908 Punnett
1909 Borradaile
1910 Clive Foster Cooper

7. *Demonstrator in Paleozoology, f. 1893*
1894-1899 Henry Woods (terminated 1899)

Lectureship in Paleozoology, f. 1899
1899 Woods

Appendix VI

Cambridge University Positions in Pathology, Bacteriology, Biology, and Pharmacology, 1870-1910[1]

1. *Professorship in Pathology, f. 1883*
   - 1884 Charles Smart Roy
   - 1897 Alfredo Antunes Kanthack
   - 1899 German Sims Woodhead

2. *Demonstratorship in Pathology, f. 1887*
   - 1886 Joseph Griffiths[2]
   - 1887 Almroth Edward Wright
   - 1887 Humphry Davy Rolleston
   - 1888 John George Adami
   - 1890 Ernest Lloyd Jones
   - 1893 Louis Cobbett
   - 1894 James Lorrain Smith
   - 1895 Walter Sydney Lazarus-Barlow
   - 1897 Thomas Strangeways-Pigg[3]

3. *Lectureship in Bacteriology and Preventive Medicine, f. 1900*
   - 1900-1906 George Henry Falkiner Nuttall (terminated 1906)

   Quick Professorship in Biology, f. 1906
   - 1906 Nuttall

4. *Lectureship in Special Pathology (Huddersfield Lectureship), f. 1905*
   - 1905 Thomas Strangeways Pigg Strangeways[3]

5. *Lectureship in Pathology, f. 1907*
   - 1907 Cobbett

6. *Lectureship in Pharmacology, f. 1908*
   - 1908 Walter Ernest Dixon

7. *Professorship in Biology, f. 1908*
   - 1908 William Bateson
   - 1910 Reginald Crundall Punnett

[1] Adapted from J. R. Tanner, *Historical Register of the University of Cambridge* (1917).

[2] Tanner fails to explain how Griffiths could have held a position that did not yet officially exist.

[3] Sometime between 1897 and 1905, Strangeways-Pigg added another Strangeways to the end of his name. After his death in 1926, the Strangeways Laboratory at Cambridge was named in his honor. *Cambridge and Its Contribution to Medicine,* ed. Arthur Rook (London, 1971), pp. 135, 265-266.

Appendix VII

Cambridge University Positions in Botany, 1870-1910[1]

1. Professorship in Botany, f. 1724
1861 Charles Cardale Babington
1895 Harry Marshall Ward
1906 Albert Charles Seward

2. Readership in Botany, f. 1883
1884 Sydney Howard Vines
1888 Francis Darwin
1904 Frederick Frost Blackman

3. Lectureship in Botany, f. 1883
1883 Darwin
1888 Walter Gardiner
1897 Blackman
1904 Arthur William Hill
1907 Reginald Philip Gregory

4. Additional Lectureship in Botany, f. 1890
1890 Seward
1906 Arthur George Tansley

5. (Senior) Demonstratorship in Botany, f. 1883
1884-88 Gardiner
1889-91 Charles Alfred Barber
1891-97 Blackman
1898-1900 Rowland Henry Biffen
1899-1905 Hill
1902-07 Gregory
1905 Frederick Tom Brook
1907-09 Albert Malins Smith

6. Additional (Junior) Demonstratorship in Botany, f. 1894
1895-98 William George Pharoe Ellis
1900-04 Laurence Lawton-Brain
1909 David Thoday

7. Demonstratorship in Paleobotany, f. 1891
1892 Henry Woods
1899 Edward Alexander Newell Arber

1 Adapted from J. R. Tanner, ed., *Historical Register of the University of Cambridge* (1917).

Appendix VIII

Cambridge University Positions in Other Related Fields: Medicine, Anatomy, Agriculture, Ethnology, Anthropology, and Experimental Psychology, 1870-1910[1]

1. *Regius Professorship in Physic, 1540*
   1851 Henry John Hayles Bond
   1872 George Edward Paget
   1892 Thomas Clifford Allbutt

2. *Professorship in Anatomy, f. 1707*
   1866 George Murray Humphry
   1883 Alexander Macalister

3. *Downing Professorship in Medicine, f. 1800*
   1841 William Webster Fisher
   1874 Peter Wallwork Latham
   1894 John Buckley Bradbury

4. *Drapers Professorship in Agriculture, f. 1899*
   1899 William Somerville
   1902 Thomas Hudson Middleton
   1907 Thomas Barlow Wood

5. *Lectureship in Physical Anthropology, f. 1899*
   1899 Wynfrid Laurence Henry Duckworth

6. *Lectureship in Ethnology, f. 1900*
   1900-09 Alfred Cort Haddon (terminated 1909)

   Readership in Ethnology, f. 1909
   1909 Haddon

7. *Lectureship in Agricultural Chemistry, f. 1900*
   1900-02 Wood (terminated 1902)

   Readership in Agricultural Chemistry, f. 1902
   1902-07 Wood (terminated 1907)

8. *Professorship in Agricultural Botany, f. 1908*
   1908 Rowland Harry Biffen

[1] Adapted from J. R. Tanner, *Historical Register of the University of Cambridge* (1917). Among still other possibly related fields, I have deliberately excluded human anatomy, surgery, and organic chemistry.

9. Lectureship in Agricultural Physiology, f. 1908
1908 Francis Hugh Adam Marshall

10. Demonstratorship in Experimental Psychology, f. 1904
1904 Charles Samuel Myers
1909 Edmund Oliver Lewis

Index of Authors Cited in Footnotes

Page numbers refer to first citation only. For the works of Bernard, Foster, Gaskell, and Langley, short titles are given.

General Index

Library of Congress Cataloging in Publication Data

Geison, Gerald L 1943-
Michael Foster and the Cambridge School of Physiology.

Based on the author's thesis, Yale, 1970.
Includes bibliographical references and index.
1. Foster, Michael, Sir, 1836-1907. 2. Physiologists—England—Biography. 3. Physiology—England—History. 4. Heart beat—History. I. Title.
QP26.F66G44 591.1'092'4 77-85539
ISBN 0-691-08197-2